T0321153

CHARACTER THEORY AND THE MCKAY CONJECTURE

The McKay conjecture is the origin of the counting conjectures in the representation theory of finite groups. This book gives a comprehensive introduction to these conjectures, while assuming minimal background knowledge. Character theory is explored in detail along the way, from the very basics to the state of the art. This includes not only older theorems but some brand new ones too. New, elegant proofs bring the reader up to date on progress in the field, leading to the final proof that if all finite simple groups satisfy the inductive McKay condition, then the McKay conjecture is true.

Open questions are presented throughout the book, and each chapter ends with a list of problems of varying degrees of difficulty.

Gabriel Navarro is Professor in the Department of Mathematics at the University of Valencia. He has published over 170 papers, and is the author of the widely cited volume *Characters and Blocks of Finite Groups* (Cambridge University Press, 1998). He is a leading researcher in character theory and in the McKay conjecture.

Character Theory and the McKay Conjecture

GABRIEL NAVARRO
University of Valencia

CAMBRIDGE
UNIVERSITY PRESS

University Printing House, Cambridge CB2 8BS, United Kingdom

One Liberty Plaza, 20th Floor, New York, NY 10006, USA

477 Williamstown Road, Port Melbourne, VIC 3207, Australia

314-321, 3rd Floor, Plot 3, Splendor Forum, Jasola District Centre, New Delhi - 110025, India

79 Anson Road, #06-04/06, Singapore 079906

Cambridge University Press is part of the University of Cambridge.

It furthers the University's mission by disseminating knowledge in the pursuit of education, learning and research at the highest international levels of excellence.

www.cambridge.org
Information on this title: www.cambridge.org/9781108428446
DOI: 10.1017/9781108552790

© Gabriel Navarro 2018

First published 2018

A catalogue record for this publication is available from the British Library

ISBN 978-1-108-42844-6 Hardback

For Isabel, Javier, Gabo, Nacho, and Vito

Contents

Preface

It is no exaggeration to say that a new era began in the representation theory of finite groups with the formulation of the McKay conjecture. From the perspective of finite group theory itself, it had been known a long time before the McKay conjecture that the interplay between a finite group and its local subgroups was the key to, among other things, the central problem of the theory: the classification of finite simple groups. From the point of view of modular representation theory of finite groups, there were fundamental connections already established (or conjectured) between local and global structures, such as Brauer's first main theorem, Dade's cyclic defect theory, the Green correspondence, and Brauer's height zero conjecture.

In 1972, in a paper dedicated to Richard Brauer, J. McKay observed that in some simple groups the number of irreducible complex characters of odd degree was equal to the same number calculated for the group $\mathbf{N}_G(P)$, the normalizer of a Sylow 2-subgroup P of G. Of course, the groups G and $\mathbf{N}_G(P)$ are very different in general, and yet they seemed to share a fundamental invariant. If true, this was an unexplained and astonishing discovery.

A year later, I. M. Isaacs proved that McKay's observation was true for solvable groups (in some sense, the "opposite" of the class of simple groups) and for groups of odd order for every prime. Although never formally formulated in its full generality, the *McKay Conjecture* took form.

The McKay Conjecture *If G is a finite group, p is a prime and $m_p(G)$ is the number of irreducible complex characters of G of degree not divisible by p, then*

$$m_p(G) = m_p(\mathbf{N}_G(P)),$$

where $P \in \mathrm{Syl}_p(G)$.

In 1975, J. L. Alperin gave a generalization of the McKay conjecture, which involved Brauer blocks. This was also the first formal statement of a conjecture that implied the still unformulated McKay conjecture. In 1976, J. B. Olsson proved the Alperin–McKay conjecture for symmetric groups, and in 1979, T. Okuyama and M. Wajima proved it for p-solvable groups. From this point on, different families of groups were checked by many mathematicians, and soon the idea that the McKay conjecture was going to prove correct was generally accepted. As Alperin used to say informally, if instead of mathematics this were physics, the McKay model would have been adopted a long time ago.

But how could the McKay conjecture be proven? During the course of the years, several deep generalizations of the conjecture have provided hints about the ingredients of a possible proof: blocks, isometries, simplicial complexes, characters over the p-adics, derived categories; some, or all, of these should or could be involved in the proof . . . but nobody can figure out exactly how.

While searching for a general conceptual proof of the McKay conjecture, assuming such a proof really existed, the Classification of Finite Simple Groups (CFSG) was achieved, or at least announced, in 1982. It soon became clear that a possible way of proving McKay could be to use the Classification. W. Feit, in his survey on the main problems in group theory after the CFSG, listed the McKay conjecture as the first problem. At the same time, he acknowledged that this conjecture did not seem to follow from the CFSG. As is well known, many statements in group theory admit a reduction to simple groups, in the sense that they are true provided that they are checked for every simple group. It is sometimes rightly criticized, perhaps, that a proof that uses the Classification does not fully explain why the result is actually true. We agree that it is not completely satisfactory to know that a theorem holds because simple groups have a certain property, but as long as no other proof or insight is found, can one really argue against the use of the Classification? In view of so many fundamental theorems in group theory that rely on the CFSG, this is hardly debatable. As we have said, however, the McKay conjecture does not reduce to simple groups, in the sense that it is not sufficient to check the McKay conjecture for simple groups in order to have a proof for every finite group. Something more complicated is going on.

In 1987, another central conjecture in representation theory, which seemed not to be connected with the McKay conjecture, was proposed. The celebrated *Alperin's weight conjecture* (AWC) claimed that the number of non-similar irreducible representations of a finite group over an algebraically closed field of characteristic p was the same as its number of p-*weights*, which is the number of irreducible projective characters of the p-local subgroups. This wonderful conjecture by J. L. Alperin was inspired by the representation theory of groups of Lie type and followed previous related work of T. Okuyama in p-solvable groups. Two years later, a reformulation of AWC by R. Knörr and G. R.

Robinson changed the whole perspective of the problem, and a wealth of reformulations of AWC by R. Boltje, J. Thévenaz and many others showed not only that AWC lies very deep in representation theory, but that it has connections with homotopy, K-theory, and other parts of mathematics.

In 1990, inspired by the Knorr–Robinson formulation, E. C. Dade announced various versions of a far-reaching generalization of the McKay conjecture (the so-called *Dade's counting conjectures*) that he expected to reduce to simple groups. If p^e is an arbitrary power of p, Dade's conjectures predict locally how many irreducible complex characters of a finite group have degree divisible by p^e but not by p^{e+1}. Dade's conjectures imply simultaneously both the McKay conjecture and AWC, which are now part of a single statement. Unfortunately, such a reduction to simple groups never appeared in print.

Also in 1990, M. Broué proposed a deep structural categorical explanation for the Alperin–McKay conjecture, but only for abelian *defect groups*. (Broué's conjecture explains McKay's, for instance, but only for groups with abelian Sylow p-subgroups.) This conjecture generalized the celebrated cyclic defect theory, and provided new and powerful insights, even at the level of characters. Broué's conjecture has led to a great deal of research in representation theory, and at the same time has opened new areas which are far removed from character theory. It is also the only one of these global–local counting conjectures that proposes an explanation for why some global numbers can be calculated locally. It remains a challenge to find a theory that generalizes Broué's conjecture to non-abelian defect groups.

In 2007, G. Malle, I. M. Isaacs and I published a reduction of the McKay conjecture to a problem on simple groups. This reduction was specially tailored for the McKay problem, and not for the other generalizations that had already been made in the course of the years (which were making the reductions much harder). Moreover, the reduction was conducted jointly with a specialist in simple groups, Malle, in order to guarantee that what was eventually going to be asked from the simple group specialists was, conceivably, achievable. This is how the so-called *inductive McKay condition* was born. It was proved that the McKay conjecture was true for every finite group if every simple group satisfies this inductive McKay condition. This condition requires the existence of a bijection between the irreducible characters of degree not divisible by p of any quasisimple group X (a perfect central extension of a simple group) and those of the normalizer of a Sylow p-subgroup of X, that additionally is equivariant with respect to some group automorphisms and that preserves certain cohomological properties associated with the characters. Since the appearance of this paper, many simple groups have been proved to satisfy the inductive

McKay condition, and the road to proving the McKay conjecture is, in theory, now built. Also, this paper soon led to reductions of all the other counting conjectures mentioned before to statements about simple groups.

But the most spectacular success occurred in 2015, when G. Malle and B. Späth managed to check the inductive McKay condition for the prime $p = 2$, thereby proving the McKay conjecture for this prime (which was McKay's original observation). This was a confirmation that what we embarked upon in 2007 was indeed a successful path.

The inductive McKay condition has inspired research on the character theory of the groups of Lie type, putting the focus on essential open problems so as to understand how automorphisms of groups of Lie type act on their irreducible characters, and the Clifford theory associated with these.

Since 2007, there have been significant simplifications on the formulation of the inductive McKay condition on quasisimple groups, and a general theory on this, with implications for the representation theory of general finite groups, has been developed (mainly by B. Späth). We shall dedicate the last chapter of this book to this topic. We wish not only to publicize the beauty of this reduction but also to engage students in the exciting tasks that remain ahead of us.

But, of course, this book is not only about the McKay and the global–local counting conjectures. Although our main goal has been to introduce the reader to these conjectures, to explain how they are related and to show how the reduction of the McKay conjecture to a question on simple groups is conducted, in order to achieve all this we need to review many of the essential and remarkable theorems of character theory.

After we have established our basics in Chapter 1, in Chapter 2 we introduce the Glauberman correspondence, which lies at the heart of the global–local counting conjectures. In Chapter 3, we analyze Galois action on characters and we digress to discuss a conjecture of W. Feit and a related theorem of R. Brauer. In Chapter 4, we allow ourselves some more digressions, in order to present results on zeros of characters, character values, and character identities, some of which appear in book form for the first time. (Among them, is a new proof of the Brauer–Nesbitt theorem on zeros of p-defect zero characters and Knörr's characterization of them; a theorem of Strunkov on the existence of p-defect zero characters; and a theorem of Robinson on characters taking roots of unity values on 2-singular elements.) Chapter 5 on normal subgroups is fundamental, since it will give us some of the techniques needed to conduct the reductions to simple groups in several later theorems. In Chapter 6, we discuss essential criteria to extend characters, and in the first of our reduction theorems, we prove that a group of even order possesses a nontrivial rational-valued irreducible character using the CFSG. In Chapter 7, we analyze degrees of characters, complex group algebras and character tables, and also give a generalization of the Itô–Michler theorem, again relying on the CFSG. Chapter 8 is devoted to presenting a proof the Howlett–Isaacs theorem on the solution of the Iwahori–Matsumoto conjecture on the solvability of groups of central type. This is the third theorem in this book that uses the CFSG. The Howlett–Isaacs theorem also relies on more delicate properties of the Glauberman correspondence, which we develop in this chapter, and that will be used later in the reduction of the McKay conjecture. In Chapter 9 we introduce the McKay conjecture, many of its generalizations and refinements, their consequences, and the interconnections between them and the block-free forms of Alperin's weight conjecture and Dade's ordinary conjecture. In Chapter 10 we finally prove that if all finite simple groups satisfy the inductive McKay condition, then the McKay conjecture is true.

At the end of each chapter, we include a section in which we comment on some related results and open questions. Each chapter concludes with a list of problems, of varying degrees of difficulty. In the Bibliographic Notes at the end of the book, we give explicit references to some of the theorems that we have covered in the text and of relevant comments made. References to all the works mentioned in this Preface can be found in the last section of Chapter 9.

It is now time to thank some colleagues, friends, and frequent collaborators, without whom this book would not have been possible. First, I would like to thank Marty Isaacs, from whom I learned all the character theory that I know. Our collaboration and friendship started in 1989, when I visited him in Berkeley to write the first of many joint papers. This book owes him a lot. Special thanks are due to Gunter Malle, with whom we started on the road to the inductive conditions, for many helpful observations on this book, which he read from beginning to end. Also, thanks to Geoff Robinson, who has helped me in this and other projects, always providing the cleverest insights. Of course, thanks to Pham Huu Tiep for his friendship and a lasting and fruitful collaboration. Benjamin Sambale has read the whole manuscript and has given me many useful comments and corrections, as has Noelia Rizo. To both I am very grateful.

Finally, I would also like to thank Silvio Dolfi, Lucía Sanus, Britta Späth, Joan Tent, and Carolina Vallejo, who have traveled with me on parts of this journey.

Notation

$\mathrm{Char}(G)$	the set of complex characters of G
$\mathrm{Irr}(G)$	the set of irreducible complex characters of G
I_n	the identity $n \times n$ matrix
1_G	the trivial or principal character of G
$\mathrm{Lin}(G)$	the group of linear characters of G
$\mathrm{cf}(G)$	the space of complex class functions on G
$\mathrm{Cl}(G)$	the set of conjugacy classes of G
$\mathrm{Irr}(\chi)$	the set of irreducible constituents of the character χ
$X(G)$	the character table of G
$[\alpha, \beta]$	the inner product of the class functions $\alpha, \beta \in \mathrm{cf}(G)$
ρ_G	the regular character of G
$\mathrm{Mat}_n(F)$	the $n \times n$ matrices over the field F
$\mathrm{diag}(a_1, \ldots, a_n)$	the diagonal $n \times n$ matrix with a_i in the (i, i)-position
$A \otimes B$	the Kronecker product of the matrices A and B
$\mathbf{Z}(A)$	the center of the algebra A
$A \oplus B$	the direct sum of the algebras A and B
RG	the group ring of the group G over the commutative ring R
$\mathbb{C}G$	the complex group algebra
$\delta_{a,b}$	the Kronecker δ symbol
e_χ	the central primitive idempotent associated with $\chi \in \mathrm{Irr}(G)$
$\bar{\chi}$	the complex conjugate of χ; in a different context, character associated with a factor group
$\ker(\chi)$	the kernel of the character χ
$\mathbf{Z}(\chi)$	the center of the character χ
$o(\chi)$	the order of the determinant of χ
ω_χ	the central character associated with χ

$\alpha \times \beta$	the direct product of the characters α and β
$\mathbb{Z}[\mathrm{Irr}(G)]$	the ring of generalized characters of G
χ_H	the restriction of the class function χ to the subgroup H
α^G	the induced class function of α to G
θ^a	if a is an isomorphism then $\theta^a(g^a) = \theta(g)$
G_θ	the stabilizer of θ in G
$\mathrm{Irr}(G|\theta)$	the irreducible characters of G whose restriction contains θ
$\mathrm{Char}(G|\theta)$	the characters of G whose restriction contains θ
$\mathrm{cf}(G|\theta)$	the $\mathbb{C}$-span of $\mathrm{Irr}(G|\theta)$
n_p	the largest power of p dividing the integer n, or p-part of n
x_p	the p-part of the group element x
$x_{p'}$	the p'-part of the group element x
G_p	the set of elements of G whose order is a power of p
$G_{p'}$	the set of elements of G whose order is not divisible by p
$\mathrm{cl}_G(x)$	the conjugacy class of $x \in G$
$o(g)$	the order of the group element g
$\mathbf{C}_G(g)$	the centralizer of $g \in G$
$\mathbf{Z}(G)$	the center of the group G
G'	the derived or commutator subgroup $[G, G]$ of G
$\mathbf{O}_p(G)$	the largest normal p-subgroup of G
$\mathbf{O}_{p'}(G)$	the largest normal subgroup of G of order not divisible by p
$\mathbf{O}^p(G)$	the smallest normal subgroup of G whose factor group is a p-group
$\mathsf{S}_n, \mathsf{A}_n$	the symmetric and the alternating group of degree n
C_n	the cyclic group of order n
$H \rtimes K$	the semidirect product of H with K
$H \wr K$	the wreath product of H by K
$\mathbf{C}_G(A)$	the subgroup of elements of G which are fixed by A
$\mathcal{C}(G)$	the set of all chains of p-subgroups of G
$\mathcal{N}(G)$	the set of all normal chains of p-subgroups of G
$\psi^{\otimes G}$	the tensor induced character
$\mathrm{Aut}(G)_H$	the subgroup of automorphisms $\alpha \in \mathrm{Aut}(G)$ with $\alpha(H) = H$
$\mathrm{Cl}_A(G)$	the set of A-invariant conjugacy classes of G
$\mathrm{Irr}_A(G)$	the set of A-invariant irreducible characters of G
$\overline{\mathbb{Q}}$	the algebraic closure of $\mathbb{Q}$
$\mathbf{R}$	the ring of algebraic integers in $\mathbb{C}$
$\mathbb{Q}_n$	the nth cyclotomic field
$\mathbb{Q}(\chi)$	the smallest field containing the values of χ
χ^σ	the Galois conjugate character of χ by σ

$\mathrm{Gal}(F/K)$ the automorphisms of F that fix every element of K.

$\mathrm{Irr}_F(G)$ the irreducible characters of G whose values are in the field F

$\mathbb{Q}(K)$ the smallest field containing the values of all the characters on K

$\widehat{X}$ if $X \subseteq G$, this is the element $\sum_{x \in X} x$ in some group algebra

(G, N, θ) means that $N \trianglelefteq G$ and $\theta \in \mathrm{Irr}(N)$ is G-invariant

$\alpha(x, y)$ the factor set α evaluated in (x, y)

$\mathrm{Irr}_{p'}(G)$ the set of irreducible characters of G of degree not divisible by p

$\mathrm{Irr}_{p'}(G|\theta)$ $\mathrm{Irr}_{p'}(G) \cap \mathrm{Irr}(G|\theta)$

$\mathrm{Ext}(G|\theta)$ the set of irreducible characters of G that extend θ

$\alpha \cdot \beta$ the central product of the characters α and β

$k_d(G)$ the set of irreducible characters $\chi \in \mathrm{Irr}(G)$ such that $\chi(1)_p = |G|_p/p^d$

1

The Basics

1.1 Characters

The main subject of this book is complex characters of finite groups, so let us first establish our notation for them. We assume that the reader is familiar with the most basic facts about the characters which we are about to review. Our main source is the book by I. M. Isaacs (see [Is06]). If G is a finite group, then Char(G) is the set of **characters** of G, which are the traces of the complex **representations** (group homomorphisms)

$$\mathcal{X} \colon G \to \mathrm{GL}_n(\mathbb{C}).$$

If $\chi \in \mathrm{Char}(G)$ is the trace of the representation $\mathcal{X}$ – that is, if

$$\chi(g) = \mathrm{trace}(\mathcal{X}(g))$$

for $g \in G$ – then we say that $\mathcal{X}$ **affords** the character χ. Since $\mathcal{X}(1) = I_n$ is the identity matrix, we say that $n = \chi(1)$ is the **degree** of χ and of $\mathcal{X}$. If $M \in \mathrm{GL}_n(\mathbb{C})$, then the map

$$\mathcal{Y} \colon G \to \mathrm{GL}_n(\mathbb{C})$$

given by $\mathcal{Y}(g) = M^{-1}\mathcal{X}(g)M$ for $g \in G$, defines another representation of G which affords the same character χ (since similar matrices have the same trace), and we say that $\mathcal{X}$ and $\mathcal{Y}$ are **similar**. A striking fact is that two complex representations $\mathcal{X}$ and $\mathcal{Y}$ are similar if and only if they afford the same character. If $\mathcal{Y}$ and $\mathcal{Z}$ are representations of G, then the diagonal sum

$$\mathcal{D} = \begin{pmatrix} \mathcal{Y} & 0 \\ 0 & \mathcal{Z} \end{pmatrix}$$

defines another representation of G. It is clear that if $\mathcal{Y}$ and $\mathcal{Z}$ afford the characters α and β, then $\mathcal{D}$ affords the character $\alpha + \beta$. This proves that the sum of characters is a character. A complex representation $\mathcal{X}$ of G is **irreducible** if it is not similar to a representation of the form

$$\begin{pmatrix} \mathcal{Y} & 0 \\ 0 & \mathcal{Z} \end{pmatrix},$$

where $\mathcal{Y}$ and $\mathcal{Z}$ are representations of G. In this case, we say the character χ afforded by $\mathcal{X}$ is **irreducible**. We denote by $\mathrm{Irr}(G)$ the set of the irreducible characters of a finite group G.

If $\mathcal{X}$ is any complex representation of G, then $\mathcal{X}$ is similar to a diagonal sum of irreducible representations, and this implies that every $\chi \in \mathrm{Char}(G)$ is a non-negative integral linear combination of $\mathrm{Irr}(G)$.

The group homomorphism

$$1_G : G \to \mathbb{C}^\times$$

given by $1_G(g) = 1$ for all $g \in G$ affords the **trivial** or **principal** character 1_G, which is, of course, irreducible. In general, the group homomorphisms

$$\lambda : G \to \mathbb{C}^\times$$

are the irreducible representations (characters) of degree 1, and are very well understood. (In fact, from a historical point of view, these were the first characters that were discovered and used in number theory.) The (irreducible) characters of G of degree 1 are called the **linear** characters of G, and we shall denote them by $\mathrm{Lin}(G)$. It is straightforward to check that $\mathrm{Lin}(G)$ is a group with multiplication given by $(\lambda \nu)(g) = \lambda(g)\nu(g)$ for $g \in G$.

At this point, we do not even know if the set $\mathrm{Irr}(G)$ is finite or not. If $\mathrm{cf}(G)$ is the complex space of **class functions** $G \to \mathbb{C}$ (that is, the complex functions which are constant on the conjugacy classes of G), it is clear that every character χ is in $\mathrm{cf}(G)$, again because similar matrices have the same trace. If $\mathrm{Cl}(G)$ is the set of conjugacy classes of G, $K \in \mathrm{Cl}(G)$ and $\delta_K : G \to \mathbb{C}$ is the **characteristic function** on K (that takes the value 1 on the elements of K and 0 otherwise), then it is trivial to prove that the functions δ_K for $K \in \mathrm{Cl}(G)$ form a basis of $\mathrm{cf}(G)$. In particular, the dimension of $\mathrm{cf}(G)$ is $|\mathrm{Cl}(G)|$, the number of conjugacy classes of G. What is definitely nontrivial is to show that $\mathrm{Irr}(G)$ is a basis of $\mathrm{cf}(G)$. In particular, we have the fundamental equality

$$|\mathrm{Irr}(G)| = |\mathrm{Cl}(G)|.$$

Also, if $\chi \in \mathrm{Char}(G)$, and we write

$$\chi = n_1 \chi_1 + \cdots + n_t \chi_t,$$

where $\chi_i \in \mathrm{Irr}(G)$ and $n_i > 0$, then the set

$$\mathrm{Irr}(\chi) = \{\chi_1, \ldots, \chi_t\}$$

is uniquely determined by χ. We say that these are the **irreducible constituents** of χ. Also, the uniquely determined integer n_i is the **multiplicity** of χ_i in χ. We deduce that a character χ is irreducible if and only if it is not the sum of two characters.

Another fundamental consequence of the fact that $\mathrm{Irr}(G)$ is a basis of $\mathrm{cf}(G)$ is that the *character table* of G is an invertible matrix. If we arbitrarily order $\mathrm{Irr}(G) = \{\chi_1, \ldots, \chi_k\}$ and take representatives $g_j \in K_j$, where $\mathrm{Cl}(G) = \{K_1, \ldots, K_k\}$, then the square matrix

$$X(G) = (\chi_i(g_j))$$

is the **character table** of G. (Of course, the character table is only uniquely defined up to row and column permutations. It is customary to choose $\chi_1 = 1_G$ and $g_1 = 1$, and to order the characters by degrees, and the conjugacy classes by the orders of their elements.)

But more is going on. The complex space $\mathrm{cf}(G)$ is a *hermitian* space. If $\alpha, \beta \in \mathrm{cf}(G)$, then

$$[\alpha, \beta] = \frac{1}{|G|} \sum_{x \in G} \alpha(x)\overline{\beta(x)}$$

defines a hermitian inner product on $\mathrm{cf}(G)$. The following is often called the fundamental theorem of character theory, and we formally state it next.

Theorem 1.1 *Let G be a finite group. Then $\mathrm{Irr}(G)$ is an orthonormal basis of $\mathrm{cf}(G)$ with respect to the previous inner product.*

In particular, if $\psi \in \mathrm{Char}(G)$, then

$$\psi = \sum_{\chi \in \mathrm{Irr}(G)} [\psi, \chi]\chi \, .$$

Hence, ψ is irreducible if and only if $[\psi, \psi] = 1$. Also, $\alpha \in \mathrm{Irr}(G)$ is an irreducible constituent of ψ if and only if $[\psi, \alpha] \neq 0$, and $[\psi, \alpha]$ is the multiplicity of α in χ.

The fact that

$$[\chi, \psi] = \delta_{\chi, \psi}$$

for $\chi, \psi \in \mathrm{Irr}(G)$ is called the **first orthogonality relation**. (In this book, $\delta_{a,b}$ is the Kronecker delta symbol, which is 1 if $a = b$ and 0 otherwise.) From Theorem 1.1, it is not difficult to derive the following.

Theorem 1.2 (Second orthogonality relation)   *Let G be a finite group and let* $g, h \in G$. *Then*

$$\sum_{\chi \in \mathrm{Irr}(G)} \chi(g)\overline{\chi(h)} = 0$$

if g and h are not conjugate in G. Otherwise, this sum is $|\mathbf{C}_G(g)|$.

The formula

$$\sum_{\chi \in \mathrm{Irr}(G)} \chi(1)^2 = |G|$$

can therefore be viewed as a particular case of the second orthogonality relation. As we shall see, however, it has a deeper structural explanation. Notice that from this formula we can easily deduce that a finite group G is abelian if and only if all the irreducible characters of G are linear. Also, we see that the **regular character** of G, which is defined as

$$\rho_G = \sum_{\chi \in \mathrm{Irr}(G)} \chi(1)\chi$$

has the value $|G|$ on the identity, and 0 elsewhere.

All the proofs of the results that we have mentioned so far can be found in Chapter 2 of [Is06], and in every textbook in character theory. The key object to prove all these results is the *complex group algebra* $\mathbb{C}G$, which we shall consider next.

1.2 Group Algebras

The foundations of character theory are ring-theoretical since they rely on the classification of certain algebras over the complex numbers (the *group algebras*). Since characters are often used to prove deep theorems on finite groups, there is a certain concern among group theorists whenever a purely group-theoretical result is proved using characters. In some sense, it is desirable that a theory solves the problems that it generates. However, there are many results in group theory for which there is no known character-free proof. The most famous is the Feit–Thompson theorem on the solvability of groups of odd order. But there are many others, such as the fact that a non-abelian simple group does not possess a nontrivial conjugacy class of prime power size. Some others, as Burnside's $p^a q^b$ theorem on the solvability of groups of order divisible by at most two primes, had taken many years to be proven by group-theoretical methods. Character theory started as a powerful tool to prove theorems on finite groups but it soon became a marvelous theory on its own.

In this book, we shall sometimes need to work with other fields different from the complex numbers. Let F be a field, and let $\mathcal{A}$ be a ring (with 1) which is also a vector space of finite dimension over F. If

$$\lambda(ab) = a(\lambda b) = (\lambda a)b$$

for all $a, b \in \mathcal{A}$ and $\lambda \in F$, then we say that $\mathcal{A}$ is an F-**algebra**. The canonical example of an F-algebra to keep in mind is

$$\mathrm{Mat}_n(F),$$

the algebra of $n \times n$ matrices over F. If $\mathcal{A}$ is an F-algebra, then

$$\mathbf{Z}(\mathcal{A}) = \{a \in \mathcal{A} \mid ab = ba \text{ for all } b \in \mathcal{A}\}$$

is also an F-algebra, which is called the **center** of $\mathcal{A}$. As is well known,

$$\mathbf{Z}(\mathrm{Mat}_n(F)) = \{\lambda I_n \mid \lambda \in F\}.$$

We can easily construct algebras by using direct sums. Recall that if $\mathcal{A}_i$ are F-algebras, then the vector space

$$\mathcal{A}_1 \oplus \cdots \oplus \mathcal{A}_k = \{(a_1, \ldots, a_k) \mid a_i \in \mathcal{A}_i\}$$

is an F-algebra with multiplication defined component-wise, and is called the **direct sum** of the algebras $\mathcal{A}_i$.

Our main interest here is the **group algebra**: if F is a field and G is a finite group, then

$$FG = \{\sum_{g \in G} a_g g \mid a_g \in F\},$$

the F-vector space with basis G, is an F-algebra with multiplication defined by

$$\left(\sum_{g \in G} a_g g\right)\left(\sum_{g \in G} b_g g\right) = \sum_{g,h \in G} a_g b_h (gh).$$

(In the same way, we can define the **group ring** RG for any commutative ring R.)

Since $z = \sum_{g \in G} a_g g \in \mathbf{Z}(FG)$ if and only if $z^y = z$ for all $y \in G$, we can easily check that the set of elements of the form

$$\hat{K} = \sum_{x \in K} x,$$

where $K \in \mathrm{Cl}(G)$, constitutes an F-basis of $\mathbf{Z}(FG)$. Hence

$$\dim_F(\mathbf{Z}(FG)) = |\mathrm{Cl}(G)|$$

for any field F. We shall later use that if $K, L, M \in \mathrm{Cl}(G)$ and we fix $x_M \in M$ arbitrarily, then

$$\hat{K}\hat{L} = \sum_{M \in \mathrm{Cl}(G)} a_{KLM}\hat{M} ,$$

where

$$a_{KLM} = |\{(x, y) \in K \times L \mid xy = x_M\}|.$$

Usually, an arbitrary algebra $\mathcal{A}$ is studied by analyzing the *algebra homomorphisms* $\mathcal{A} \to \mathrm{Mat}_n(F)$. Recall that an **algebra homomorphism** between two F-algebras $\mathcal{A}$ and $\mathcal{B}$ is an F-linear, multiplicative map

$$\tau : \mathcal{A} \to \mathcal{B}$$

with $\tau(1) = 1$. If τ is bijective, then $\mathcal{A}$ and $\mathcal{B}$ are **isomorphic**, which is written $\mathcal{A} \cong \mathcal{B}$.

Notice that in order to show that an F-linear map τ with $\tau(1) = 1$ is an algebra homomorphism, it is enough to check that it is multiplicative on some F-basis of $\mathcal{A}$. Hence, an algebra homomorphism

$$\mathcal{X} : FG \to \mathrm{Mat}_n(F)$$

is simply a group homomorphism $\mathcal{X} : G \to \mathrm{GL}_n(F)$ extended F-linearly.

At this point, the representation theory of finite groups splits into two different vast territories: when the characteristic of the field divides the order of the group and when it does not. This book is mainly about the latter situation. (In fact, most, but not all, of what we have to say here is about complex characters and characteristic zero fields.)

If F is any field of characteristic zero, then we can define F-representations of G, similarity and irreducibility of F-representations, and the F-characters of G, in the same way as we did for $\mathbb{C}$.

Theorem 1.3 (Wedderburn) *Let G be a finite group, and let F be an algebraically closed field of characteristic zero. Let k be the number of conjugacy classes of G.*

(a) *There are exactly k irreducible non-similar F-representations $\{\mathcal{X}_1, \ldots, \mathcal{X}_k\}$ of G.*

(b) *If χ_i is the F-character afforded by $\mathcal{X}_i$, then $\{\chi_1, \ldots, \chi_k\}$ are F-linearly independent. Also, $\mathcal{X}_i(FG) = \mathrm{Mat}_{\chi_i(1)}(F)$.*

(c) *We have that*

$$FG \cong \bigoplus_{i=1}^{k} \mathrm{Mat}_{\chi_i(1)}(F) .$$

In particular,

$$|G| = \sum_{i=1}^{k} \chi_i(1)^2.$$

An ingredient in the proof of Theorem 1.3 is Maschke's theorem, which asserts that the algebra FG is *semisimple.*

We shall use the following fact later in Chapter 3. Here $\overline{\mathbb{Q}}$ denotes the algebraic closure of $\mathbb{Q}$ in $\mathbb{C}$.

Theorem 1.4 *Let G be a finite group, and let $\chi \in \mathrm{Irr}(G)$. Then there exists a representation*

$$\mathcal{X} \colon G \to \mathrm{GL}_n(\overline{\mathbb{Q}})$$

affording χ.

Proof Let k be the number of conjugacy classes of G. Let $F = \overline{\mathbb{Q}}$. By Theorem 1.3, let $\{\psi_1, \dots, \psi_k\}$ be the set of irreducible F-characters of G. Now, each F-character is a complex character and therefore we can write

$$\psi_i = \sum_{j=1}^{k} a_{ij} \chi_j,$$

where $\mathrm{Irr}(G) = \{\chi_1, \dots, \chi_k\}$, and the a_{ij} are nonnegative integers. Since the F-irreducible characters $\{\psi_1, \dots, \psi_k\}$ are F-linearly independent, we have that the matrix $(\psi_i(g_j))$ is invertible, where g_j are representatives of the conjugacy classes of G. In particular, $\{\psi_1, \dots, \psi_k\}$ is a basis of $\mathrm{cf}(G)$, and the matrix (a_{ij}) is also invertible, because it is the matrix of a change of bases. In particular, it cannot have a column of zeros: given $1 \le s \le k$ there is i such that $a_{is} \ne 0$. Thus $\sum_{i=1}^{k} (a_{is})^2 \ge 1$ for all s. Now,

$$|G| = \sum_{i=1}^{k} \psi_i(1)^2 = \sum_{i=1}^{k} \left(\sum_{j=1}^{k} a_{ij} \chi_j(1) \right)^2 = \sum_{s,t=1}^{k} \chi_s(1)\chi_t(1) \left(\sum_{i=1}^{k} a_{is} a_{it} \right).$$

Since

$$\sum_{s=1}^{k} \chi_s(1)^2 \left(\sum_{i=1}^{k} (a_{is})^2 \right) \ge |G|,$$

we deduce that

$$\sum_{i=1}^{k} a_{is} a_{it} = \delta_{s,t}.$$

Hence, for each index s there exists a unique j such that $a_{js} \neq 0$, and in fact $a_{js} = 1$. Thus $\psi_j = \chi_s \in \mathrm{Irr}(G)$. This easily implies that $\mathrm{Irr}(G) = \{\psi_1, \ldots, \psi_k\}$. □

Complex group algebras are not fully understood without the **central primitive idempotents**, which are defined as the elements

$$e_\chi = \frac{\chi(1)}{|G|} \sum_{g \in G} \chi(g^{-1}) g \in \mathbf{Z}(\mathbb{C}G)$$

for $\chi \in \mathrm{Irr}(G)$.

Theorem 1.5 *Let G be a finite group. We have that*

$$e_\chi e_\psi = \delta_{\chi,\psi} e_\chi$$

for $\chi, \psi \in \mathrm{Irr}(G)$. In particular, $\{e_\chi \mid \chi \in \mathrm{Irr}(G)\}$ is a basis of $\mathbf{Z}(\mathbb{C}G)$.

Corollary 1.6 (Generalized orthogonality relations) *Let G be a finite group and let $g, h \in G$. Let $\chi, \psi \in \mathrm{Irr}(G)$. Then*

$$\frac{1}{|G|} \sum_{g \in G} \chi(gh)\psi(g^{-1}) = \delta_{\chi,\psi} \frac{\chi(h)}{\chi(1)}.$$

Proof The coefficient of h^{-1} in $e_\chi e_\psi$ is

$$\left(\frac{\chi(1)\psi(1)}{|G|^2} \right) \sum_{\substack{x,y \in G \\ xy = h^{-1}}} \chi(x^{-1})\psi(y^{-1}) = \left(\frac{\chi(1)\psi(1)}{|G|^2} \right) \sum_{g \in G} \chi(gh)\psi(g^{-1}),$$

by writing $g = y$. By Theorem 1.5, this equals

$$\delta_{\chi,\psi} \frac{\chi(1)}{|G|} \chi(h),$$

and the result easily follows. □

From Theorem 1.5, we can easily deduce that $e_\chi \mathbb{C}G$ is an algebra (with identity e_χ) which is isomorphic to $\mathrm{Mat}_{\chi(1)}(\mathbb{C})$. We shall not need this fact, however.

We end this section on group algebras with a quite elementary and useful result.

Theorem 1.7 (Schur's lemma) *Suppose that $\mathcal{X} \colon G \to \mathrm{GL}_n(\mathbb{C})$ is an irreducible representation of G. If $M \in \mathrm{Mat}_n(\mathbb{C})$ is such that $M\mathcal{X}(g) = \mathcal{X}(g)M$ for all $g \in G$, then $M = \lambda I_n$ for some $\lambda \in \mathbb{C}$.*

Proof See Lemma 2.25 of [Is06]. □

All the results that we have mentioned in this section are contained in Chapters 1 and 2 of [Is06]. (The statement that we have used for Theorem 1.3 can also be found in Chapter 1 of [Na98], for example.)

1.3 Character Values, Kernel, Center, and Determinant

Suppose that $\chi \in \mathrm{Char}(G)$ is afforded by the representation $\mathcal{X}$ of degree n. If $g \in G$, then the matrix $\mathcal{X}(g)$ is similar to a diagonal matrix $\mathrm{diag}(\epsilon_1, \ldots, \epsilon_n)$, where $\epsilon_j^{o(g)} = 1$, by elementary linear algebra. (We use $o(g)$ to denote the order of the element g.) Hence

$$\chi(g) = \epsilon_1 + \cdots + \epsilon_n,$$

and $\chi(g)$ is an algebraic integer in the cyclotomic field $\mathbb{Q}_{o(g)}$. (In this book, $\mathbb{Q}_m$ will represent the field $\mathbb{Q}(\xi)$, where $\xi \in \mathbb{C}$ is a primitive mth root of unity. An **algebraic integer** is a root of any monic polynomial in $\mathbb{Z}[x]$, and the set **R** of algebraic integers forms a ring in $\mathbb{C}$, by elementary number theory. It is also a well-known elementary fact that $\mathbf{R} \cap \mathbb{Q} = \mathbb{Z}$.) We deduce that $\mathcal{X}(g^{-1})$ is similar to $\mathrm{diag}(\bar{\epsilon}_1, \ldots, \bar{\epsilon}_n)$ and

$$\chi(g^{-1}) = \overline{\chi(g)},$$

where here $\bar{\epsilon}$ is the complex conjugate of $\epsilon \in \mathbb{C}$. Furthermore, if we define $\bar{\mathcal{X}}(g) = \overline{\mathcal{X}(g)}$ (the matrix in which we complex-conjugate every entry), then we see that $\bar{\mathcal{X}}$ is a representation affording the character

$$\bar{\chi}(g) = \overline{\chi(g)}.$$

Furthermore, $[\bar{\chi}, \bar{\chi}] = [\chi, \chi]$, and therefore $\bar{\chi} \in \mathrm{Irr}(G)$ if and only if $\chi \in \mathrm{Irr}(G)$. The character $\bar{\chi}$ is the **complex conjugate** of χ.

Notice that if $\chi(1) = 1$, then $\chi(g)$ is a root of unity for every $g \in G$, and in particular $\chi(g) \neq 0$ for all $g \in G$. (The converse of this is true, and it is a theorem of Burnside, which we shall consider later.)

Since $|\epsilon_j| = 1$, by using the triangle inequality for complex numbers we have that

$$|\chi(g)| \le \chi(1).$$

(The well-known triangle inequality asserts that if $\alpha_i \in \mathbb{C}$, then

$$|\alpha_1 + \cdots + \alpha_n| \le |\alpha_1| + \cdots + |\alpha_n|,$$

with equality if and only if there are nonnegative real numbers λ_i and some $\alpha \in \mathbb{C}$ such that $\alpha_i = \lambda_i \alpha$ for all i.) Using again the triangle inequality, we also deduce that $|\chi(g)| = \chi(1)$ if and only if $\mathcal{X}(g)$ is a scalar matrix, and that $\chi(g) = \chi(1)$ if and only if $\mathcal{X}(g) = I_n$. Hence the subgroup $\ker(\mathcal{X})$ is uniquely determined by the character χ that it affords. This subgroup, denoted by $\ker(\chi)$, is the **kernel** of the character χ. The same happens with the subgroup $\mathbf{Z}(\chi) = \{g \in G \mid \mathcal{X}(g) \text{ is scalar}\}$, which is called the **center** of the character χ. A character χ is **faithful** if $\ker(\chi) = 1$. Notice that

$$\bigcap_{\chi \in \mathrm{Irr}(G)} \ker(\chi) = 1$$

by the second orthogonality relation.

Since the determinant map

$$\det : \mathrm{GL}_n(\mathbb{C}) \to \mathbb{C}^\times$$

is a group homomorphism, we have that χ has associated a linear character $\det(\chi)$ given by

$$\det(\chi)(g) = \det(\mathcal{X}(g)).$$

Notice that, again, $\det(\chi)$ only depends on χ, since two representations affording χ are similar. We write $o(\chi)$ to denote the order of $\det(\chi)$ in the group $\mathrm{Lin}(G)$. This is called the **determinantal order** of χ.

1.4 More Algebraic Integers

If $\chi \in \mathrm{Irr}(G)$ and $g \in G$, we already know that $\chi(g) \in \mathbb{R}$. But there are some other algebraic integers associated with χ of the utmost importance. In order to introduce them, we need to come back, again, to representations. If $\mathcal{X}$ affords $\chi \in \mathrm{Irr}(G)$, let us extend $\mathcal{X}$ linearly to a homomorphism of algebras

$$\mathcal{X} : \mathbb{C}G \to \mathrm{Mat}_n(\mathbb{C}).$$

Recall that Schur's lemma (Theorem 1.7) tells us that the only matrices A satisfying $A\mathcal{X}(g) = \mathcal{X}(g)A$ for all $g \in G$ are the scalar matrices. Now, if $z \in \mathbf{Z}(\mathbb{C}G)$, then

$$\mathcal{X}(z) = \alpha_z I_n$$

for some scalar $\alpha_z \in \mathbb{C}$. Since $\mathcal{X}$ is an algebra homomorphism, it is also clear that the map

$$\omega_\chi : \mathbf{Z}(\mathbb{C}G) \to \mathbb{C}$$

given by $\omega_\chi(z) = \alpha_z$ is an algebra homomorphism. (These are called the **central characters**.) If $x \in G$, $K = \mathrm{cl}_G(x)$ is the G-conjugacy class of x, and $\hat{K} = \sum_{x \in K} x$, then, by taking traces, we see that

$$\omega_\chi(\hat{K}) = \frac{|K|\chi(x)}{\chi(1)}.$$

Using that $\hat{K}\hat{L}$ is an integer linear combination of $\{\hat{M} \mid M \in \mathrm{Cl}(G)\}$, where $K, L \in \mathrm{Cl}(G)$, and that ω_χ is multiplicative, it is not difficult to prove the following essential fact.

Theorem 1.8 *If $\chi \in \mathrm{Irr}(G)$, $x \in G$ and K is the conjugacy class of x in G, then*

$$\frac{|K|\chi(x)}{\chi(1)} \in \mathbf{R}.$$

Proof See Theorem (3.7) of [Is06].   □

We recall the following consequence.

Corollary 1.9 *Let G be a finite group, and let $\chi \in \mathrm{Irr}(G)$. Then $\chi(1)$ divides $|G|$.*

Proof If $K \in \mathrm{Cl}(G)$ and $x_K \in K$, we have that

$$\frac{|G|}{\chi(1)} = \frac{|G|}{\chi(1)}[\chi, \chi] = \sum_{K \in \mathrm{Cl}(G)} \frac{|K|\chi(x_K)}{\chi(1)} \chi(x_K^{-1}) \in \mathbf{R}.$$

Therefore

$$\frac{|G|}{\chi(1)} \in \mathbf{R} \cap \mathbb{Q} = \mathbb{Z},$$

as desired.   □

With slightly more effort, it is elementary to prove the following extension of Corollary 1.9, which turns out to be quite useful.

Theorem 1.10 *Let G be a finite group, and let $\chi \in \mathrm{Irr}(G)$. Then $\chi(1)$ divides $|G : \mathbf{Z}(G)|$.*

Proof This follows from Theorem 3.12 of [Is06]. □

1.5 Characters and Factor Groups

Since we are going to use it frequently, let us record our convention for the characters of G/N, where N is a normal subgroup of G.

Theorem 1.11 *Let G be a finite group and suppose that $N \trianglelefteq G$.*

(a) *If $\chi \in \mathrm{Char}(G)$ with $N \subseteq \ker(\chi)$, then $\chi(ng) = \chi(g)$ for every $n \in N$ and $g \in G$.*

(b) *Let $\mathcal{A} = \{\chi \in \mathrm{Char}(G)$ such that $N \subseteq \ker(\chi)\}$. If $\chi \in \mathcal{A}$, then the function $\tilde{\chi} : G/N \to \mathbb{C}$ given by*

$$\tilde{\chi}(Ng) = \chi(g)$$

is a well-defined character of G/N.

(c) *The map $\chi \mapsto \tilde{\chi}$ is a bijection $\mathcal{A} \to \mathrm{Char}(G/N)$ satisfying $[\alpha, \beta] = [\tilde{\alpha}, \tilde{\beta}]$ for $\alpha, \beta \in \mathcal{A}$.*

Proof Let $\chi \in \mathcal{A}$ and let $\mathcal{X} : G \to \mathrm{GL}_n(\mathbb{C})$ be a representation of G affording χ. As we have already mentioned, we know that $N \subseteq \ker(\mathcal{X})$. Hence, the map $\tilde{\mathcal{X}} : G/N \to \mathrm{GL}_n(\mathbb{C})$ given by $\tilde{\mathcal{X}}(Ng) = \mathcal{X}(g)$ is a well-defined representation of G/N affording $\tilde{\chi}$. In particular, notice that $\chi(g) = \chi(ng)$ for $n \in N$ and $g \in G$. Now, let T be a complete set of representatives of cosets of N in G. If $\alpha, \beta \in \mathcal{A}$, then

$$[\alpha, \beta] = \frac{1}{|G|} \sum_{n \in N, t \in T} \alpha(nt)\overline{\beta(nt)} = \frac{1}{|G|}|N| \sum_{t \in T} \alpha(t)\overline{\beta(t)}$$

$$= \frac{1}{|G/N|} \sum_{t \in T} \tilde{\alpha}(Nt)\overline{\tilde{\beta}(Nt)} = [\tilde{\alpha}, \tilde{\beta}].$$

Finally, if $\mathcal{Y}$ is a representation of G/N affording $\gamma \in \mathrm{Char}(G/N)$, then the map $\mathcal{X}(g) = \mathcal{Y}(Ng)$ defines a representation of G. If χ is the character afforded by $\mathcal{X}$, then $N \subseteq \ker(\chi)$ and $\tilde{\chi} = \gamma$. This completes the proof of the theorem. □

In this book, we shall identify the characters χ and $\tilde{\chi}$ in Theorem 1.11 when there is no risk of confusion, and we shall view $\mathrm{Irr}(G/N)$ as a subset of Irr(G), namely, as the subset of the irreducible characters χ of G such that $N \subseteq \ker(\chi)$. Some authors call this "inflation of characters."

Coming back to the set Lin(G) of linear characters of G, with our established convention we have that

$$\mathrm{Lin}(G) = \mathrm{Irr}(G/G')\,.$$

(In this book, $G' = [G, G]$ is the commutator or derived subgroup of G.) This is because any linear character $\lambda \colon G \to \mathbb{C}^\times$ satisfies that $G/\ker(\lambda) \cong \lambda(G)$ is abelian.

As we have already mentioned, if $\lambda, \nu \in \mathrm{Lin}(G)$, then the product $\lambda\nu \in \mathrm{Lin}(G)$ and $\mathrm{Lin}(G)$ is a group. (In fact, $\mathrm{Lin}(G)$ is isomorphic to G/G', by Problem 2.7 of [Is06].) Sometimes, the following is useful.

Lemma 1.12 *If $\lambda \in Lin(G)$, then $o(\lambda) = |G : \ker(\lambda)|$.*

Proof We have that $G/\ker(\lambda)$ is isomorphic to $\lambda(G)$, which is a subgroup of $\mathbb{C}^\times$. Hence $\lambda(G)$ is cyclic, and the lemma easily follows. $\qquad\square$

We have more to say about products of characters in the next section.

1.6 Products of Characters

We have seen how easy it is to prove that the sum of two characters is a character. It is also easy to check that the product of a linear character and a character is a character. Indeed, if $\lambda \in \mathrm{Lin}(G)$ and $\mathcal{X}$ affords $\chi \in \mathrm{Char}(G)$, then

$$(\lambda\mathcal{X})(g) = \lambda(g)\mathcal{X}(g)$$

is a representation of G with character $\lambda\chi$. Furthermore, since $[\lambda\chi, \lambda\chi] = [\chi, \chi]$, we have that $\lambda\chi \in \mathrm{Irr}(G)$ if and only if $\chi \in \mathrm{Irr}(G)$. A little more difficult is the task of showing that the product of two arbitrary characters is also a character.

In order to prove this fact, we shall need to do a bit more work. We only require a straightforward fact about matrix multiplication. Recall that if $A \in \mathrm{Mat}_n(\mathbb{C})$ and $B \in \mathrm{Mat}_m(\mathbb{C})$, then $A \otimes B \in \mathrm{Mat}_{nm}(\mathbb{C})$ is defined as

$$A \otimes B = \begin{pmatrix} a_{11}B & \ldots & a_{1n}B \\ \vdots & \ddots & \vdots \\ a_{n1}B & \ldots & a_{nn}B \end{pmatrix},$$

where $A = (a_{ij})$. This is sometimes called the **Kronecker product** of matrices. It is easy, but somewhat tedious, to check that

$$(A \otimes B)(C \otimes D) = AC \otimes BD$$

for $C \in \mathrm{Mat}_n(\mathbb{C})$ and $D \in \mathrm{Mat}_m(\mathbb{C})$. In fact, $A \otimes B$ is invertible if and only if A and B are invertible, and in this case

$$(A \otimes B)^{-1} = A^{-1} \otimes B^{-1}.$$

In other words, we have that the map

$$\mathrm{GL}_n(\mathbb{C}) \times \mathrm{GL}_m(\mathbb{C}) \to \mathrm{GL}_{nm}(\mathbb{C})$$

given by $(A, B) \mapsto A \otimes B$ is a group homomorphism.

Theorem 1.13 *Suppose that $\mathcal{X}$ and $\mathcal{Y}$ are complex representations of G of degrees n and m, affording the characters α and β, respectively. Then $\mathcal{X} \otimes \mathcal{Y} \colon G \to \mathrm{GL}_{nm}(\mathbb{C})$ defined by $(\mathcal{X} \otimes \mathcal{Y})(g) = \mathcal{X}(g) \otimes \mathcal{Y}(g)$ is a representation of G affording the character $\alpha\beta$.*

Proof We only need to check that

$$\mathrm{trace}(A \otimes B) = \mathrm{trace}(A)\mathrm{trace}(B),$$

but this is straightforward. □

In general, if $\alpha, \beta \in \mathrm{Irr}(G)$, then $\alpha\beta$ needs not be an irreducible character. In fact, if α and β are not linear, it is very rare that $\alpha\beta \in \mathrm{Irr}(G)$. However, we shall study some cases where this does happen. The following, if one thinks about it, is one such case.

Theorem 1.14 *Suppose that G and H are finite groups. Let $\alpha \in \mathrm{Char}(G)$ and $\beta \in \mathrm{Char}(H)$. Then the function $\alpha \times \beta \colon G \times H \to \mathbb{C}$ defined by*

$$(\alpha \times \beta)(g, h) = \alpha(g)\beta(h)$$

is a character of $G \times H$. Moreover, the map $f \colon \mathrm{Irr}(G) \times \mathrm{Irr}(H) \to \mathrm{Irr}(G \times H)$ given by $(\alpha, \beta) \mapsto \alpha \times \beta$ is a well-defined bijection.

Proof Let $\mathcal{X}$ be a representation of G affording α and let $\mathcal{Y}$ be a representation of H affording β. Now $\mathcal{Z}(g, h) = \mathcal{X}(g)$ defines a representation of $G \times H$ affording $\alpha \times 1_H$, which is therefore a character of $G \times H$. In the same way, $1_G \times \beta$ is a character of $G \times H$, and we deduce that $\alpha \times \beta = (\alpha \times 1_H)(1_G \times \beta)$

is a character of $G \times H$ by Theorem 1.13. By using the definition of inner products, we check that

$$[\alpha_1 \times \beta_1, \alpha_2 \times \beta_2] = [\alpha_1, \alpha_2][\beta_1, \beta_2]$$

for $\alpha_i \in \mathrm{Char}(G)$ and $\beta_i \in \mathrm{Char}(H)$. We deduce that f is well-defined and injective. (If $\alpha \in \mathrm{Irr}(G)$ and $\beta \in \mathrm{Irr}(H)$, then $[\alpha \times \beta, \alpha \times \beta] = 1$, and therefore $\alpha \times \beta \in \mathrm{Irr}(G \times H)$.) Since the number of conjugacy classes of $G \times H$ is $|\mathrm{Irr}(G)||\mathrm{Irr}(H)|$, we also have that f is onto. $\square$

As is clear, Theorem 1.14 can be stated for n factors: if G_i is a finite group for $i = 1, \ldots, n$, then $\mathrm{Irr}(G_1 \times \cdots \times G_n)$ consists of the functions $\psi_1 \times \cdots \times \psi_n$ defined by

$$(\psi_1 \times \cdots \times \psi_n)(g_1, \ldots, g_n) = \prod_{i=1}^{n} \psi_i(g_i),$$

for $\psi_i \in \mathrm{Irr}(G_i)$.

To end this section, we introduce the abelian group $\mathbb{Z}[\mathrm{Irr}(G)]$, which is the set of $\mathbb{Z}$-linear combinations of the irreducible characters of a finite group G. (These are called **generalized** or **virtual** characters of G.) Now that we can multiply characters, we have that $\mathbb{Z}[\mathrm{Irr}(G)]$ is a commutative ring with identity.

1.7 Restriction and Induction of Characters

The interplay between the characters of a group and those of its subgroups is capital. If G is a finite group, H is a subgroup of G, and $\chi \in \mathrm{Char}(G)$, then the **restricted** function $\chi_H : H \to \mathbb{C}$ given by

$$\chi_H(h) = \chi(h)$$

is a character of H. To prove this, we only need to choose a representation $\mathcal{X}$ affording χ, and consider the restriction of $\mathcal{X}$ to H. We call χ_H the character χ **restricted to** H. We may write

$$\chi_H = \sum_{\alpha \in \mathrm{Irr}(H)} [\chi_H, \alpha]\alpha.$$

In the case that $\chi \in \mathrm{Irr}(G)$, $\alpha \in \mathrm{Irr}(H)$ and $[\chi_H, \alpha] \neq 0$, that is, if α is an irreducible constituent of χ_H, it is customary to say that χ **lies over** α, or that α **lies below** χ.

There is also a way to go from characters of H to characters of G. Given $\alpha \in \mathrm{cf}(H)$, we define $\dot{\alpha} : G \to \mathbb{C}$ as $\dot{\alpha}(x) = \alpha(x)$ if $x \in H$, and $\dot{\alpha}(x) = 0$, if

$x \in G - H$. (The function $\dot{\alpha}$ does not need to be a class function of G.) The **induced** function $\alpha^G : G \to \mathbb{C}$ is defined as

$$\alpha^G(x) = \frac{1}{|H|} \sum_{g \in G} \dot{\alpha}(gxg^{-1}).$$

It easily follows from the definition that $\alpha^G \in \mathrm{cf}(G)$. Notice that

$$\alpha^G(1) = |G : H|\alpha(1).$$

The following essential result has an easy proof.

Theorem 1.15 (Frobenius reciprocity) *Suppose that G is a finite group and H is a subgroup of G. Let $\alpha \in \mathrm{cf}(H)$ and $\chi \in \mathrm{cf}(G)$. Then*

$$[\chi, \alpha^G] = [\chi_H, \alpha].$$

In particular, if $\alpha \in \mathrm{Char}(H)$, then $\alpha^G \in \mathrm{Char}(G)$.

Proof This is Lemma 5.2 of [Is06]. □

The next theorem is used rather frequently. It is easily generalized to the case where H and K are arbitrary subgroups of G. (See Problem 1.7(b).)

Theorem 1.16 (Mackey) *Suppose that G is a finite group, and assume that $G = HK$ for subgroups H, K of G. If $\alpha \in \mathrm{cf}(H)$, then*

$$(\alpha^G)_K = (\alpha_{H \cap K})^K.$$

Proof Let T be a complete set of representatives of left cosets of $H \cap K$ in H. Then T is a complete set of representatives of left cosets of K in G, using that $HK = G$. If $x \in K$, then

$$\alpha^G(x) = \frac{1}{|H|} \sum_{t \in T, k \in K} \dot{\alpha}(tkxk^{-1}t^{-1}) = \frac{|T|}{|H|} \sum_{k \in K} \dot{\alpha}(kxk^{-1}) = (\alpha_{H \cap K})^K(x),$$

where in the second equality we have used that $\dot{\alpha}(tkxk^{-1}t^{-1}) = \dot{\alpha}(kxk^{-1})$, because $T \subseteq H$ and $\alpha \in \mathrm{cf}(H)$. □

It does not often occur that $\chi_H \in \mathrm{Irr}(H)$ if $\chi \in \mathrm{Irr}(G)$, or that $\alpha^G \in \mathrm{Irr}(G)$ if $\alpha \in \mathrm{Irr}(H)$. To analyze when this happens is an interesting problem in character theory, and we shall study some cases in this book.

Next we prove that in abelian groups, the characters of subgroups *extend* to the group.

Corollary 1.17 *Suppose that G is abelian. If $H \leq G$ and $\lambda \in \mathrm{Irr}(H)$, then there exists $\nu \in \mathrm{Irr}(G)$ such that $\nu_H = \lambda$.*

Proof Let $\nu \in \mathrm{Irr}(G)$ be any irreducible constituent of λ^G. Then $0 \neq [\lambda^G, \nu] = [\lambda, \nu_H]$. Since $\nu(1) = 1$, then $\nu_H \in \mathrm{Irr}(H)$, and thus $\lambda = \nu_H$, by the first orthogonality relation. □

The following will be useful later on.

Theorem 1.18 *Suppose that G is a finite group, $N \trianglelefteq G$, and $H \leq G$ is such that $G = NH$. Let $M = N \cap H$. Then the restriction map $\mathrm{Char}(G/N) \to \mathrm{Char}(H/M)$ is a bijection satisfying*

$$[\alpha, \beta] = [\alpha_H, \beta_H]$$

for $\alpha, \beta \in \mathrm{Char}(G/N)$. Hence, restriction defines a bijection $\mathrm{Irr}(G/N) \to \mathrm{Irr}(H/M)$.

Proof Let $\chi \in \mathrm{Char}(G/N)$. Thus $\chi \in \mathrm{Char}(G)$ with $N \subseteq \ker(\chi)$, and $\chi(ng) = \chi(g)$ for $n \in N$ and $g \in G$, by Theorem 1.11(a). Now $M = N \cap H \subseteq \ker(\chi) \cap H = \ker(\chi_H)$ and $\chi_H \in \mathrm{Char}(H/M)$. If T is a complete set of representatives of cosets of M in H, note that T is a complete set of representatives of cosets of N in G. If $\alpha, \beta \in \mathrm{Char}(G/N)$, then

$$[\alpha, \beta] = \frac{1}{|G|} \sum_{n \in N, t \in T} \alpha(nt)\overline{\beta(nt)} = \frac{1}{|G/N|} \sum_{t \in T} \alpha(t)\overline{\beta(t)}$$

$$= \frac{1}{|H|} \sum_{m \in M, t \in T} \alpha(mt)\overline{\beta(mt)} = [\alpha_H, \beta_H].$$

In particular, this proves that restriction defines an injection $\mathrm{Irr}(G/N) \to \mathrm{Irr}(H/M)$. Since G/N and H/M are isomorphic then $|\mathrm{Irr}(G/N)| = |\mathrm{Irr}(H/M)|$, and necessarily restriction is a bijection. □

If H is a subgroup of G, then we can define two natural homomorphisms of free abelian groups, namely, the restriction

$$\mathrm{Res}_H^G : \mathbb{Z}[\mathrm{Irr}(G)] \to \mathbb{Z}[\mathrm{Irr}(H)]$$

and the induction

$$\mathrm{Ind}_H^G : \mathbb{Z}[\mathrm{Irr}(H)] \to \mathbb{Z}[\mathrm{Irr}(G)].$$

Frobenius reciprocity is telling us that the matrices of these linear maps with respect to the canonical bases $\mathrm{Irr}(G)$ and $\mathrm{Irr}(H)$ are transposes of one another.

1.8 Characters and Normal Subgroups

Let G be a finite group and let $N \trianglelefteq G$. If $\mathcal{X}$ is a representation of N and $g \in G$, then, using that $n \mapsto gng^{-1}$ is an automorphism of N, we have that

$$\mathcal{X}^g(n) = \mathcal{X}(gng^{-1})$$

defines a representation of N. If $\mathcal{X}$ affords the character θ, then $\mathcal{X}^g$ affords the character θ^g, where

$$\theta^g(n) = \theta(gng^{-1})$$

for $n \in N$. We say that θ and θ^g are **conjugate** characters in G by g. Since it is clear that

$$(\theta^g)^h = \theta^{gh}$$

for $g, h \in G$ and $\theta^1 = \theta$, we have that G acts on $\mathrm{Char}(N)$. The **inertia subgroup** of θ in G, which is

$$G_\theta = \{g \in G \mid \theta^g = \theta\},$$

is the stabilizer of θ in G, and therefore it is a subgroup of G. Furthermore, by the orbit-stabilizer theorem we have that $|G : G_\theta| = t$ is the number of different conjugates of θ in G. If we write $T = G_\theta$ and

$$G = Tx_1 \cup \cdots \cup Tx_t$$

as a union of disjoint right cosets of T in G, then it is clear that $\{\theta_1, \ldots, \theta_t\}$ is the set of all the different G-conjugates of θ, where $\theta_i = \theta^{x_i}$. Since θ is a class function of N, we have that $N \subseteq T$.

Notice too that

$$[\alpha^g, \beta^g] = [\alpha, \beta]$$

for $\alpha, \beta \in \mathrm{Char}(N)$ and $g \in G$, so we conclude that $\theta^g \in \mathrm{Irr}(N)$ if and only if $\theta \in \mathrm{Irr}(N)$.

If we induce a character θ of a normal subgroup, then the induction formula simplifies. We have that $\theta^G(g) = 0$ if $g \in G - N$, and

$$(\theta^G)_N = |T : N| \sum_{j=1}^{t} \theta_j.$$

From this latter equation, Clifford's theorem has an immediate proof.

Theorem 1.19 (Clifford) *Let G be a finite group, and let $N \trianglelefteq G$. Suppose that $\chi \in \mathrm{Irr}(G)$, and let $\theta \in \mathrm{Irr}(N)$ be an irreducible constituent of the restriction χ_N. If $\{\theta_1, \ldots, \theta_t\}$ are the different G-conjugates of θ, then*

$$\chi_N = e \sum_{j=1}^{t} \theta_j,$$

where $e = [\chi_N, \theta]$. In particular, e, t and $\theta(1)$ divide $\chi(1)$.

Proof See Theorem (6.2) of [Is06]. □

The number e appearing in Clifford's theorem is quite important in character theory, and we have more to say about it later on.

If $N \trianglelefteq G$ and $\theta \in \mathrm{Irr}(N)$, it has become usual to denote by

$$\mathrm{Irr}(G|\theta)$$

the set of $\chi \in \mathrm{Irr}(G)$ such that $[\chi_N, \theta] \neq 0$. By Frobenius reciprocity, this is the set of the irreducible constituents of the induced character θ^G. Hence

$$\mathrm{Irr}(G|\theta) = \mathrm{Irr}(\theta^G),$$

and we can write

$$\theta^G = \sum_{\chi \in \mathrm{Irr}(G|\theta)} e_\chi \chi,$$

where $e_\chi = [\theta^G, \chi] = [\theta, \chi_N]$. If θ is G-invariant, then $(\theta^G)_N = |G : N|\theta$ and we conclude that

$$|G : N| = \sum_{\chi \in \mathrm{Irr}(G|\theta)} e_\chi^2.$$

Sometimes, we use $\mathrm{Char}(G|\theta)$ to denote the set of characters of G all of whose irreducible constituents are in $\mathrm{Irr}(G|\theta)$.

Suppose that we wish to study some property of the set $\mathrm{Irr}(G|\theta)$. Let us say, for instance, that we wish to count $|\mathrm{Irr}(G|\theta)|$, or prove that the integer e in Clifford's theorem divides $|G : N|$. (We shall discuss these two problems further in Chapter 5.) There is a remarkably useful result called the **Clifford correspondence**, which relates $\mathrm{Irr}(G_\theta|\theta)$ with $\mathrm{Irr}(G|\theta)$ via induction of characters. Using this, it is often the case that there is no loss to assume that θ is G-invariant.

Theorem 1.20 (Clifford correspondence) *Suppose that G is a finite group, $N \trianglelefteq G$, $\theta \in \mathrm{Irr}(N)$, and $T = G_\theta$ is the stabilizer of θ in G. Then the following hold.*

(a) *The map $\psi \mapsto \psi^G$ defines a bijection* $\mathrm{Irr}(T|\theta) \to \mathrm{Irr}(G|\theta)$.
(b) *If $\psi \in \mathrm{Irr}(T|\theta)$, then* $[(\psi^G)_N, \theta] = [\psi_N, \theta]$.
(c) *If $\psi \in \mathrm{Irr}(T|\theta)$ and $\chi = \psi^G$, then*

$$\chi_T = \psi + \Delta,$$

where Δ is a character of T, or zero, such that no irreducible constituent of Δ lies over θ.

Proof See Theorem (6.11) of [Is06].   □

In the Clifford correspondence, we say that $\psi \in \mathrm{Irr}(T|\theta)$ and $\psi^G \in \mathrm{Irr}(G|\theta)$ are **Clifford correspondents** over θ.

As an application of the Clifford correspondence, we extend Corollary 1.10 using some of the theorems that we have been stating so far.

Corollary 1.21 (Itô) *Suppose that G is a finite group, and $A \trianglelefteq G$ is abelian. If $\chi \in \mathrm{Irr}(G)$, then $\chi(1)$ divides $|G : A|$.*

Proof We argue by induction on $|G|$. Let $\lambda \in \mathrm{Irr}(A)$ be under χ and let $T = G_\lambda$ be the stabilizer of λ in G. Let $\psi \in \mathrm{Irr}(T|\lambda)$ be the Clifford correspondent of χ over λ. Then $\chi = \psi^G$ by the Clifford correspondence. If $T < G$, then by induction $\psi(1)$ divides $|T : A|$ and thus $\chi(1) = |G : T|\psi(1)$ divides $|G : A|$. Therefore, we may assume that $T = G$. Since λ is now G-invariant, notice that $K = \ker(\lambda) \trianglelefteq G$. Since χ lies over λ, it follows that χ lies over $\lambda_K = 1_K$. Then, by Clifford's theorem (Theorem 1.19) we have that $\chi_K = \chi(1)1_K$, and therefore $K \subseteq \ker(\chi)$. By working in the group G/K, and using the associated characters $\tilde{\chi} \in \mathrm{Irr}(G/K)$ and $\tilde{\lambda} \in \mathrm{Irr}(A/K)$ of Theorem 1.11, we may assume that $K = 1$. In this case, if $g \in G$ and $a \in A$, we have that $\lambda^g(a) = \lambda(a)$. Then $\lambda(gag^{-1}a^{-1}) = 1$, and using that λ is faithful, we deduce that $A \subseteq \mathbf{Z}(G)$. We apply now Theorem 1.10.   □

There is a case in which we have absolute control on the set $\mathrm{Irr}(G|\theta)$: if there is some $\chi \in \mathrm{Irr}(G)$ such that $\chi_N = \theta$. Then it is said that θ **extends** to G and that χ is an **extension** of θ to G. It is a classical subject in character theory to find conditions that guarantee that a character θ of N extends, and we shall dedicate some parts of this book to investigate this. (We already know that this always happens in abelian groups, by Corollary 1.17.) We derive Gallagher's theorem below from a useful slightly more general result that we shall use in Chapter 5.

Theorem 1.22 *Suppose that $N \trianglelefteq G$. Assume that $\theta \in \mathrm{Irr}(N)$ extends to some $\chi \in \mathrm{Irr}(G)$. If $\varphi \in \mathrm{Irr}(N)$ is G-invariant and $\varphi\theta \in \mathrm{Irr}(N)$, then the map $\beta \mapsto \beta\chi$ is a well-defined bijection $\mathrm{Irr}(G|\varphi) \to \mathrm{Irr}(G|\varphi\theta)$.*

Proof See Theorem (6.16) of [Is06]. ☐

Corollary 1.23 (Gallagher correspondence) *Suppose that $N \trianglelefteq G$. Assume that $\chi \in \mathrm{Irr}(G)$ is such that $\chi_N = \theta \in \mathrm{Irr}(N)$. Then the map $\mathrm{Irr}(G/N) \to \mathrm{Irr}(G|\theta)$ given by $\beta \mapsto \beta\chi$ is a well-defined bijection.*

Proof Set $\varphi = 1_N$ in Theorem 1.22. ☐

We have said before that it is not usually the case that induction or products of irreducible characters become irreducible. The Clifford and the Gallagher correspondences are two important exceptions.

Later, the following will be useful.

Corollary 1.24 *Suppose that $N \trianglelefteq G$ with G/N abelian. Let $\theta \in \mathrm{Irr}(N)$ be G-invariant.*

(a) Then θ extends to G if and only if $|\mathrm{Irr}(G|\theta)| = |G/N|$.
(b) If $\chi \in \mathrm{Irr}(G|\theta)$, then $\mathrm{Irr}(G|\theta) = \{\lambda\chi \mid \lambda \in \mathrm{Lin}(G/N)\}$.

Proof First, we prove (a). If θ extends to G, then $|\mathrm{Irr}(G|\theta)| = |\mathrm{Irr}(G/N)|$ by Corollary 1.23. Since G/N is abelian, then $|\mathrm{Irr}(G/N)| = |G/N|$. Conversely, assume that $|\mathrm{Irr}(G|\theta)| = |G/N|$. If $\chi \in \mathrm{Irr}(G|\theta)$ and $\chi_N = e_\chi\theta$, then we know that

$$|G/N| = \sum_{\chi \in \mathrm{Irr}(G|\theta)} e_\chi^2 \geq |G/N|.$$

We conclude that $e_\chi = 1$ for all $\chi \in \mathrm{Irr}(G|\theta)$.

Next, we prove (b). Write $\chi_N = e\theta$. Notice that

$$\sum_{\lambda \in \mathrm{Lin}(G/N)} \lambda$$

is zero on the elements of $G - N$, by the second orthogonality relation. Hence, we have that the character

$$\psi = \sum_{\lambda \in \mathrm{Lin}(G/N)} \lambda\chi$$

has the value zero on the elements of $G - N$, and $\psi_N = e|G : N|\theta$. Hence

$$e\theta^G = \sum_{\lambda \in \mathrm{Lin}(G/N)} \lambda\chi,$$

because these $e\theta^G$ and ψ have the same value on every element of G. Since $\mathrm{Irr}(\theta^G) = \mathrm{Irr}(e\theta^G)$, the proof of part (b) is complete. □

To finish this introductory chapter, let us prove a useful result which is a direct consequence of the Gallagher correspondence.

Theorem 1.25 *Let $N \trianglelefteq G$ and let $\chi \in \mathrm{Irr}(G)$ be such that $\chi_N = \theta \in \mathrm{Irr}(N)$. Let $\psi \in \mathrm{Char}(G)$. For each $g \in G$, let*

$$S(Ng) = \sum_{x \in Ng} \chi(x)\overline{\psi(x)}.$$

If $\psi = \chi$, then $S(Ng) = |N|$ and if no irreducible constituent of ψ lies over θ, then $S(Ng) = 0$.

Proof Notice that S is a class function of G/N. Let $\tau \in \mathrm{Irr}(G/N)$, and view it as an irreducible character of G with $N \subseteq \ker(\tau)$. Now

$$[S, \tau] = \frac{1}{|G/N|} \sum_{Ng \in G/N} S(Ng)\overline{\tau(Ng)} = \frac{1}{|G/N|} \sum_{Ng \in G/N} \left(\sum_{x \in Ng} \chi(x)\overline{\psi(x)}\tau(x) \right)$$

$$= |N|[\chi\bar{\tau}, \psi],$$

where here $\bar{\tau}$ is the complex-conjugate of τ. We see that $[S, \tau] = 0$ if no irreducible constituent of ψ lies over θ (because $(\chi\bar{\tau})_N$ is a multiple of θ). Hence $S(Ng) = 0$ for all $g \in G$ in this case. If $\psi = \chi$, then $[S, 1_{G/N}] = |N|$ and $[S, \tau] = 0$ if $\tau \neq 1_{G/N}$, using the Gallagher correspondence. Therefore $S = |N|1_{G/N}$, as desired. □

Problems

(1.1) Let G be a finite group, and let $\Psi = \sum_{\chi \in \mathrm{Irr}(G)} \chi^2$. Prove that $\Psi(g) = 0$ if and only if g is not G-conjugate to g^{-1}.

(1.2) Suppose that $\chi \in \mathrm{Char}(G)$.
   (a) Prove that

$$\ker(\chi) = \bigcap_{\psi \in \mathrm{Irr}(G), [\chi, \psi] \neq 0} \ker(\psi).$$

(b) Prove that

$$\det(\chi) = \prod_{\psi \in \mathrm{Irr}(G)} \det(\psi)^{[\chi, \psi]}.$$

(1.3) (*Knörr*) Let G be a finite group, let $g_1, \ldots, g_k$ be representatives of the conjugacy classes of G, and write $\mathrm{Irr}(G) = \{\chi_1, \ldots, \chi_k\}$. Suppose that $\chi \in \mathbb{Z}[\mathrm{Irr}(G)]$.

(a) Define the $\mathbb{C}$-linear map $f : \mathrm{cf}(G) \to \mathrm{cf}(G)$ by $f(\alpha) = \chi\alpha$. Let δ_{K_i} be the class function that has the value 1 on the conjugacy class K_i of g_i and is 0 elsewhere. Show that the matrices of f with respect to the bases $\{\delta_{K_i}\}$ and $\mathrm{Irr}(G)$ are, respectively, $D = \mathrm{diag}(\chi(g_1), \ldots, \chi(g_k))$ and $M = (m_{ij})$, where $m_{ij} = [\chi \chi_i, \chi_j]$.

(b) Deduce that $\sum_{i=1}^{k} \chi(g_i)$ and $\prod_{i=1}^{k} \chi(g_i)$ are integers. More generally, show that the polynomial

$$p(x) = \prod_{i=1}^{k} (x - \chi(g_i)) \in \mathbb{Z}[x].$$

(c) Suppose that $z = \prod_{i=1}^{k} \chi(g_i) \neq 0$, and consider the class function φ defined by

$$\varphi(g) = \frac{z}{\chi(g)}$$

for $g \in G$. Define the matrix $N = (n_{ij})$ by setting $n_{ij} = [\varphi\chi_i, \chi_j]$. Show that $N = zM^{-1} = \det(M)M^{-1}$. Deduce that $n_{ij} \in \mathbb{Z}$, and that φ is a generalized character.

(d) Suppose that $z \neq 0$ and that $\gcd(z, |G|) = 1$. If $\psi \in \mathbb{Z}[\mathrm{Irr}(G)]$ is such that $\psi(g)/\chi(g)$ is an algebraic integer for every $g \in G$, then show that $\psi/\chi \in \mathbb{Z}[\mathrm{Irr}(G)]$.

(*Note*: Some of these results were independently obtained by D. Chillag [Ch86], P. Ferguson and I. M. Isaacs [FI89], and G. Mason [Mas86]. From part (a), we see that the values of any character χ are the eigenvalues of an integer matrix M. This was noted by D. Chillag.)

(1.4) (*Chillag*) Assume the hypothesis and notation of Problem 1.3. Let $X = (\chi_i(g_j))$ be the character table of G. Show that

$$X^{-1}MX = D.$$

(*Note*: In particular, this shows again that character values are algebraic integers.)

(1.5) Let G be a finite group. Consider the basis $\{e_\chi \mid \chi \in \mathrm{Irr}(G)\}$ of $\mathbf{Z}(\mathbb{C}G)$, consisting of the central primitive idempotents. Show that $\omega_\psi(e_\chi) = \delta_{\chi,\psi}$ for $\chi, \psi \in \mathrm{Irr}(G)$, and

$$z = \sum_{\chi \in \mathrm{Irr}(G)} \omega_\chi(z) e_\chi$$

for $z \in \mathbf{Z}(\mathbb{C}G)$.

(1.6) Prove the following properties of the induction of class functions (which we shall use frequently).
(a) If $\alpha \in \mathrm{cf}(H)$, and $H \le K \le G$, then $(\alpha^K)^G = \alpha^G$.
(b) If $\alpha \in \mathrm{cf}(H)$, $H \le G$, and $\beta \in \mathrm{cf}(G)$, then

$$(\alpha \beta_H)^G = (\alpha^G)\beta .$$

(1.7) Suppose that $H \le G$, let $\varphi \in \mathrm{cf}(H)$, and let $g \in G$. Define φ^g as the class function of H^g defined by $\varphi^g(h^g) = \varphi(h)$ for $h \in H$.
(a) If φ is a character of H, then prove that φ^g is a character of H^g. Also, prove that φ^g is irreducible if and only if φ is irreducible.
(b) (*Mackey*) Suppose that $H, K \le G$, and write

$$G = \bigcup_{t \in T} H t K$$

as a disjoint union. Let $\varphi \in \mathrm{cf}(H)$. Prove that

$$(\varphi^G)_K = \sum_{t \in T} ((\varphi^t)_{H^t \cap K})^K .$$

(1.8) Let H be a subgroup of G, and let $\theta \in \mathrm{Irr}(H)$. Prove that θ^G is irreducible if and only if

$$[(\theta^g)_{H \cap H^g}, \theta_{H \cap H^g}] = 0$$

for all $g \in G - H$.

(1.9) Let G be a finite group and let H be a subgroup of G. If the induction map $\mathbb{Z}[\mathrm{Irr}(H)] \to \mathbb{Z}[\mathrm{Irr}(G)]$ is surjective, then prove that $H = G$. (*Note*: I. M. Isaacs and I showed that if the restriction map $\mathbb{Z}[\mathrm{Irr}(G)] \to \mathbb{Z}[\mathrm{Irr}(H)]$ is injective, then $H = G$.)

(1.10) Prove that the ring $\mathbb{Z}[\mathrm{Irr}(G)]$ does not need to be an integral domain.

(1.11) Suppose that G is a finite group that acts on a finite set $\Omega = \{\alpha_1, \ldots, \alpha_n\}$. For $g \in G$, let $\chi(g) = |\{\alpha \in \Omega \mid \alpha \cdot g = \alpha\}|$, and let $\mathcal{X}(g)$ be the $n \times n$ matrix (a_{ij}), where $a_{ij} = 1$ if $\alpha_i \cdot g = \alpha_j$ and $a_{ij} = 0$, otherwise.

(a) Prove that $\mathcal{X}$ is a representation of G that affords the character χ. (The character χ is called the **permutation character** of the action.)

(b) If $\mathcal{O}_1, \ldots, \mathcal{O}_k$ are the different G-orbits on Ω, $\beta_i \in \mathcal{O}_i$ and $H_i = G_{\beta_i}$, show that

$$\chi = (1_{H_1})^G + \cdots + (1_{H_k})^G.$$

(c) Prove that $[\chi, 1_G]$ is the number of G-orbits on Ω.

(d) Suppose that the action of G on Ω is transitive. Let $\alpha \in \Omega$. Show that $[\chi, \chi]$ is the number of orbits of G_α on Ω. Deduce that the action of G on Ω is **2-transitive** (that is, G is transitive on Ω, and G_α on $\Omega - \{\alpha\}$) if and only if $\chi = 1 + \psi$ for some $\psi \in \mathrm{Irr}(G)$.

(1.12) (*L. Solomon*) Let G be a finite group, and let Ψ be the permutation character of the action of G on itself by conjugation, so that $\Psi(g) = |\mathbf{C}_G(g)|$. Let $g_1, \ldots, g_k$ be representatives of the conjugacy classes of G, and write $\mathrm{Irr}(G) = \{\chi_1, \ldots, \chi_k\}$.

(a) Prove that $c_i = \sum_{j=1}^{k} \chi_i(g_j) = [\Psi, \chi_i] \in \mathbb{Z}$.

(b) Show that the sum of all the entries in the character table of G is a positive integer $s \le |G|$.

(*Note*: Since $\mathbf{Z}(G) = \ker(\Psi)$, it follows that $c_i = 0$ if $\mathbf{Z}(G)$ is not contained in $\ker(\chi_i)$. The converse is not true. Another proof of the first part is in Problem 1.3(b).)

(1.13) (*Knörr*) Suppose that a finite group G acts on a finite set Ω and assume that $\gcd(|\Omega|, |G|) = 1$. Let π be the permutation character of this action and assume that $\pi(g)$ divides $|\Omega|$ for every $g \in G$. Show that the function $\frac{|\Omega|}{\pi}$ is a generalized character of G.
(*Hint*: Use Problem 1.3(d).)

(1.14) Suppose that $N \le H \le G$, where $N \trianglelefteq G$. Let $\theta \in \mathrm{Irr}(N)$ and let $\chi \in \mathrm{Irr}(G|\theta)$. Let $T = G_\theta$, and let $\psi \in \mathrm{Irr}(T)$ be the Clifford correspondent of χ over θ.

(a) Show that

$$\chi_H = (\psi_{T \cap H})^H + \Delta,$$

where Δ is a character of H (or zero) having no irreducible constituent over θ.

(b) If $\eta \in \mathrm{Irr}(H|\theta)$ and $\nu \in \mathrm{Irr}(T \cap H|\theta)$ is the Clifford correspondent of η over θ, show that

$$[\chi_H, \eta] = [\psi_{T \cap H}, \nu].$$

(1.15) (*G. Higman*) Suppose that G is a finite abelian group. For $\chi \in \text{Irr}(G)$, let $e_\chi = \frac{1}{|G|} \sum_{g \in G} \chi(g^{-1})g$ be the central primitive idempotent.

(a) If $g \in G$, prove that

$$g = \sum_{\chi \in \text{Irr}(G)} \chi(g)e_\chi .$$

(b) If $u \in \mathbb{C}G$ and $u = \sum_{g \in G} a_g g = \sum_{\chi \in \text{Irr}(G)} b_\chi e_\chi$, show that

$$b_\chi = \sum_{g \in G} a_g \chi(g) \quad \text{and} \quad a_g = \frac{1}{|G|} \sum_{\chi \in \text{Irr}(G)} b_\chi \chi(g^{-1}).$$

(c) Now suppose that $u \in \mathbb{Z}G$ and that $u^m = 1$ for some positive integer m. Show that $u = \pm g$, for some $g \in G$.
(*Hint:* Show that $b_\chi^m = 1$. If $a_g \neq 0$, use the triangle inequality to deduce that $b_\chi \chi(g^{-1}) = b_{1_G}$ for all $\chi \in \text{Irr}(G)$, and $|a_g| = |b_{1_G}| = 1$.)

2

Action on Characters by Automorphisms

If G is a finite group and A is any group of automorphisms of G, then A acts naturally on the characters of G. In the case where the orders of A and G are coprime, something fundamental happens. The aim of this chapter is to construct the Glauberman correspondence.

2.1 Actions by Automorphisms

If $a: G \to H$ is a group homomorphism, we shall often denote by $g^a \in H$ the image of $g \in G$ under a. Since we shall be using it rather frequently, let us record the following trivial fact.

Theorem 2.1 *Suppose that $a : G \to H$ is an isomorphism of finite groups. For $\chi \in \mathrm{Char}(G)$, we define $\chi^a: H \to \mathbb{C}$ by $\chi^a(h) = \chi(h^{a^{-1}})$ for $h \in H$. Then the map $\chi \mapsto \chi^a$ is a bijection from $\mathrm{Char}(G)$ to $\mathrm{Char}(H)$ which sends $\mathrm{Irr}(G)$ onto $\mathrm{Irr}(H)$. Also, $[\alpha^a, \beta^a] = [\alpha, \beta]$ for $\alpha, \beta \in \mathrm{Char}(G)$.*

Proof If $\mathcal{X}$ is a representation of G affording χ, then $\mathcal{X}^a$ defined by

$$\mathcal{X}^a(h) = \mathcal{X}(h^{a^{-1}})$$

is a representation of H affording the character χ^a. It is trivial to check that $(\chi^a)^{a^{-1}} = \chi$ and $(\psi^{a^{-1}})^a = \psi$, from which it follows that the map $\chi \mapsto \chi^a$ is bijective. Now, one can easily check that

$$[\alpha^a, \beta^a] = [\alpha, \beta]$$

for $\alpha, \beta \in \mathrm{Char}(G)$, and this implies that $\mathrm{Irr}(G)$ is sent onto $\mathrm{Irr}(H)$. $\square$

The more general situation where a is simply a group homomorphism, which also includes Theorem 1.11, is discussed in Problem 2.1.

If $K \leq G$ and $g \in G$, then conjugation by g is an isomorphism $K \rightarrow K^g$. If $\alpha \in \text{Char}(K)$ then α^g is the character of K^g such that $\alpha^g(k^g) = \alpha(k)$ for $k \in K$. This is a particular case of Theorem 2.1. (See also Problem 1.7.)

If G is a finite group, then we denote by $\text{Aut}(G)$ the group of automorphisms of G. (In this book, when we compose maps, ab means first a and then b. Hence if $a, b \in \text{Aut}(G)$ and $g \in G$, notice that g^{ab} means $b(a(g))$.) It follows from its definition that

$$(\chi^a)^b = \chi^{ab}$$

for $\chi \in \text{Irr}(G)$ and $a, b \in \text{Aut}(G)$. Since the identity automorphism fixes all the irreducible characters, we have that the group $\text{Aut}(G)$ (and therefore any of its subgroups) acts on the set $\text{Irr}(G)$. In fact, we shall consider a slightly more general setting.

One of the most basic and important situations in group theory is when a group A *acts by automorphisms* on another group G. This is based on the following. Suppose that we have $N \trianglelefteq G$ and $H \leq G$. Then, by using the conjugation formula

$$n^h = h^{-1}nh,$$

we have that $(nm)^h = n^h m^h$ and $(n^h)^k = n^{hk}$ for all $n, m \in N$, $h, k \in H$. In other words, if $\alpha_h \colon N \rightarrow N$ is the conjugation map defined by $\alpha_h(n) = n^h$, then the map

$$\alpha \colon H \rightarrow \text{Aut}(N)$$

given by $h \mapsto \alpha_h$ is a group homomorphism.

We say that a group A **acts by automorphisms** on a group G if there is a group homomorphism

$$\alpha \colon A \rightarrow \text{Aut}(G).$$

To simplify notation, we simply write

$$\alpha(a)(g) = g^a.$$

(This notation *forgets* α, but, as we shall see, once α is given, there is no need to remember it again.) Using this notation, the familiar equations $(gh)^a = g^a h^a$ and $(g^a)^b = g^{ab}$ hold for all $g, h \in G$, $a, b \in A$.

There is an equivalent way of defining action by automorphisms, if we do not wish to mention α at all. Suppose that A acts on G as a set. That is, for every $g \in G$ and $a \in A$, we have a uniquely defined $g^a \in G$ such $g^1 = g$ and $(g^a)^b = g^{ab}$ for $g \in G$, and $a, b \in A$. Then A acts by automorphisms on G if we also have that $(gh)^a = g^a h^a$ for $g, h \in G$ and $a \in A$.

If $\alpha : A \to \text{Aut}(G)$ is a group homomorphism, then the subgroup

$$\ker(\alpha) = \{a \in A \mid g^a = g \text{ for all } g \in G\} = \mathbf{C}_A(G)$$

is called the **kernel** of the action. Notice that $\mathbf{C}_A(G) \trianglelefteq A$, and that $A/\mathbf{C}_A(G)$ again acts on G. If $A = \mathbf{C}_A(G)$, then we say that the action is **trivial**. If $\mathbf{C}_A(G) = 1$, then we say that the action is **faithful**. In this latter case, A is isomorphic to a subgroup of $\text{Aut}(G)$.

Let us consider some examples.

Example 2.2 (i) If $A \leq \text{Aut}(G)$, then we have that A acts on G by automorphisms by setting

$$g^a = a(g)$$

for $a \in A$ and $g \in G$.

(ii) If $G = (\mathbb{Z}/p\mathbb{Z})^n$, where p is a prime, then it is well known that $\text{Aut}(G) = \text{GL}_n(p)$. If $A \leq \text{GL}_n(p)$, then A acts on G with

$$v^a = va$$

for $v \in G$ and $a \in A$. For instance, if $G = (\mathbb{Z}/3\mathbb{Z})^2$,

$$a = \begin{pmatrix} 1 & 0 \\ 0 & -1 \end{pmatrix} \quad \text{and} \quad b = \begin{pmatrix} 1 & 1 \\ 0 & 1 \end{pmatrix},$$

then $A = \langle a, b \rangle = S_3$, the symmetric group, acts on G via

$$(x, y)^a = (x, -y) \quad \text{and} \quad (x, y)^b = (x, x + y).$$

Whenever we have that A acts by automorphisms on G, then we can place G and A inside a group Γ that contains G and A in such a way that $G \trianglelefteq \Gamma$, $A \leq \Gamma$, and the action given is simply the conjugation action inside Γ. To see this we need to remember about the **semidirect product** of G and A. This is the group

$$\Gamma = \{(g, a) \mid g \in G, a \in A\}$$

with multiplication

$$(g, a)(h, b) = (gh^{a^{-1}}, ab).$$

If we identify G with $G \times 1$, and A with $1 \times A$, then $G \trianglelefteq \Gamma$, $A \leq \Gamma$, $\Gamma = GA$ and $G \cap A = 1$.

When A acts on G by automorphisms, then A permutes the group theoretical objects of G in a natural way. For instance, if $H \leq G$ and $a \in A$, then $H^a = \{h^a \mid h \in H\}$ is another subgroup of G. If K is the conjugacy class of $x \in G$, then K^a is the conjugacy class of $x^a \in G$. If $P \in \mathrm{Syl}_p(G)$, then $P^a \in \mathrm{Syl}_p(G)$, etc. If χ is a complex character of G afforded by the complex representation $\mathcal{X}$, then we already know by Theorem 2.1 that $\mathcal{X}^a$ defined by $\mathcal{X}^a(g^a) = \mathcal{X}(g)$ is another representation of G that affords the character χ^a, where

$$\chi^a(g^a) = \chi(g).$$

Also,

$$[\chi^a, \psi^a] = [\chi, \psi]$$

for characters χ, ψ of G, and $\chi^a \in \mathrm{Irr}(G)$ if and only if $\chi \in \mathrm{Irr}(G)$.

When we have an action of A on G by automorphisms, it is interesting to study the "A-structure of G" that is, the group theoretical objects of G that A *can see*. In other words, we are interested in the A-invariant objects of G. For instance, the elements of G that A *sees* form the **fixed point subgroup**

$$\mathbf{C}_G(A) = \{g \in G \mid g^a = g \text{ for all } a \in A\} \leq G.$$

A subgroup $H \leq G$ is A-**invariant** if $H^a = H$ for all $a \in A$ and, of course, A acts on H in this case. For instance, if H is characteristic in G, then H is automatically A-invariant. Also, if $N \trianglelefteq G$ is A-invariant, then A also acts by automorphisms on G/N by letting

$$(Ng)^a = Ng^a$$

for $g \in G$ and $a \in A$. The conjugacy classes of G that A *sees* are the A-invariant conjugacy classes, and we shall denote them by

$$\mathrm{Cl}_A(G) = \{K \in \mathrm{Cl}(G) \mid K^a = K \text{ for all } a \in A\}.$$

The A-invariant irreducible complex characters are denoted by

$$\mathrm{Irr}_A(G) = \{\chi \in \mathrm{Irr}(G) \mid \chi^a = \chi \text{ for all } a \in A\}.$$

As we already know, the fundamental theorem in character theory tells us that

$$|\mathrm{Irr}(G)| = |\mathrm{Cl}(G)|$$

for every finite group G. If now we have an action of A on G, then it is natural to ask: is it true that

$$|\mathrm{Irr}_A(G)| = |\mathrm{Cl}_A(G)|?$$

Let us start with the easy case where G is abelian. In this case, it is trivial to check that

$$\mathrm{Irr}_A(G) = \mathrm{Irr}(G/[G, A]),$$

where $[G, A]$ is the subgroup of G generated by $g^{-1}g^a$ for $g \in G$, and $a \in A$. Also, if G is abelian then it is clear that $|\mathrm{Cl}_A(G)| = |\mathbf{C}_G(A)|$. Hence, in the case where G is abelian, our question can be reformulated: is it true that

$$|G/[G, A]| = |\mathbf{C}_G(A)|?$$

The answer is, in general, negative. For instance, Example 2.2(ii) above is a counterexample.

2.2 Brauer's Lemma on Character Tables

There are two cases where the equality

$$|\mathrm{Irr}_A(G)| = |\mathrm{Cl}_A(G)|$$

always holds. If A is cyclic, this will follow from *Brauer's Lemma on Character Tables*, whose proof we shall explore next. If $\gcd(|A|, |G|) = 1$ and A is solvable, then the equality will follow from a fundamental result in character theory: the Glauberman correspondence.

Theorem 2.3 (Brauer's Lemma on Character Tables) *Let A and G be finite groups, and suppose that A acts on the sets $\mathrm{Irr}(G)$ and on $\mathrm{Cl}(G)$. Assume that*

$$\chi^a(g^a) = \chi(g)$$

for $\chi \in \mathrm{Irr}(G)$, $a \in A$ and $g \in G$, where $g^a \in K^a$, and K is the conjugacy class of g. If $a \in A$, then a fixes the same number of irreducible characters of G as conjugacy classes of G.

Proof Write $\mathrm{Irr}(G) = \{\chi_1, \ldots, \chi_k\}$, $\mathrm{Cl}(G) = \{K_1, \ldots, K_k\}$ and $K_i = \mathrm{cl}_G(g_i)$, where $g_i \in G$. Let $X = (\chi_i(g_j))$ be the character table of G, which we know is an invertible $k \times k$ matrix.

Let $a \in A$. Let $P = (p_{ij})$ be the $k \times k$ matrix defined by $p_{ij} = 0$ if $(\chi_i)^a \neq \chi_j$, and $p_{ij} = 1$, otherwise. Also, let $Q = (q_{ij})$, where $q_{ij} = 0$ if $(K_i)^a \neq K_j$ and $q_{ij} = 1$, otherwise. We claim that $PX = XQ$. The (i, j)-entry of PX is

$$\sum_{t=1}^{k} p_{it}\chi_t(g_j) = (\chi_i)^a(g_j).$$

The (i, j)-entry of XQ is

$$\sum_{t=1}^{k} \chi_i(g_t)q_{tj} = \chi_i((g_j)^{a^{-1}}).$$

By hypothesis, $(\chi_i)^a(g_j) = \chi_i((g_j)^{a^{-1}})$, and this proves the claim. Now, $P = XQX^{-1}$ and by taking traces, we get the result. $\qquad\qquad\qquad$ $\square$

We shall use (several times in this chapter) that if a p-group P acts on a set Ω, then

$$|\Omega| \equiv |\Omega_0| \bmod p,$$

where Ω_0 is the set of P-fixed points of Ω. This easily follows from the orbit-stabilizer theorem. In particular, if $|\Omega|$ is not divisible by p, then Ω_0 is not empty.

Another elementary fact of group theory is used in this book: if p is a prime, then every element x of a finite group can be written in a unique way as $x = x_p x_{p'}$, where x_p has order a power of p, $x_{p'}$ has order not divisible by p, and x_p and $x_{p'}$ commute. Furthermore, x_p and $x_{p'}$ are powers of x. We say that x_p is the p-**part** of x and $x_{p'}$ is the p'-**part** of x.

The two facts mentioned above are used in the proof of the following classical application of Brauer's lemma on character tables.

Theorem 2.4 *Suppose that A is a finite group that acts by automorphisms on a finite group G. Assume that A fixes every $\chi \in \mathrm{Irr}(G)$. If $\gcd(|A|, |G|) = 1$, then A acts trivially on G.*

Proof We argue by induction on $|A|$. Since we wish to prove that every element of A acts trivially on G, then we may assume that $A = \langle a \rangle$ is cyclic. If p divides $o(a)$, then it suffices to show that a_p acts trivially on G. Therefore, we may assume that A is a cyclic p-group. By Theorem 2.3, we have that A fixes every conjugacy class K of G. Let $C = \mathbf{C}_G(A)$. Since $|K|$ divides $|G|$, then p does not divide $|K|$ by hypothesis, and we have that $K \cap C \neq \emptyset$ by the orbit-stabilizer theorem. Then

$$G = \bigcup_{g \in G} C^g.$$

By elementary group theory, we have that $C = G$, as desired. $\qquad\qquad$ $\square$

Of course, the hypothesis that $\gcd(|A|, |G|) = 1$ is necessary in Theorem 2.4, as shown by any non-abelian group acting on itself by conjugation.

2.3 The Glauberman Correspondence

We start our approach to the Glauberman correspondence with a group theoretical lemma.

Lemma 2.5 *Suppose that A is a p-group that acts by automorphisms on a group G of order not divisible by p. Let $C = \mathbf{C}_G(A)$. Then the map $K \mapsto K \cap C$ is a well-defined bijection between $\mathrm{Cl}_A(G)$ and $\mathrm{Cl}(C)$.*

Proof We work in $\Gamma = GA$, the semidirect product. Thus $A \in \mathrm{Syl}_p(\Gamma)$. Suppose that $K = \mathrm{cl}_G(g)$ is A-invariant. Then A acts on the set K. Since $|K|$ is not divisible by p, then A fixes an element in K. Thus, we may assume that $g \in C$. Now, suppose that $g^h \in K \cap C$ for some $h \in G$. Then $A \subseteq \mathbf{C}_\Gamma(g^h)$ and $A, A^{h^{-1}}$ are Sylow p-subgroups of $\mathbf{C}_\Gamma(g)$. Thus there exists $x \in \mathbf{C}_\Gamma(g) = A\mathbf{C}_G(g)$ such that $A^{h^{-1}} = A^x$. Writing $x = ay$ for $y \in \mathbf{C}_G(g)$, we have that $c = yh \in \mathbf{N}_G(A)$. However, $[\mathbf{N}_G(A), A] \subseteq G \cap A = 1$, and thus $\mathbf{N}_G(A) = C$. Now, $g^c = g^h$, and this proves that $K \cap C$ is a conjugacy class of C. Of course, the map $K \mapsto K \cap C$ is one-to-one, since distinct conjugacy classes are disjoint. The map is onto since given $c \in C$, then $K = \mathrm{cl}_G(c)$ is A-invariant, and necessarily $K \cap C = \mathrm{cl}_C(c)$ by the first part. $\qquad\square$

Lemma 2.5 is also true only assuming that $\gcd(|G|, |A|) = 1$. In this case, we have to use the Schur–Zassenhaus theorem instead of Sylow theory. In this book, however, our main interest is when the group acting is a p-group.

We see from Lemma 2.5 that there is a canonical bijection between $\mathrm{Cl}_A(G)$ and $\mathrm{Cl}(C)$. The Glauberman correspondence establishes that there is a canonical bijection between $\mathrm{Irr}_A(G)$ and $\mathrm{Irr}(C)$. It is an amazing fact that many results on characters have analogs on conjugacy classes and vice versa.

We will construct the Glauberman correspondence by using some algebra homomorphisms. The following lemma is standard.

Lemma 2.6 *Let $\mathcal{A}$ be an F-algebra and let $\omega_1, \dots, \omega_k$ be distinct algebra homomorphisms $\mathcal{A} \to F$. Then $\omega_1, \dots, \omega_k$ are F-linearly independent (in the dual space $\mathcal{A}^*$). In particular, $k \le \dim_F(\mathcal{A})$.*

Proof We prove it by induction on k and certainly we may assume that $k > 1$, using that $\omega_i(1) = 1$. Now, suppose that

$$\sum_{i=1}^{k} \alpha_i \omega_i = 0$$

for some $\alpha_i \in F$. By induction, we may assume that $\alpha_i \ne 0$ for all i. If $a, b \in \mathcal{A}$, notice that

$$\sum_{i=1}^{k} \alpha_i \omega_i(a) \omega_i(b) = \sum_{i=1}^{k} \alpha_i \omega_i(ab) = 0.$$

Also,

$$\left(\sum_{i=1}^{k} \alpha_i \omega_i(a)\right) \omega_k(b) = 0.$$

Hence, we deduce that

$$\sum_{i=1}^{k-1} \alpha_i(\omega_i(b) - \omega_k(b))\omega_i(a) = 0$$

for all $a, b \in A$. Thus

$$\sum_{i=1}^{k-1} \alpha_i(\omega_i(b) - \omega_k(b))\omega_i = 0,$$

for all $b \in A$. By induction, $\alpha_1(\omega_1(b) - \omega_k(b)) = 0$ for all $b \in A$, and then $\omega_1 = \omega_k$, a contradiction. The last part follows since $\dim(A) = \dim(A^*)$. $\square$

In this book we will use many times the following. Let p be a prime and let M be any maximal ideal of $\mathbf{R}$ (the ring of algebraic integers in $\mathbb{C}$) containing $p\mathbf{R}$. Write $F = \mathbf{R}/M$, a field of characteristic p, and let $*: \mathbf{R} \to F$ be the natural ring homomorphism. Notice that if $n \in \mathbb{Z}$, then $n^* = 0$ if and only if n is divisible by p. (If $n \in M$ is coprime to p, then $1 = up + vn$ for some integers u and v, and $1 \in M$, which is not possible.)

If $a, b \in \mathbf{R}$, then we write $a \equiv b$ mod M if and only if $a^* = b^*$, which happens if and only if $a - b \in M$. If $a, b \in \mathbb{Z}$, then notice that $a \equiv b$ mod M if and only if $a \equiv b$ mod p.

If $\chi \in \mathrm{Irr}(G)$ and K is a conjugacy class of G, we know by Theorem 1.8 that

$$\omega_\chi(\hat{K}) = \frac{|K|\chi(x_K)}{\chi(1)} \in \mathbf{R}$$

for any $x_K \in K$, and that ω_χ is an algebra homomorphism $\mathbf{Z}(\mathbb{C}G) \to \mathbb{C}$. It is then clear that

$$\lambda_\chi(\hat{K}) = \omega_\chi(\hat{K})^*$$

defines an algebra homomorphism

$$\lambda_\chi: \mathbf{Z}(FG) \to F,$$

by extending F-linearly. (Recall that the set of class sums $\hat{K} = \sum_{x \in K} x$ forms an F-basis of $\mathbf{Z}(FG)$.)

If p does not divide the order of G and $\chi \in \mathrm{Irr}(G)$, then we show next that λ_χ uniquely determines χ.

Lemma 2.7 *Suppose that p does not divide $|G|$. Then the maps λ_χ with $\chi \in$ Irr(G) are distinct and they are all the algebra homomorphisms $\mathbf{Z}(FG) \to F$.*

Proof Since the F-dimension of $\mathbf{Z}(FG)$ is $|\mathrm{Cl}(G)| = |\mathrm{Irr}(G)|$, by Lemma 2.6 it suffices to show that the λ_χ are distinct. Now, for $\psi \in \mathrm{Irr}(G)$, let

$$f_\psi^* = \sum_{K \in \mathrm{Cl}(G)} \psi(x_K^{-1})^* \hat{K} \in \mathbf{Z}(FG).$$

Now we use the first orthogonality relation of characters. Let $\chi, \psi \in \mathrm{Irr}(G)$. Hence

$$|G|\delta_{\chi,\psi} = \sum_{K \in \mathrm{Cl}(G)} |K|\chi(x_K)\psi(x_K^{-1}) = \chi(1) \sum_{K \in \mathrm{Cl}(G)} \omega_\chi(\hat{K})\psi(x_K^{-1}).$$

Then

$$|G|^* \delta_{\chi,\psi} = \chi(1)^* \sum_{K \in \mathrm{Cl}(G)} \omega_\chi(\hat{K})^* \psi(x_K^{-1})^* = \chi(1)^* \lambda_\chi(f_\psi^*).$$

Using that $|G|$ and $\chi(1)$ are not divisible by p (by Corollary 1.9), we have that $\lambda_\chi(f_\chi^*) \neq 0$ and $\lambda_\psi(f_\chi^*) = 0$ for $\psi \neq \chi$. This proves the lemma. □

The following lemma is the key to our construction of the Glauberman correspondence.

Lemma 2.8 *Suppose that a p-group A acts by automorphisms on a finite group G of order not divisible by p. Let $C = \mathbf{C}_G(A)$. For $\chi \in \mathrm{Irr}_A(G)$, we define an F-linear map $\delta_\chi : \mathbf{Z}(FC) \to F$ by setting*

$$\delta_\chi(\widehat{K \cap C}) = \lambda_\chi(\hat{K}),$$

for every A-invariant class K of G, and extending linearly. Then δ_χ is an algebra homomorphism.

Proof Notice that $\delta_\chi(1) = 1$. By Lemma 2.5, the map

$$K \mapsto K \cap C$$

is a bijection between the A-invariant conjugacy classes of G and the conjugacy classes of C. It is enough then to show that δ_χ is multiplicative on a basis of $\mathbf{Z}(FC)$. Since $\lambda_\chi(\hat{K}\hat{L}) = \lambda_\chi(\hat{K})\lambda_\chi(\hat{L})$, it suffices to show that

$$\delta_\chi(\widehat{K \cap C}\,\widehat{L \cap C}) = \lambda_\chi(\hat{K}\hat{L})$$

for every pair of A-invariant classes K and L of G.

Now, A acts on the set $\mathrm{Cl}(G)$, and we can write

$$\mathrm{Cl}(G) - \mathrm{Cl}_A(G) = \mathcal{O}(K_1) \cup \cdots \cup \mathcal{O}(K_s),$$

where $\mathcal{O}(K_i)$ is the A-orbit of the conjugacy class K_i, which has length $p^{b_i} > 1$, and $\mathrm{Cl}_A(G)$ is the set of A-invariant classes of G.

We know that if K and L are conjugacy classes of G, working in $\mathbf{Z}(FG)$, we have that

$$\hat{K}\hat{L} = \sum_{M \in \mathrm{Cl}(G)} (a_{KLM})^* \hat{M}$$

where $a_{KLM} = |\{(x, y) \in K \times L \mid xy = x_M\}|$ and x_M is a fixed (but arbitrary) element of M. Notice that $a_{KLM} = a_{K^a L^a M^a}$ for any $a \in A$. Suppose now that K and L are A-invariant conjugacy classes of G. Then, using that $a_{KLM} = a_{KLM^a}$, we may write

$$\hat{K}\hat{L} = \sum_{M \in \mathrm{Cl}_A(G)} (a_{KLM})^* \hat{M} + \sum_{j=1}^{s} c_j \left(\sum_{M \in \mathcal{O}(K_j)} \hat{M} \right)$$

for some $c_j \in F$.

Suppose now that $M \in \mathrm{Cl}_A(G)$ and fix $x_M \in M \cap C$. Since A naturally acts on the set

$$\{(x, y) \in K \times L \mid xy = x_M\}$$

with fixed points

$$\{(x, y) \in (K \cap C) \times (L \cap C) \mid xy = x_M\},$$

it follows that

$$a_{KLM} = |\{(x, y) \in K \times L \mid xy = x_M\}|$$
$$\equiv |\{(x, y) \in (K \cap C) \times (L \cap C) \mid xy = x_M\}| = a_{K \cap C L \cap C M \cap C} \pmod{p}.$$

Hence

$$\widehat{K \cap C}\,\widehat{L \cap C} = \sum_{M \in \mathrm{Cl}_A(G)} (a_{KLM})^* \widehat{M \cap C}$$

in $\mathbf{Z}(FC)$. Now, since λ_χ is constant on each A-orbit (because χ is A-invariant), we have that

$$\lambda_\chi(\hat{K}\hat{L}) = \sum_{M \in \mathrm{Cl}_A(G)} (a_{KLM})^* \lambda_\chi(\hat{M}) + \sum_{j=1}^{s} c_j \lambda_\chi \left(\sum_{M \in \mathcal{O}(K_j)} \hat{M} \right)$$

$$= \sum_{M \in \mathrm{Cl}_A(G)} (a_{KLM})^* \lambda_\chi(\hat{M}) + \sum_{j=1}^{s} c_j p^{b_j} \lambda_\chi(\hat{K}_j)$$

$$= \sum_{M \in \mathrm{Cl}_A(G)} (a_{KLM})^* \lambda_\chi(\hat{M})$$

$$= \sum_{M \in \mathrm{Cl}_A(G)} (a_{KLM})^* \delta_\chi(\widehat{M \cap C}) = \delta_\chi \left(\sum_{M \in \mathrm{Cl}_A(G)} (a_{KLM})^* \widehat{M \cap C} \right)$$

$$= \delta_\chi(\widehat{K \cap C} \widehat{L \cap C}),$$

as required. $\qquad\square$

Now, we are ready to prove the existence of the Glauberman correspondence. Assume again that A is a p-group that acts by automorphisms on a finite group G of order not divisible by p, and let $C = \mathbf{C}_G(A)$. If $\chi \in \mathrm{Irr}_A(G)$, since δ_χ is an algebra homomorphism $\mathbf{Z}(FC) \to F$ by Lemma 2.8, then we know by Lemma 2.7 that there is a unique $\chi' \in \mathrm{Irr}(C)$ such that $\delta_\chi = \lambda_{\chi'}$. Hence, for every A-invariant conjugacy class K in G, we have that

$$\lambda_\chi(\hat{K}) = \lambda_{\chi'}(\widehat{K \cap C}).$$

In other words, for $x \in C$, we have

$$\left(\frac{\chi(x)|K|}{\chi(1)} \right)^* = \left(\frac{\chi'(x)|K \cap C|}{\chi'(1)} \right)^*$$

where $K = \mathrm{cl}_G(x)$ is the class of x in G. Notice that $|K| \equiv |K \cap C| \not\equiv 0 \bmod p$, since, again, A acts on K with fixed points $K \cap C$. Then, by multiplying by $\chi(1)^* \chi'(1)^*$ on both sides and using that $*$ is a ring homomorphism, we obtain

$$\chi'(1)\chi(x) \equiv \chi(1)\chi'(x) \bmod M$$

for all $x \in C$.

Theorem 2.9 (The Glauberman correspondence) *Suppose that A is a p-group that acts by automorphisms on a finite group G of order not divisible by p. Let $C = \mathbf{C}_G(A)$. Then the map $\mathrm{Irr}_A(G) \to \mathrm{Irr}(C)$ defined by $\chi \mapsto \chi'$ is a bijection. Furthermore, $[\chi_C, \chi'] \equiv \pm 1 \bmod p$ and $[\chi_C, \theta] \equiv 0 \bmod p$ for $\chi' \neq \theta \in \mathrm{Irr}(C)$.*

Proof Let $\chi \in \mathrm{Irr}_A(G)$ and let $\theta \in \mathrm{Irr}(C)$. Then

$$\chi'(1)|C|[\chi_C, \theta] = \chi'(1) \sum_{x \in C} \chi(x)\theta(x^{-1})$$

$$\equiv \chi(1) \sum_{x \in C} \chi'(x)\theta(x^{-1}) = \chi(1)|C|[\chi', \theta] \bmod M.$$

Since p does not divide $|C|$, this implies that

$$\chi'(1)[\chi_C, \theta] \equiv \chi(1)[\chi', \theta] \bmod p.$$

Notice that since p does not divide $\chi'(1)$, for $\theta \neq \chi'$ we have that

$$[\chi_C, \theta] \equiv 0 \bmod p.$$

Also, if $\theta = \chi'$, we conclude that

$$\chi'(1)[\chi_C, \chi'] \equiv \chi(1) \not\equiv 0 \bmod p.$$

Hence, we see that χ' is the unique irreducible constituent of χ_C with multiplicity not divisible by p.

Now, let $\chi, \varphi \in \mathrm{Irr}_A(G)$. For each A-invariant class K of G, we choose $x_K \in K \cap C$. Then

$$
\begin{aligned}
|G|[\chi, \varphi] = \sum_{x \in G} \chi(x)\varphi(x^{-1}) &= \sum_{K \in \mathrm{Cl}(G)} |K|\chi(x_K)\varphi(x_K^{-1}) \\
&\equiv \sum_{K \in \mathrm{Cl}_A(G)} |K|\chi(x_K)\varphi(x_K^{-1}) \equiv \sum_{K \in \mathrm{Cl}_A(G)} |K \cap C|\chi(x_K)\varphi(x_K^{-1}) \\
&= \sum_{x \in C} \chi(x)\varphi(x^{-1}) = |C|[\chi_C, \varphi_C] \bmod M,
\end{aligned}
$$

where in the first congruence we are using that χ and φ are A-invariant, and therefore constant on the A-orbits of conjugacy classes of G, and in the second that $|K| \equiv |K \cap C| \bmod p$. Furthermore, since $|G| \equiv |C| \bmod p$, we deduce that

$$[\chi, \varphi] \equiv [\chi_C, \varphi_C] \bmod p.$$

Write

$$\chi_C = [\chi_C, \chi']\chi' + p\Delta \quad \text{and} \quad \varphi_C = [\varphi_C, \varphi']\varphi' + p\Xi,$$

where $[\chi_C, \chi']$ and $[\varphi_C, \varphi']$ are not divisible by p, and Δ and Ξ are characters of C or zero. Then we have that

$$[\chi, \varphi] \equiv [\chi_C, \chi'][\varphi_C, \varphi'][\chi', \varphi'] \bmod p. \tag{1}$$

This easily proves that our map $\chi \to \chi'$ is injective.

On the other hand, if $\alpha \in \mathrm{Irr}(C)$ is not in the image of our map, then $[\chi_C, \alpha] = [\chi, \alpha^G]$ is divisible by p for every $\chi \in \mathrm{Irr}_A(G)$. Also, by the induction formula, notice that

$$(\alpha^G)^a = (\alpha^a)^G = \alpha^G.$$

Then we have that $[\alpha^G, \mu] = [\alpha^G, \mu^a]$ for all $a \in A$ and $\mu \in \mathrm{Irr}(G)$. Hence, if we write α^G as linear combination of $\mathrm{Irr}(G)$ and evaluate in 1, we deduce that p divides $\alpha^G(1) = |G : C|\alpha(1)$. This is not possible.

Finally, taking $\chi = \varphi$ in (1), we obtain

$$1 \equiv [\chi_C, \chi']^2 \bmod p$$

and then

$$[\chi_C, \chi'] \equiv \pm 1 \bmod p,$$

as required.   □

We see that if A is a p-group acting on a group G of order not divisible by p, then it is very easy to determine the Glauberman correspondent of $\chi \in \mathrm{Irr}_A(G)$: $\chi' \in \mathrm{Irr}(C)$ is simply the unique irreducible constituent of χ_C with multiplicity not divisible by p. This uniqueness is very important in later applications.

Lemma 2.10 *Suppose that $G \trianglelefteq GA$, where G has order not divisible by p, and A is a p-subgroup of GA. Let $': \mathrm{Irr}_A(G) \to \mathrm{Irr}(C)$ be the A-Glauberman correspondence, where $C = \mathbf{C}_G(A)$. Suppose that X acts by automorphisms on GA fixing A setwise. Then X acts on $\mathrm{Irr}_A(G)$ and if $x \in X$, then*

$$(\chi^x)' = (\chi')^x$$

for $\chi \in \mathrm{Irr}_A(G)$.

Proof By elementary group theory, we have that G is invariant under any automorphism of GA. Notice that X permutes $\mathrm{Irr}_A(G)$. Indeed, if $\chi \in \mathrm{Irr}(G)$ is A-invariant and $x \in X$, then $(\chi^x)^a = \chi^x$ because $xax^{-1} \in A$ for $a \in A$. Also, $C^x = C$. Write

$$\chi_C = e\chi' + p\Delta,$$

where e is not divisible by p and Δ is a character of C or zero. Then

$$(\chi^x)_C = (\chi_C)^x = e(\chi')^x + p\Delta^x.$$

Now, $(\chi')^x$ is the unique irreducible constituent of $(\chi^x)_C$ with multiplicity not divisible by p, and then $(\chi^x)' = (\chi')^x$.   □

Theorem 2.11 *If A is a solvable group acting coprimely on G by automorphisms, then*

$$|\mathrm{Irr}_A(G)| = |\mathrm{Irr}(\mathbf{C}_G(A))|.$$

Proof We argue by induction on $|A|$. Since A is solvable, we may find $B \trianglelefteq A$ a nontrivial normal p-subgroup. We have therefore defined the B-Glauberman correspondence $': \mathrm{Irr}_B(G) \to \mathrm{Irr}(\mathbf{C}_G(B))$. Since A acts on GB, by Lemma 2.10, notice that this correspondence commutes with the action of A. Therefore it maps $\mathrm{Irr}_A(G)$ onto $\mathrm{Irr}_A(\mathbf{C}_G(B))$. In particular, $|\mathrm{Irr}_A(G)| = |\mathrm{Irr}_A(\mathbf{C}_G(B))|$. Now, A/B acts coprimely on $\mathbf{C}_G(B)$ and by induction, we have that

$$|\mathrm{Irr}_{A/B}(\mathbf{C}_G(B))| = |\mathrm{Irr}(\mathbf{C}_{\mathbf{C}_G(B)}(A/B))|.$$

Since

$$\mathrm{Irr}_{A/B}(\mathbf{C}_G(B)) = \mathrm{Irr}_A(\mathbf{C}_G(B))$$

and $\mathbf{C}_{\mathbf{C}_G(B)}(A/B) = \mathbf{C}_G(A)$, the theorem is proven. $\square$

If a group A acts on two sets Ω_1 and Ω_2, recall that the actions are **permutation isomorphic** if there exists a bijection $\alpha \colon \Omega_1 \to \Omega_2$ such that $\alpha(\omega \cdot a) = \alpha(\omega) \cdot a$ for $\omega \in \Omega_1$ and $a \in A$. In this case, we say that the bijection α is A-**equivariant**. It is well known that this happens if and only if, for every subgroup B of A, the number of fixed points of B on Ω_1 equals the number on Ω_2. (See for instance Lemma (13.23) of [Is06].)

Corollary 2.12   *Suppose that A acts coprimely on G by automorphisms, and assume that A is solvable. Then A fixes the same number of irreducible characters of G as conjugacy classes of G. In particular, the actions of A on $\mathrm{Irr}(G)$ and $\mathrm{Cl}(G)$ are permutation isomorphic.*

Proof   The first part follows from Theorem 2.11 and Lemma 2.5 (in its general coprime version). The second part now follows from what we have said on permutation isomorphic actions. $\square$

2.4 More Properties of the Glauberman Correspondence

We shall need the following properties of the Glauberman correspondence in Chapter 8.

Theorem 2.13   *Suppose that A is a p-group that acts by automorphisms on a finite group G of order not divisible by p. Assume that $N \trianglelefteq G$ is A-invariant and write $C = \mathbf{C}_G(A)$. Let $\theta \in \mathrm{Irr}_A(N)$ and $\chi \in \mathrm{Irr}_A(G)$, and let $\theta' \in \mathrm{Irr}(C \cap N)$ and $\chi' \in \mathrm{Irr}(C)$ be the Glauberman correspondents of θ and χ, respectively.*

(a) Then χ lies over θ if and only if χ' lies over θ'.
(b) If G/N is abelian and $G = NC$, then $\chi_N = \theta$ if and only if $\chi'_{C \cap N} = \theta'$.

Proof Since χ_N is A-invariant, we have that A acts on the set $\operatorname{Irr}(\chi_N)$. Let $\mathcal{O}_1, \ldots, \mathcal{O}_t$ be the orbits of this action. By Clifford's theorem, we can write

$$\chi_N = e\left(\sum_{\gamma \in \mathcal{O}_1} \gamma + \cdots + \sum_{\gamma \in \mathcal{O}_t} \gamma\right),$$

where e is not divisible by p. (This last part follows since e divides $\chi(1)$ and p does not divide $|G|$.) Since $\gamma_{C \cap N} = (\gamma^a)_{C \cap N}$ for $a \in A$ and $\gamma \in \operatorname{Irr}(N)$, we have that

$$[\chi_{C \cap N}, \theta'] \equiv e \sum_{\substack{\gamma \in \operatorname{Irr}_A(N) \\ [\chi_N, \gamma] \neq 0}} [\gamma_{C \cap N}, \theta'] \bmod p.$$

Since $[\gamma_{C \cap N}, \theta'] \equiv 0 \bmod p$ unless $\gamma = \theta$ by Theorem 2.9, we conclude that $[\chi_{C \cap N}, \theta'] \not\equiv 0 \bmod p$ if and only if $[\chi_N, \theta] \neq 0$. Now, again by Theorem 2.9, we can write $\chi_C = f\chi' + p\Delta$, where f is not divisible by p, and Δ is a character of C or zero. Then

$$[\chi_{C \cap N}, \theta'] \equiv f[\chi'_{C \cap N}, \theta'] \bmod p.$$

Hence $[\chi_{C \cap N}, \theta'] \not\equiv 0 \bmod p$ if and only if $[\chi'_{C \cap N}, \theta'] \not\equiv 0 \bmod p$. Since C is a p'-group, using Clifford's theorem we conclude that happens if and only if $[\chi'_{C \cap N}, \theta'] \neq 0$. This proves part (a).

For part (b), assume now that $NC = G$ and G/N is abelian. Since C acts on NA normalizing A, notice that θ is C-invariant if and only of θ' is C-invariant, by Lemma 2.10. So we may assume that both θ and θ' are C-invariant. Next, we claim that every $\chi \in \operatorname{Irr}(G|\theta)$ is A-invariant. Since θ is A-invariant, we have that A acts on $\operatorname{Irr}(G|\theta)$. We also know by Lemma 1.24(b) that $\operatorname{Lin}(G/N)$ acts transitively on $\operatorname{Irr}(G|\theta)$ by multiplication. Thus $|\operatorname{Irr}(G|\theta)|$ divides $|G : N|$, and therefore $|\operatorname{Irr}(G|\theta)|$ is not divisible by p. Since A is a p-group, it follows that A fixes some $\chi \in \operatorname{Irr}(G|\theta)$. Since A acts trivially on G/N, then every $\lambda \in \operatorname{Lin}(G/N)$ is A-invariant, and therefore all characters in $\operatorname{Irr}(G/N)$ are A-invariant. Now, by part (a) we have that $|\operatorname{Irr}(G|\theta)| = |\operatorname{Irr}(C|\theta')|$. By Lemma 1.24(a), we have that θ extends to G if and only if $|\operatorname{Irr}(G|\theta)| = |G : N|$ and θ' extends to C if and only if $|\operatorname{Irr}(C|\theta')| = |C : C \cap N|$. Since $|G : N| = |CN : N| = |C : C \cap N|$, the proof is finished. $\square$

In several parts of this book, we shall use a nearly trivial fact. If A is a p-group acting coprimely on G and $B \trianglelefteq A$ acts trivially on G, then there is no essential difference between the action of A on G and the action of A/B on G. For instance, it is immediate to check that the A-Glauberman correspondence on G coincides with the A/B-Glauberman correspondence.

Theorem 2.14 *Suppose that A is a p-group that acts by automorphisms on a group G of order not divisible by p. Assume that $B \trianglelefteq A$, and let $D = \mathbf{C}_G(B)$ and $C = \mathbf{C}_G(A)$. Let $f_B \colon \mathrm{Irr}_B(G) \to \mathrm{Irr}(D)$ be the B-Glauberman correspondence, let $f_A \colon \mathrm{Irr}_A(G) \to \mathrm{Irr}(C)$ be the A-Glauberman correspondence, and let $f_{A/B} \colon \mathrm{Irr}_{A/B}(D) \to \mathrm{Irr}(\mathbf{C}_D(A/B))$ be the A/B-Glauberman correspondence. Then $f_B(\mathrm{Irr}_A(G)) = \mathrm{Irr}_A(D)$ and $f_A(\chi) = f_{A/B}(f_B(\chi))$ for $\chi \in \mathrm{Irr}_A(G)$.*

Proof We have that A acts on GB normalizing B. By Lemma 2.10, we have that $f_B \colon \mathrm{Irr}_B(G) \to \mathrm{Irr}(D)$ commutes with the action of A, and in particular $f_B(\mathrm{Irr}_A(G)) = \mathrm{Irr}_A(D)$. Suppose now that $\chi \in \mathrm{Irr}_A(G)$ and let $\eta = f_B(\chi) \in \mathrm{Irr}(D)$ be the B-Glauberman correspondent of χ. Thus $\chi_D = e\eta + p\Delta$, where e is not divisible by p and Δ is a character of D or zero, by Theorem 2.9. Now, notice that A acts on D and that B is in the kernel of the action. Also, $\mathbf{C}_D(A/B) = \mathbf{C}_D(A) = \mathbf{C}_G(A) = C$. Let $\eta' = f_{A/B}(\eta)$. By Theorem 2.9, we have $\eta_C = d\eta' + p\Xi$, where p does not divide d and Ξ is a character of C or zero. Then

$$\chi_C = e\eta_C + p\Delta_C = de\eta' + p(e\Xi + \Delta_C).$$

Since p does not divide de, we conclude that $\chi' = \eta'$, by Theorem 2.9. $\qquad\square$

2.5 Notes

G. Glauberman constructed a canonical bijection $\mathrm{Irr}_A(G) \to \mathrm{Irr}(\mathbf{C}_G(A))$ whenever a solvable group A acts coprimely by automorphisms on G in [Gl68]. Using the same ideas as in Theorem 2.11, we can also construct canonical bijections $\mathrm{Irr}_A(G) \to \mathrm{Irr}(\mathbf{C}_G(A))$ by induction. (There are ways to choose canonically the subgroup B in the proof of Theorem 2.11, and this gives a way to construct a canonical map. Glauberman showed, however, that one gets the same map for every choice of B.) As the reader must be aware, we are using the words *natural* or *canonical* to refer to the Glauberman correspondence. We do not attempt to make a formal definition here of what this means, although we understand that a map is canonical or natural if it is independent of any choice that is used to construct it. We shall try to illustrate the importance of the existence of canonical bijections in several parts of this book.

What happens if A is non-solvable? Then, by the Feit–Thompson theorem we have that $|A|$ has even order, and therefore that $|G|$ is odd. In this case, I. M. Isaacs constructed in 1973 a canonical bijection $\mathrm{Irr}_A(G) \to \mathrm{Irr}(\mathbf{C}_G(A))$ (see [Is73]). Isaacs's result is too complicated to present here. Finally, T. R. Wolf proved in his PhD thesis in 1978 that the Glauberman and the Isaacs correspondences coincide whenever A is solvable and $|G|$ is odd (see [Wo78a]). Consequently, there exists a uniquely defined canonical map

$$\mathrm{Irr}_A(G) \to \mathrm{Irr}(\mathbf{C}_G(A))$$

(called the **Glauberman–Isaacs correspondence**) which is defined whenever a group A acts coprimely by automorphisms on a finite group G. As a consequence, if a group A acts coprimely by automorphisms on another group G, then the actions of A on $\mathrm{Irr}(G)$ and $\mathrm{Cl}(G)$ are always permutation isomorphic.

The importance of the Glauberman correspondence in representation theory cannot be overstated. It is fundamental in the proofs of the global–local counting conjectures (which we shall discuss later in this book) for p-solvable groups, and it is used in some parts of the Langlands program. In fact, the reductions to simple groups of these global–local counting conjectures are modeled upon properties of the Glauberman correspondence.

Beside the cases where A is cyclic, or $|A|$ and $|G|$ are coprime, there are no general theorems that guarantee that the actions of A on $\mathrm{Irr}(G)$ and $\mathrm{Cl}(G)$ are permutation isomorphic. As we know, this does not happen in general, even if G and A are abelian (see Problem 2.2). It is not even true if G is simple, as pointed out by Thomas Breuer: the simple group $O_8^+(3)$ is a counterexample.

Not everything is known about the Glauberman correspondence at the time of this writing. For instance, it remains an open problem to show that $\chi'(1)$ divides $\chi(1)$. This problem was reduced to quasisimple groups by B. Hartley and A. Turull (see [HT94]).

Problems

(2.1) Suppose that $a \colon G \to H$ is a group homomorphism. If $\chi \in \mathrm{Char}(H)$, we define $\chi^{a^{-1}} \colon G \to \mathbb{C}$ as $\chi^{a^{-1}}(g) = \chi(g^a)$ for $g \in G$.
   (a) Prove that $\chi^{a^{-1}}$ is a character of G with kernel $a^{-1}(\ker(\chi))$.
   (b) If $\chi, \psi \in \mathrm{Char}(H)$, prove that $[\chi^{a^{-1}}, \psi^{a^{-1}}] = [\chi_K, \psi_K]$, where $K = a(G) \le H$.
   (c) Prove that $\chi^{a^{-1}}$ is irreducible if and only if $\chi_K \in \mathrm{Irr}(K)$.

(2.2) If an abelian group A acts by automorphisms on another abelian group G, then it is in general not true that $|\mathrm{Irr}_A(G)| = |\mathrm{Irr}(\mathbf{C}_G(A))|$. Check that the following is an example: set $G = (\mathbb{Z}/2\mathbb{Z})^3$ and let $A = C_2 \times C_2$ given by

$$A = \left\{ I_3, \begin{pmatrix} 0 & 0 & 1 \\ 0 & 1 & 0 \\ 1 & 0 & 0 \end{pmatrix}, \begin{pmatrix} 0 & 0 & 1 \\ 1 & 1 & 1 \\ 1 & 0 & 0 \end{pmatrix}, \begin{pmatrix} 1 & 0 & 0 \\ 1 & 1 & 1 \\ 0 & 0 & 1 \end{pmatrix} \right\} \le \mathrm{GL}_3(2).$$

(2.3) Suppose that $N \trianglelefteq G$. Show that the actions of G on $\mathrm{Irr}(N)$ and $\mathrm{Cl}(N)$ have the same number of orbits.
(*Hint:* Problem 1.11(c) is relevant.)

(2.4) (*Glauberman's lemma*) Suppose that A is a finite group that acts by automorphisms on a finite group G. Assume that both A and G act on a set Ω such that G acts transitively on Ω and $(\alpha \cdot g) \cdot a = (\alpha \cdot a) \cdot g^a$ for all $g \in G$, $a \in A$, and $\alpha \in \Omega$. If $\gcd(|A|, |G|) = 1$, then show that A fixes an element of Ω. Also, if Ω_0 is the set of A-fixed points of Ω, then prove that $\mathbf{C}_G(A)$ acts transitively on Ω_0.
(*Hint:* Show that the semidirect product GA acts transitively on Ω, and use the Schur–Zassenhaus theorem. If A is a p-group, one can use Sylow theory instead.)

(2.5) Suppose that A is a finite group that acts by automorphisms on a finite group G. Let $N \trianglelefteq G$ be A-invariant with $\gcd(|A|, |G : N|) = 1$. Let $\chi \in \mathrm{Irr}(G)$ be A-invariant. Prove that χ_N has an A-invariant irreducible constituent and that all of them are C-conjugate, where $C/N = \mathbf{C}_{G/N}(A)$.
(*Hint:* Use Glauberman's lemma.)

(2.6) Suppose that A is a p-group that acts by automorphisms on a finite group G of order not divisible by p. Let $C = \mathbf{C}_G(A)$ and let $N \trianglelefteq G$ be A-invariant. Let $\chi \in \mathrm{Irr}_A(G)$, $\theta \in \mathrm{Irr}_A(N)$, and let $\chi' \in \mathrm{Irr}(C)$ and $\theta' \in \mathrm{Irr}(C \cap N)$ be the Glauberman correspondents of χ and θ, respectively.
  (a) If $T = G_\theta$, prove that $T \cap C = C_{\theta'}$.
  (b) If $\psi \in \mathrm{Irr}(T | \theta)$ is the Clifford correspondent of χ over θ, prove that $\psi' \in \mathrm{Irr}(T \cap C)$ is the Clifford correspondent of χ' over θ'.

(2.7) Suppose that A is a p-group that acts by automorphisms on a finite group G of order not divisible by p. Assume that $C \subseteq H \leq G$ is an A-invariant subgroup, where $C = \mathbf{C}_G(A)$. If $\chi \in \mathrm{Irr}_A(G)$, then prove that the restriction χ_H contains a unique A-invariant irreducible constituent μ such that $[\chi_H, \mu]$ is not divisible by p. Furthermore, show that $\mu' = \chi'$.
(*Hint:* If $\eta \in \mathrm{Irr}(H)$ and $a \in A$, use that $[\chi_H, \eta] = [\chi_H, \eta^a]$ and $\eta_C = (\eta^a)_C$.)

(2.8) (*Hawkes*) Suppose that A is a p-group that acts by automorphisms on a finite group G of order not divisible by p. Suppose that $\mathrm{Irr}_A(G) = \{\chi_1, \ldots, \chi_k\}$ and $\mathrm{Cl}_A(G) = \{K_1, \ldots, K_k\}$. If $x_i \in K_i$, show that the matrix

$$X = (\chi_i(x_j))$$

is invertible. Deduce that the restriction map $\mathbb{C}[\text{Irr}_A(G)] \to \text{cf}(C)$ is an isomorphism, where $\mathbb{C}[\text{Irr}_A(G)]$ is the $\mathbb{C}$-span of $\text{Irr}_A(G)$ and $C = \mathbf{C}_G(A)$.

(*Hint:* Show that X^* is invertible in F.)

(*Note:* If A is not a p-group, then it is not known if the conclusion of this problem still holds.)

3

Galois Action on Characters

Some of the deep results in character theory have to do with the Galois action on the irreducible characters of finite groups.

3.1 Galois Action on Characters

Suppose that G is a finite group and let χ be a character of G. The **field of values** of χ is

$$\mathbb{Q}(\chi) = \mathbb{Q}(\chi(g) \mid g \in G),$$

that is, the smallest subfield of $\mathbb{C}$ containing the values of χ. If n is the exponent of G, then we know that $\mathbb{Q}(\chi) \subseteq \mathbb{Q}_n$, the nth cyclotomic field, because $\chi(g)$ is a sum of nth roots of unity. In particular, $\mathbb{Q}(\chi)/\mathbb{Q}$ is a normal (abelian) extension. Now, if $\mathbb{Q}(\chi) \subseteq F \subseteq \mathbb{C}$ is any field and $\sigma : F \to F$ is a field automorphism, then $\sigma(\mathbb{Q}(\chi)) = \mathbb{Q}(\chi)$, by elementary Galois theory. Thus we may define the **Galois conjugate** function $\chi^\sigma : G \to \mathbb{C}$ by letting

$$\chi^\sigma(g) = \chi(g)^\sigma$$

for $g \in G$. It is clear that χ^σ is a class function of G and that

$$[\chi^\sigma, \chi^\sigma] = \frac{1}{|G|} \sum_{g \in G} \chi^\sigma(g)\overline{\chi^\sigma(g)} = \left(\frac{1}{|G|} \sum_{g \in G} \chi(g)\overline{\chi(g)} \right)^\sigma = [\chi, \chi],$$

using that σ and complex conjugation commute. In fact, χ^σ is also a character of G, but this is a little more delicate to show. The problem, of course, is that it is not necessarily true that there exists a representation $G \to \mathrm{GL}_m(F)$ affording χ. (The smallest example of this occurs in the quaternion group of order 8.) If F/K is a field extension, in this book $\mathrm{Gal}(F/K)$ is the group of all field automorphisms $\sigma : F \to F$ such that $\sigma(k) = k$ for all $k \in K$.

Theorem 3.1 *Let G be a finite group and let $\chi \in \mathrm{Char}(G)$. Let $\mathbb{Q}(\chi) \subseteq F \subseteq \mathbb{C}$ by any field.*

(a) *If $\sigma \in \mathrm{Gal}(F/\mathbb{Q})$, then $\chi^\sigma \in \mathrm{Char}(G)$. Also, $\mathbb{Q}(\chi) = \mathbb{Q}(\chi^\sigma)$, and $\chi \in \mathrm{Irr}(G)$ if and only if $\chi^\sigma \in \mathrm{Irr}(G)$.*

(b) *The set of Galois conjugates of χ is $\Omega = \{\chi^\sigma \mid \sigma \in \mathrm{Gal}(\mathbb{Q}(\chi)/\mathbb{Q})\}$. Also, $|\Omega| = |\mathbb{Q}(\chi) : \mathbb{Q}|$. Furthermore, $\mathrm{Gal}(F/\mathbb{Q})$ acts on Ω, and this action is transitive if $F/\mathbb{Q}$ is a normal extension.*

Proof By Theorem 1.4 applied to each of the irreducible constituents of χ, there exists a representation $\mathcal{X}$ of G with entries in $\overline{\mathbb{Q}}$ affording χ. Now, all the entries of all the matrices $\mathcal{X}(g)$ for $g \in G$ are algebraic numbers, so all of them lie in some finite extension of the rationals. By taking its normal closure, we then find a finite normal extension $K/\mathbb{Q}$ and a representation $\mathcal{X} : G \to \mathrm{GL}_n(K)$ affording χ. Then $\mathbb{Q}(\chi) \subseteq K \cap F$. Since $K/\mathbb{Q}$ is normal, then the isomorphism $\sigma|_{\mathbb{Q}(\chi)}$ extends to some $\tau \in \mathrm{Gal}(K/\mathbb{Q})$. Now, the representation

$$\mathcal{X}^\tau(g) = \mathcal{X}(g)^\tau,$$

where $\mathcal{X}(g)^\tau$ is the matrix that results from applying τ to every entry of $\mathcal{X}(g)$, affords the character χ^τ. Since $\chi^\tau = \chi^\sigma$, we conclude that $\chi^\sigma \in \mathrm{Char}(G)$. Also,

$$\mathbb{Q}(\chi^\sigma) = \mathbb{Q}(\chi(g)^\sigma \mid g \in G) = \sigma(\mathbb{Q}(\chi)) = \mathbb{Q}(\chi).$$

Since $[\chi^\sigma, \chi^\sigma] = [\chi, \chi]$, the proof of part (a) is complete.

We claim next that the set of Galois conjugate characters of χ is

$$\Omega = \{\chi^\sigma \mid \sigma \in \mathrm{Gal}(\mathbb{Q}(\chi)/\mathbb{Q})\}.$$

This is true because whenever $\mathbb{Q}(\chi) \subseteq E \subseteq \mathbb{C}$ and $\tau \in \mathrm{Gal}(E/\mathbb{Q})$, then $\chi^\tau = \chi^{\tau'}$, where $\tau' \in \mathrm{Gal}(\mathbb{Q}(\chi)/\mathbb{Q})$ is the restriction of τ to $\mathbb{Q}(\chi)$. Notice furthermore that

$$\chi^{\sigma\tau} = (\chi^\sigma)^\tau$$

for $\sigma, \tau \in \mathrm{Gal}(F/\mathbb{Q})$. (Recall that $\sigma\tau$ means that we apply first σ and then τ.) Since the trivial automorphism 1_F fixes χ, we have that $\mathrm{Gal}(F/\mathbb{Q})$ acts on Ω. This action is transitive if $F/\mathbb{Q}$ is normal, because every isomorphism $\mathbb{Q}(\chi) \to \mathbb{Q}(\chi)$ extends to F. In particular, $\mathrm{Gal}(\mathbb{Q}(\chi)/\mathbb{Q})$ acts transitively on Ω. Finally, we have that $\chi^\sigma = \chi$ if and only if $\sigma = 1_{\mathbb{Q}(\chi)}$ for $\sigma \in \mathrm{Gal}(\mathbb{Q}(\chi)/\mathbb{Q})$, and therefore

$$|\Omega| = |\mathrm{Gal}(\mathbb{Q}(\chi)/\mathbb{Q})| = |\mathbb{Q}(\chi) : \mathbb{Q}|,$$

as desired. $\qquad\square$

If $F \subseteq \mathbb{C}$ is any field containing $\mathbb{Q}(\chi)$ for every $\chi \in \text{Irr}(G)$, then we have that $\text{Gal}(F/\mathbb{Q})$ acts on $\text{Irr}(G)$. Of course, we obtain the smallest Galois group that acts on $\text{Irr}(G)$ if we let F be the smallest field containing the values of all the irreducible characters of G. As we know, $F \subseteq \mathbb{Q}_n$ in this case, where n is the exponent of G. But it is not necessary to use the smallest possible Galois group. For most purposes, it is enough to consider $\mathcal{G} = \text{Gal}(\mathbb{Q}_{|G|}/\mathbb{Q})$, which does act on the characters of every subgroup of G and of every quotient of G. (If we want a group that simultaneously acts on the characters of all finite groups, then we can use $\text{Gal}(\overline{\mathbb{Q}}/\mathbb{Q})$, the absolute Galois group, or $\text{Gal}(\mathbb{Q}^{\text{ab}}/\mathbb{Q})$, the Galois group of the smallest field that contains all roots of unity; but we shall not need that.)

3.2 Galois Action on Classes

Suppose that G is a finite group of exponent dividing n, where n is a positive integer. As we know now, the group $\mathcal{G} = \text{Gal}(\mathbb{Q}_n/\mathbb{Q})$ acts on $\text{Irr}(G)$. We show next that $\mathcal{G}$ also acts on $\text{Cl}(G)$, the set of conjugacy classes of G. In fact, $\mathcal{G}$ acts on G as a set. Indeed, if $\sigma \in \mathcal{G}$, then there exists a unique element $m + n\mathbb{Z}$ in the ring $\mathbb{Z}/n\mathbb{Z}$, with m coprime with n, such that $\sigma(\xi) = \xi^m$ for every root of unity in $\mathbb{Q}_n$. (In fact, $\mathcal{G}$ is isomorphic to the multiplicative group of units of $\mathbb{Z}/n\mathbb{Z}$.) Now, if $x \in G$, then we let $x^\sigma = x^m$, and we observe that this is well-defined since $x^n = 1$. Also, $(x^\sigma)^\tau = x^{\sigma\tau}$, the identity of $\mathcal{G}$ fixes every element of G, and therefore $\mathcal{G}$ does act on G. Notice too that $(x^y)^\sigma = (x^y)^m = (x^m)^y = (x^\sigma)^y$ for $x, y \in G$. If $K = \text{cl}_G(x)$ is the conjugacy class of $x \in G$, then

$$K^\sigma = \{k^\sigma \mid k \in K\} = \text{cl}_G(x^\sigma).$$

It is clear that this defines an action of $\mathcal{G}$ on $\text{Cl}(G)$.

Lemma 3.2 *Suppose that G is a finite group of exponent dividing n. Let $\mathcal{G} = \text{Gal}(\mathbb{Q}_n/\mathbb{Q})$. If $\sigma \in \mathcal{G}$, then*

$$\chi^\sigma(g) = \chi(g^\sigma)$$

for $g \in G$ and $\chi \in \text{Char}(G)$.

Proof If $\mathcal{X}$ affords χ, then $\mathcal{X}(g)$ is similar to a diagonal matrix $\text{diag}(\xi_1, \ldots, \xi_t)$ for some nth roots of unity ξ_i. Suppose that $\sigma(\xi) = \xi^m$ for any nth root of unity ξ in $\mathbb{Q}_n$. Then $\mathcal{X}(g^\sigma) = \mathcal{X}(g^m) = \mathcal{X}(g)^m$ is similar to $\text{diag}(\xi_1, \ldots, \xi_t)^m = \text{diag}(\xi_1^m, \ldots, \xi_t^m)$. Thus

$$\chi(g^\sigma) = \xi_1^m + \cdots + \xi_t^m = (\xi_1 + \cdots + \xi_t)^\sigma = \chi(g)^\sigma = \chi^\sigma(g),$$

as required. □

Now that we have actions of $\mathcal{G}$ on $\mathrm{Irr}(G)$ and on $\mathrm{Cl}(G)$, it is natural to ask if they are permutation isomorphic. The following shows that this is the case if $\mathcal{G}$ is cyclic.

Theorem 3.3 *Assume that G is a finite group of exponent dividing n, and let $\mathcal{G} = \mathrm{Gal}(\mathbb{Q}_n/\mathbb{Q})$. Let $\sigma \in \mathcal{G}$. Then σ fixes the same number of conjugacy classes of G as irreducible characters of G. In particular, if $\mathcal{H} \leq \mathcal{G}$ is cyclic, then the actions of $\mathcal{H}$ on $\mathrm{Irr}(G)$ and $\mathrm{Cl}(G)$ are permutation isomorphic.*

Proof We wish to use Brauer's lemma on character tables (Theorem 2.3). Notice that

$$\chi^\sigma(g^{\sigma^{-1}}) = \chi(g)$$

by Lemma 3.2. If $K \in \mathrm{Cl}(G)$ and $\sigma \in \mathcal{G}$, let us define $K \cdot \sigma = K^{\sigma^{-1}}$. Since $\mathcal{G}$ is abelian, this defines an action on $\mathrm{Cl}(G)$ that satisfies the hypotheses of Theorem 2.3. Since K is invariant by σ if and only if it is invariant by σ^{-1}, the first part of the theorem follows. If $\mathcal{H} \leq \mathcal{G}$ is cyclic, then every element of $\mathcal{H}$ fixes the same number of characters as classes, and the conclusion follows from the result quoted before Corollary 2.12. □

Corollary 3.4 *Let G be a finite p-group of order p^a, where p is an odd prime. Then the actions of $\mathrm{Gal}(\mathbb{Q}_{p^a}/\mathbb{Q})$ on the irreducible characters and classes of G are permutation isomorphic.*

Proof It is well known that the group $\mathrm{Gal}(\mathbb{Q}_{p^a}/\mathbb{Q})$ is cyclic if p is odd. □

Later, we shall give an example of a 2-group for which the conclusion of the previous corollary does not hold.

A classic theme in group theory is *reality*. An element $g \in G$ is **real** in G if $\chi(g) \in \mathbb{R}$ for every $\chi \in \mathrm{Irr}(G)$. Note that if g is real in G then every G-conjugate of g is real. The class $K = \mathrm{cl}_G(g)$ is **real** if g is real in G, and we write $\mathrm{Cl}_{\mathbb{R}}(G)$ for the set of real classes of G. Notice that $g \in G$ is real if and only if $\chi(g) = \chi(g^{-1})$ for all $\chi \in \mathrm{Irr}(G)$, by using that $\chi(g^{-1}) = \overline{\chi(g)}$. This happens if and only if g and g^{-1} are G-conjugate by the second orthogonality relation. An irreducible character $\chi \in \mathrm{Irr}(G)$ is **real** or **real valued** if $\chi(g) \in \mathbb{R}$ for all $g \in G$. We write $\mathrm{Irr}_{\mathbb{R}}(G)$ for the set of irreducible real characters of G.

Corollary 3.5 (Burnside) *Let G be a finite group. Then*

$$|\mathrm{Irr}_{\mathbb{R}}(G)| = |\mathrm{Cl}_{\mathbb{R}}(G)|.$$

In particular, $|\mathrm{Irr}_{\mathbb{R}}(G)| = 1$ *if and only if G has odd order.*

Proof Let $\mathcal{G} = \mathrm{Gal}(\mathbb{Q}_{|G|}/\mathbb{Q})$, and let $\sigma \in \mathcal{G}$ be the complex conjugation automorphism of $\mathbb{Q}_{|G|}$. Then $\chi^{\sigma} = \bar{\chi}$, the complex conjugate character of χ, and $x^{\sigma} = x^{-1}$ for $x \in G$. Now we apply Theorem 3.3 to obtain

$$|\mathrm{Irr}_{\mathbb{R}}(G)| = |\mathrm{Cl}_{\mathbb{R}}(G)|.$$

If G has odd order and $x^g = x^{-1}$ for some $x, g \in G$, then $x^{g^2} = x$, and therefore all powers of g^2 centralize x. Since $\langle g \rangle = \langle g^2 \rangle$ because $\langle g \rangle$ has odd order, we conclude that $x = x^g = x^{-1}$, and thus $x = 1$. Therefore $|\mathrm{Irr}_{\mathbb{R}}(G)| = 1$ if $|G|$ is odd. If G has even order and $g \in G$ is an element of order 2, then g is real, and so G has at least two different real classes. Therefore G has at least two different irreducible real characters. $\qquad\square$

Even more subtle than *reality* is *rationality*. An element $g \in G$ is **rational** in G if $\chi(g) \in \mathbb{Q}$ for all $\chi \in \mathrm{Irr}(G)$, and a conjugacy class of G is **rational** in G if consists of rational elements. An irreducible character $\chi \in \mathrm{Irr}(G)$ is **rational** or **rational valued** if $\chi(g) \in \mathbb{Q}$ for all $g \in G$. The definition of when an element of a finite group is rational can also be given purely group theoretically.

Lemma 3.6 *Let* $x, y \in G$, *where G is a finite group.*

(a) *We have that* $\langle x \rangle = \langle y \rangle$ *if and only if* $x = y^m$ *where m is an integer coprime with* $|G|$.
(b) *If* $\langle x \rangle = \langle y \rangle$, *then there is* $\sigma \in \mathrm{Gal}(\mathbb{Q}_{|G|}/\mathbb{Q})$ *such that* $\chi(x) = \chi(y)^{\sigma}$ *for every* $\chi \in \mathrm{Char}(G)$.

Proof (a) Assume that $\langle y \rangle = \langle x \rangle$. Then $x = y^n$, where n is coprime with $o(y)$. Now, write $|G| = vw$, where v and w are integers such that the set of primes dividing $o(y)$ is the set of primes dividing v, and $\gcd(o(y), w) = 1$. Now, it is enough to let $m \equiv n \bmod o(y)$ and $m \equiv 1 \bmod w$, using the Chinese Remainder Theorem. It is clear that m is coprime with $|G|$ and $y^n = y^m$. The converse is clear.

(b) By part (a), there is m coprime with $|G|$ such that $x = y^m$. Let $\sigma \in \mathrm{Gal}(\mathbb{Q}_{|G|}/\mathbb{Q})$ be such that $\xi^{\sigma} = \xi^m$ for $|G|$th roots of unity ξ. Now

$$\chi(x) = \chi(y^m) = \chi^{\sigma}(y) = \chi(y)^{\sigma}$$

by Lemma 3.2. $\qquad\square$

Theorem 3.7 *Let G be a finite group. Then $g \in G$ is rational in G if and only if whenever $\langle x \rangle = \langle g \rangle$, then g and x are G-conjugate.*

Proof Notice that g is rational in G if and only if $\chi(g)^\sigma = \chi(g)$ for all $\chi \in \mathrm{Irr}(G)$ and all $\sigma \in \mathrm{Gal}(\mathbb{Q}_{|G|}/\mathbb{Q})$. This happens if and only if $\chi(g^m) = \chi(g)$ for all m coprime with $|G|$ and all $\chi \in \mathrm{Irr}(G)$, using Lemma 3.2. By the second orthogonality relation, this happens if and only if g^m and g are G-conjugate for all m coprime with $|G|$. By Lemma 3.6(a) the elements g^m with m coprime with $|G|$ are exactly the generators of $\langle g \rangle$. $\qquad\qquad\square$

So far we have defined real and rational characters and classes of a finite group. But there is a more general concept: if $\mathbb{Q} \subseteq F \subseteq \mathbb{C}$ is any field, we say that $g \in G$ is an *F*-**element** if $\chi(g) \in F$ for all $\chi \in \mathrm{Irr}(G)$. A conjugacy class of G is an *F*-**class** if it consists of *F*-elements, and it is said that $\chi \in \mathrm{Irr}(G)$ is an *F*-**character** if $\chi(g) \in F$ for all $g \in G$. If $|G| = n$, notice that $\chi \in \mathrm{Irr}(G)$ is an *F*-character if $\chi^\sigma = \chi$ for all $\sigma \in \mathrm{Gal}(\mathbb{Q}_n/(\mathbb{Q}_n \cap F))$, by Galois theory. By using Lemma 3.2, we also have that a conjugacy class K is an *F*-class if $K^\sigma = K$ for all $\sigma \in \mathrm{Gal}(\mathbb{Q}_n/(\mathbb{Q}_n \cap F))$

In general, the Galois actions on characters and classes are not permutation isomorphic. If this is the case for a finite group G, then there would exist a bijection $f \colon \mathrm{Irr}(G) \to \mathrm{Cl}(G)$ commuting with the action of $\mathcal{G} = \mathrm{Gal}(\mathbb{Q}_{|G|}/\mathbb{Q})$. By Problem 3.1, this is equivalent to assuming that for every field $\mathbb{Q} \subseteq F \subseteq \mathbb{Q}_{|G|}$, the number of *F*-classes is the number of *F*-characters. J. Thompson and E. C. Dade found examples where this is not true, for 2-groups and for groups of odd order, respectively. Nowadays, with the use of computers, this is an easy check. The smallest group for which these actions are not permutation isomorphic is SmallGroup(32,9), while SmallGroup(567,7) is the smallest for groups of odd order. (The computer programs **GAP** and **Magma** have a database of all groups of order up to 2000, except for order 1024. The command SmallGroup(*x*, *y*) produces the *y*th group of order *x* in this database.) The SmallGroup(32,9) has six irreducible rational characters and eight rational classes, for instance. Interestingly enough, we only know of one simple group where the Galois actions on classes and characters are not permutation isomorphic: the Tits group $^2F_4(2)'$.

As we are seeing, one of the main ideas in character theory is to discover analogies between characters and classes; that is, between rows and columns of the character table. In fact, the aim is to detect as many group theoretical properties from the character table, and the other way around, as possible. In the following section we discuss yet another one.

3.3 A Theorem of Brauer and a Conjecture of Feit

It is well known that the character table of a finite group G does not determine the orders of the elements of G. (For instance, the dihedral group D_8 and the quaternion group Q_8 have the same character table.) Nevertheless, there are non-trivial connections between orders of elements and character values, and we shall explore these.

First, let us start by reminding the reader that if m and n are positive integers, then

$$\mathbb{Q}_m \cap \mathbb{Q}_n = \mathbb{Q}_{\gcd(m,n)},$$

by Galois theory. We shall use this fact in what follows.

Now, if G is a finite group and $\chi \in \text{Irr}(G)$, then the **Feit number** f_χ of χ is the smallest positive integer f such that $\mathbb{Q}(\chi) \subseteq \mathbb{Q}_f$. Since $\mathbb{Q}(\chi) \subseteq \mathbb{Q}_n$, where n is the exponent of G, then we have that

$$\mathbb{Q}(\chi) \subseteq \mathbb{Q}_{f_\chi} \cap \mathbb{Q}_n = \mathbb{Q}_{\gcd(f_\chi, n)}.$$

By definition, $f_\chi \leq \gcd(f_\chi, n)$, and we conclude that $f_\chi = \gcd(f_\chi, n)$. In particular, f_χ divides n.

Feit's conjecture asserts that there always exists $g \in G$ such that $o(g) = f_\chi$. In this section, we prove a related theorem of Richard Brauer.

Lemma 3.8 *Let χ be a character of G, and let $g \in G$. If $\chi(g)$ is in $\mathbb{Q}_{p^a n}$ but not in $\mathbb{Q}_{p^{a-1}n}$, where p is a prime, p does not divide n and $a \geq 1$, then p^a divides the order of g.*

Proof Write $o(g) = p^b m$, where $b \geq 0$ and m is not divisible by p. Then $\chi(g) \in \mathbb{Q}_{p^b m}$. Now, we have that $\chi(g) \in \mathbb{Q}_d$, where $d = \gcd(p^b m, p^a n)$. Thus $d = p^e \gcd(n, m)$, where e is the minimum of a and b. Therefore $\chi(g) \in \mathbb{Q}_{p^e n}$. Then $\mathbb{Q}_{p^e n}$ is not contained in $\mathbb{Q}_{p^{a-1}n}$. Hence $e \geq a$. Thus $e = a$, and therefore $b \geq a$. Hence p^a divides the order of g. $\square$

Notice that Lemma 3.8 proves the Feit conjecture in the case where f_χ is a prime power.

Theorem 3.9 (Brauer) *Let G be a finite group, and let $\chi \in \text{Irr}(G)$. Let F be a field such that $\mathbb{Q}(\chi) \subseteq F \subseteq \mathbb{Q}_{|G|}$. Let $\sigma_1, \ldots, \sigma_n \in \text{Gal}(F/\mathbb{Q})$. Then one of the following cases holds:*

(a) There exists $g \in G$ such that $\chi(g) \neq \chi(g)^{\sigma_i}$ for all i.

(b) There exist an odd number k of indices $1 \leq i_1 < \cdots < i_k \leq n$ such that

$$\chi^{\sigma_{i_1} \cdots \sigma_{i_k}} = \chi.$$

Proof Let $\mathcal{G} = \mathrm{Gal}(F/\mathbb{Q})$. By Theorem 3.1, we have that $\mathcal{G}$ acts on the set $\Omega = \{\chi^\sigma \mid \sigma \in \mathrm{Gal}(\mathbb{Q}(\chi)/\mathbb{Q})\}$ of Galois conjugates of χ. We can extend this action $\mathbb{Z}$-linearly to $\mathbb{Z}[\Omega]$, the subgroup of $\mathbb{Z}[\mathrm{Irr}(G)]$ with basis Ω. Let $R = \mathbb{Z}[\mathcal{G}]$ be the group ring. Now we define a map $(r, \psi) \mapsto r \cdot \psi$ from $R \times \mathbb{Z}[\Omega] \to \mathbb{Z}[\Omega]$ in the following way: if $r = \sum_{\sigma \in \mathcal{G}} a_\sigma \sigma \in R$ and $\psi \in \mathbb{Z}[\Omega]$, then we let

$$r \cdot \psi = \sum_{\sigma \in \mathcal{G}} a_\sigma \psi^\sigma.$$

It is straightforward to check that

$$(rs) \cdot \psi = r \cdot (s \cdot \psi)$$

for $r, s \in R$ and $\psi \in \mathbb{Z}[\Omega]$. Also, $1 \cdot \psi = \psi$. (In fact, $\mathbb{Z}[\Omega]$ becomes an R-module, but we do not need that.) Notice that if $\psi \in \mathbb{Z}[\Omega]$ and $\psi(g) = 0$ for some $g \in G$, then $(r \cdot \psi)(g) = 0$ for every $r \in R$, and we shall use this fact in the next paragraph.

If $r_1, \ldots, r_k \in R$ and $(r_1 \cdots r_k) \cdot \chi \neq 0$, then we claim that there exists $g \in G$ such that $(r_i \cdot \chi)(g) \neq 0$ for all i. Otherwise, for every $g \in G$ there exists j (depending on g) such that $(r_j \cdot \chi)(g) = 0$. Then, using that R is abelian, we would have that

$$((r_1 \cdots r_k) \cdot \chi)(g) = ((r_1 \cdots r_{j-1}r_{j+1} \cdots r_k) \cdot (r_j \cdot \chi))(g) = 0,$$

for all $g \in G$, which is impossible. This proves the claim.

Now, write $r_i = 1 - \sigma_i \in R$. If $(r_1 \cdots r_n) \cdot \chi \neq 0$, then by the preceding paragraph, there exists $g \in G$ such that $(r_i \cdot \chi)(g) \neq 0$ for all i. Since

$$(r_i \cdot \chi) = \chi - \chi^{\sigma_i},$$

then we are in case (a) of the theorem. So we may assume that $(r_1 \cdots r_n) \cdot \chi = 0$. Notice that

$$r_1 \cdots r_n = (1 - \sigma_1) \cdots (1 - \sigma_n)$$
$$= 1 - \sum_{1 \leq i \leq n} \sigma_i + \sum_{1 \leq i < j \leq n} \sigma_i \sigma_j - \sum_{1 \leq i < j < k \leq n} \sigma_i \sigma_j \sigma_k + \ldots$$

Therefore

$$\chi + \sum_{1 \leq i < j \leq n} \chi^{\sigma_i \sigma_j} + \ldots = \sum_i \chi^{\sigma_i} + \sum_{1 \leq i < j < k \leq n} \chi^{\sigma_i \sigma_j \sigma_k} + \ldots$$

By using linear independence of characters, χ has to be one of the irreducible characters on the right-hand side. □

The following result of Brauer proves a case of Feit's conjecture.

Theorem 3.10 (Brauer) *Let $\chi \in \mathrm{Irr}(G)$, and write $f_\chi = m$. Let m_0 be the product of those prime power factors $p_i^{a_i}$ of m with $a_i > 1$. Then G has an element of order m_0.*

Proof Write $m = p_1^{a_1} \cdots p_n^{a_n} p_{n+1} \cdots p_k$, where the p_is are pairwise different primes, and $a_i \geq 2$ for all $i = 1, \ldots, n$. By definition, $m_0 = p_1^{a_1} \cdots p_n^{a_n}$.

Let $F = \mathbb{Q}_m$. For $i = 1, \ldots, n$, we let $F_i = \mathbb{Q}_{\frac{m}{p_i}}$. By the definition of f_χ, we have that $\mathbb{Q}(\chi) \subseteq F$ but $\mathbb{Q}(\chi)$ is not contained in F_i for any i. Notice that

$$|F : F_i| = \frac{\varphi(m)}{\varphi(\frac{m}{p_i})} = p_i,$$

where φ is the Euler function, by elementary Galois theory. Therefore F/F_i is a Galois extension of degree p_i. Let $\langle \sigma_i \rangle = \mathrm{Gal}(F/F_i)$ of order p_i. Let $\sigma = \sigma_{i_1} \cdots \sigma_{i_k}$ be any product of pairwise different σ_i's. Assume that χ is σ-fixed. Then we would have that $\sigma_{i_1} \cdots \sigma_{i_k} \in \mathrm{Gal}(F/\mathbb{Q}(\chi))$. Since the orders of the σ_is are pairwise coprime, it follows that each σ_{i_j} fixes χ. (In an abelian group, if a product g of pairwise coprime order elements lies in some subgroup, then every factor lies in the subgroup too. This is because every factor is a power of g.) But then $\mathbb{Q}(\chi)$ is in the fixed field of σ_{i_1} which is $F_{i_1} = \mathbb{Q}_{\frac{m}{p_{i_1}}}$. This is impossible.

Thus we are not in case (b) of Brauer's theorem (Theorem 3.9), and we deduce that there exists $g \in G$ such that $\chi(g)^{\sigma_i} \neq \chi(g)$ for all $i = 1, \ldots, n$. Therefore $\chi(g) \notin F_i$. By Lemma 3.8, we have that $p_i^{a_i}$ divides the order of g for all $i = 1, \ldots, n$. In particular, m_0 divides the order of g. Now, some power of g has order m_0. □

3.4 Order of Certain Elements

Next, we prove another result on character values and the order of certain elements, which is fairly elementary. It is convenient to introduce the **field of values of a conjugacy class** $K = \mathrm{cl}_G(g)$, which is $\mathbb{Q}(K) = \mathbb{Q}(\chi(g) \mid \chi \in \mathrm{Irr}(G))$, the smallest field containing the values of the characters of G evaluated in g. Since $\chi(g)$ is a sum of $o(g)$th roots of unity, we have that $\mathbb{Q}(K) \subseteq \mathbb{Q}_{o(g)}$.

Theorem 3.11 *Suppose that G is a finite group and let $g \in G$. Let $K = \mathrm{cl}_G(g)$ be the conjugacy class of g in G. Then there is a natural isomorphism between the groups $\mathbf{N}_G(\langle g \rangle)/\mathbf{C}_G(g)$ and $\mathrm{Gal}(\mathbb{Q}_{o(g)}/\mathbb{Q}(K))$. In particular,*

$$|\mathbf{N}_G(\langle g \rangle)/\mathbf{C}_G(g)| = |\mathbb{Q}_{o(g)} : \mathbb{Q}(K)|.$$

Proof Write $o(g) = n$. Let $H = \mathbf{N}_G(\langle g \rangle)$. Given $h \in H$, then $\langle g^h \rangle = \langle g \rangle$, and therefore there is an integer k coprime to n, uniquely determined modulo $n\mathbb{Z}$, such that $g^h = g^k$. This defines a natural injective homomorphism $f : H/\mathbf{C}_G(g) \to U$, where U is the group of units of the ring $\mathbb{Z}/n\mathbb{Z}$. If V is the image of this map and v is an integer coprime with n, notice that $v + n\mathbb{Z} \in V$ if and only if g^v and g are G-conjugate.

Now let ξ be any primitive nth root of unity. Given an integer k coprime with n, there is a unique $\sigma_k \in \mathrm{Gal}(\mathbb{Q}_n/\mathbb{Q})$ sending $\xi \mapsto \xi^k$ and, in fact, the map $k + n\mathbb{Z} \mapsto \sigma_k$ defines an isomorphism $\sigma : U \to \mathrm{Gal}(\mathbb{Q}_n/\mathbb{Q})$. It remains to show that $\sigma(V) = \mathrm{Gal}(\mathbb{Q}_n/\mathbb{Q}(K))$. Notice that if χ is any character of G, then

$$\chi(g)^{\sigma_k} = \chi(g^k)$$

by Lemma 3.2 applied to the character $\chi_{\langle g \rangle}$.

Now, if k is an integer coprime with n, we have that $k + n\mathbb{Z} \in V$ if and only if g^k and g are G-conjugate. By the second orthogonality relation, this happens if and only if $\chi(g^k) = \chi(g)$ for all $\chi \in \mathrm{Irr}(G)$, which happens if and only if $\chi(g)^{\sigma_k} = \chi(g)$ for all $\chi \in \mathrm{Irr}(G)$. Thus $k + n\mathbb{Z} \in V$ if and only if $\sigma_k \in \mathrm{Gal}(\mathbb{Q}_n/\mathbb{Q}(K))$. This completes the proof of the theorem. $\square$

Corollary 3.12 *Let G be a finite group and let g be a nonidentity element of the center of a Sylow p-subgroup of G. If $o(g) = p^e$, then e is the smallest positive integer such that $\mathbb{Q}(K)$ is contained in $\mathbb{Q}_{p^e}$, where $K = \mathrm{cl}_G(g)$.*

Proof We know that $\mathbb{Q}(K) \subseteq \mathbb{Q}_{p^e}$. By hypothesis, $e \geq 1$. If $e = 1$, then notice that there is nothing to prove. So we may assume that $e \geq 2$. We know by Galois theory that $|\mathbb{Q}_{p^e} : \mathbb{Q}_{p^{e-1}}| = p$. By hypothesis $|G : \mathbf{C}_G(g)|$ is not divisible by p. Therefore, by Theorem 3.11 we know that $|\mathbb{Q}_{p^e} : \mathbb{Q}(K)|$ is not divisible by p. If $\mathbb{Q}(K) \subseteq \mathbb{Q}_{p^{e-1}}$, then p would divide

$$|\mathbb{Q}_{p^e} : \mathbb{Q}(K)| = |\mathbb{Q}_{p^e} : \mathbb{Q}_{p^{e-1}}| |\mathbb{Q}_{p^{e-1}} : \mathbb{Q}(K)|$$

and this is not possible. $\square$

3.5 Notes

Feit's conjecture appeared in [Fe80]. G. Amit and D. Chillag proved Feit's conjecture for solvable groups [AC86] after R. Gow had proved it for groups of odd order (see [Go81]). Almost at the same time, P. Ferguson and A. Turull gave another proof for solvable groups of a more general result (see [FT86]). Later on Ferguson and Turull proved several cases of the Feit conjecture for non-solvable groups. Up to this day, Feit's conjecture remains open, even for simple groups.

An interesting property of the Feit number f_χ of $\chi \in \mathrm{Irr}(G)$ is that f_χ divides $|G|/\chi(1)$. This was proved by W. Feit (see [Fe82b]) after having been conjectured by J. H. Conway, but the only known proof uses modular representation theory.

Rational groups (that is, groups whose irreducible complex characters are all rational valued) is a fascinating subject. R. Gow proved that a solvable rational group has order $2^a 3^b 5^c$ for $a, b, c \geq 0$ (see [Go76]). Much later, P. Hedegüs showed in [He05] that in this case the Sylow 5-subgroup is normal and elementary abelian. J. G. Thompson studied the composition factors of rational groups of order a prime p and showed that $p \leq 11$ (see [Tho08]). It is believed that $p \leq 5$, but this very deep conjecture remains open too. The non-abelian composition factors of rational groups were classified by W. Feit and G. Seitz in [FS88]: they are the alternating groups and five specific simple groups.

It is reasonable to expect a relationship between the rationality of a finite group with rationality properties of its Sylow 2-subgroups. A longstanding conjecture asserted that if G is rational, then a Sylow 2-subgroup of G is rational, but I. M. Isaacs and I disproved it with the use of computers in [IN12]. (The group SmallGroup(1536,408632705) is a counterexample.) Other similar conjectures have had an affirmative answer, however. For instance, R. Gow conjectured that in a **real group** (a group all of whose irreducible characters are real-valued), then P/P' is elementary abelian, where $P \in \mathrm{Syl}_2(G)$. This was proved by Pham Huu Tiep and me (see [NT16b]).

We wrote above that *rationality* is somehow more subtle than *reality*. We have easily shown that a group of even order has a nontrivial irreducible real-valued character. To show that a group of even order has a nontrivial rational-valued character, however, requires the Classification of Finite Simple Groups (CFSG). This result will be proven in Theorem 6.7. As we are going to see in some parts of this book, it is a fact that some deep problems in character theory are being solved (or studied) by using the CFSG.

Problems

(3.1) Let G be a finite group, and let $\mathcal{G} = \mathrm{Gal}(\mathbb{Q}_{|G|}/\mathbb{Q})$. Prove that the actions of $\mathcal{G}$ on $\mathrm{Irr}(G)$ and $\mathrm{Cl}(G)$ are permutation isomorphic if and only if the multisets $\{\mathbb{Q}(\chi) \mid \chi \in \mathrm{Irr}(G)\}$ and $\{\mathbb{Q}(K) \mid K \in \mathrm{Cl}(G)\}$ are the same.

(3.2) Let G be a finite group, and let $y \in G$. If $\chi \in \mathrm{Char}(G)$, then prove $\{\chi(x) \mid \langle x \rangle = \langle y \rangle\}$ is the set of the Galois conjugates of $\chi(y)$.

(3.3) If a 2-group T acts nontrivially on a finite group X by automorphisms, show that there is $t \in T$ and $1 \neq x \in X$ such that $x^t = x^{-1}$.

(3.4) Let G be a finite group, and let $N \trianglelefteq G$. Prove the following properties of real elements.
  (a) If x is real in G, then there is some 2-element $y \in G$ such that $x^y = x^{-1}$.
  (b) If G/N has odd order, then the set of real elements of G is the set of real elements of N.
  (c) If x is real in G, then x^m is real in G for every integer m.
  (d) Suppose $Nx \in G/N$ is real. If $\gcd(o(x), |N|) = 1$, then x is real in G.
  (e) Suppose that $Nx \in G/N$ is real of odd order. Then $Nx = Ny$ for some real $y \in G$.

(3.5) Let G be a finite group and let $g \in G$ be rational in G. Prove that g^m is rational in G for every m.

(3.6) Let G be a finite group, and let $\chi \in \mathrm{Irr}(G)$. If $f_\chi = p^a q^b$, where p and q are odd primes, then prove Feit's conjecture for χ.
  (*Hint:* Use Brauer's theorem (Theorem 3.9).)

(3.7) Suppose that A is a p-group acting coprimely by automorphisms on a finite group G. Let $\chi \in \mathrm{Irr}_A(G)$ and let $\chi' \in \mathrm{Irr}(\mathbf{C}_G(A))$ be its Glauberman correspondent. Show that $\mathbb{Q}(\chi) = \mathbb{Q}(\chi')$.

(3.8) Suppose that χ is a character of a finite group and let $\sigma \in \mathrm{Gal}(\mathbb{Q}(\chi)/\mathbb{Q})$. Prove that
$$\det(\chi^\sigma) = \det(\chi)^\sigma.$$

(3.9) Suppose that $N \trianglelefteq G$ and let $\theta \in \mathrm{Irr}(N)$. Let T^* be the set of $g \in G$ such that $\theta^g = \theta^\sigma$ for some $\sigma \in \mathrm{Gal}(\mathbb{Q}(\theta)/\mathbb{Q})$, and let $T = G_\theta$ be the stabilizer of θ in G.
  (a) Prove that $T \leq T^* \leq G$, and that induction defines a bijection $\mathrm{Irr}(T^*|\theta) \to \mathrm{Irr}(G|\theta)$ such that $\mathbb{Q}(\psi) = \mathbb{Q}(\psi^G)$ for $\psi \in \mathrm{Irr}(T^*|\theta)$.
  (b) Prove that $T \trianglelefteq T^*$ and $T^*/T \cong \mathrm{Gal}(\mathbb{Q}(\theta)/\mathbb{Q}(\theta^G))$.
  (*Note:* The subgroup T^* is called the **semi-inertia** subgroup of θ in G.)

4

Character Values and Identities

A theorem of J. G. Thompson (which uses nonelementary number theory) asserts that if $\chi \in \mathrm{Irr}(G)$, then $\chi(g)$ is zero or a root of unity for at least a third of the elements $g \in G$. Among other topics, in this chapter we discuss zeros of characters, p-defect zero characters, a theorem of G. R. Robinson on characters taking roots of unity values on 2-singular elements, and some related character identities that are crucial in studying some open problems on products of conjugacy classes in finite groups.

4.1 Burnside's Theorem

As we know, the values of any linear character of a finite group G are roots of unity. W. Burnside proved that every nonlinear irreducible character takes the value 0. We present a slight variation of this result due to P. X. Gallagher.

Theorem 4.1 (Gallagher) *Let G be a finite group, let H be a subgroup of G and let $\chi \in \mathrm{Irr}(G)$. Then $\chi(x) \neq 0$ for all $x \in G - H$ if and only if $\chi_H \in \mathrm{Irr}(H)$ and $|\chi(x)| = 1$ for all $x \in G - H$.*

Proof For $x, y \in G$, set $x \equiv y$ if $\langle x \rangle = \langle y \rangle$, and let C be the equivalence class of $x \in G$. Notice that every C is either contained in H or disjoint to H. Let $n = |G|$ and $\mathcal{G} = \mathrm{Gal}(\mathbb{Q}_n/\mathbb{Q})$. If m is any integer coprime with n, notice that the map $g \mapsto g^m$ is a permutation of C. Now, if $\sigma \in \mathcal{G}$, by Lemma 3.2 there is m coprime with n, such that $\chi(g)^\sigma = \chi(g^m)$ for all $g \in G$. Thus we see that $\prod_{y \in C} \chi(y)$ is σ-invariant for all $\sigma \in \mathcal{G}$, and therefore it is an integer (since it is a rational algebraic integer). If $\chi(y)$ is nonzero for all $y \in C$, we conclude that

$$\prod_{y \in C} |\chi(y)|^2 = \prod_{y \in C} \chi(y) \overline{\prod_{y \in C} \chi(y)}$$

58

is a nonzero integer. By the inequality (and the equality) of the arithmetic and geometric means, we conclude that

$$\frac{1}{|C|} \sum_{y \in C} |\chi(y)|^2 \geq (\prod_{y \in C} |\chi(y)|^2)^{\frac{1}{|C|}} \geq 1 ,$$

and

$$\frac{1}{|C|} \sum_{y \in C} |\chi(y)|^2 = 1$$

if and only if $|\chi(y)| = 1$ for all $y \in C$.

Assume now that $\chi(y) \neq 0$ for all $y \in G - H$. Then summing over all the equivalence classes disjoint to H, and using that $[\chi_H, \chi_H] \geq 1$, we have that

$$|G| = \sum_{x \in G} |\chi(x)|^2 = |H|[\chi_H, \chi_H] + \sum_{C \cap H = \emptyset} \sum_{y \in C} |\chi(y)|^2$$

$$\geq |H|[\chi_H, \chi_H] + \sum_{C \cap H = \emptyset} |C| = |H|[\chi_H, \chi_H] + |G - H|$$

$$\geq |G|.$$

We deduce that $[\chi_H, \chi_H] = 1$ and that $|\chi(y)| = 1$ for all $y \in G - H$. $\qquad\square$

By elementary number theory, any algebraic integer α in a cyclotomic field with $|\alpha| = 1$ is a root of unity. (See Problem 4.1.) Hence in Theorem 4.1, we could have written the conclusion: $\chi_H \in \mathrm{Irr}(H)$ and $\chi(x)$ is a root of unity for all $x \in G - H$. The groups $G = \mathrm{GL}_2(3)$, $H = \mathrm{SL}_2(3)$ for the characters of degree 3, or $G = \mathrm{S}_5$, $H = \mathrm{A}_5$ for characters of degree 5 give nontrivial examples in which the hypotheses of Theorem 4.1 are satisfied.

Corollary 4.2 (Burnside) *Let G be a finite group and let $\chi \in \mathrm{Irr}(G)$ be nonlinear. Then there is $g \in G$ such that $\chi(g) = 0$.*

Proof Apply Theorem 4.1 to $H = 1$. $\qquad\square$

We shall see yet another variation of Burnside's theorem in Theorem 5.20.

Once we know that nonlinear characters have zeros, it is interesting to study the nature of these zeros. It is a theorem of G. Malle, J. B. Olsson and me that every nonlinear irreducible character takes a zero value on an element of prime power order ([MNO00]). Although we will see in this book a few theorems that use the CFSG, the complication of this one is excessive. As shown by the quaternion group $\mathbf{Q}_8$, zeros of characters are not necessarily taken on elements of prime order.

4.2 Defect Zero Characters

The most celebrated theorem that guarantees zero values of characters is the Brauer–Nesbitt theorem on *defect zero* characters. Recall that if p is a prime and G is a finite group, then $\chi \in \mathrm{Irr}(G)$ is said to have p-**defect zero** if $\chi(1)_p = |G|_p$. (If $n \geq 1$ is an integer and p is a prime, then n_p is the largest power of p that divides n, and we say that n_p is the p-**part** of n.) Defect zero characters are ubiquitous in the representation theory of finite groups.

Brauer and Nesbitt proved that a p-defect zero character has the value 0 on every element g whose order is divisible by p. The proof that we present here is an elementary proof discovered independently by B. Külshammer and M. Leitz.

We start with an identity of characters. If G is a group and $m \geq 1$, then we denote by G^m the direct product $G \times \cdots \times G$ (m times).

Lemma 4.3 *Suppose that G is a finite group, $\chi \in \mathrm{Irr}(G)$ and $g \in G$. If $m \geq 1$ is any integer, then*

$$\sum_{\substack{(x_1,\ldots,x_m)\in G^m \\ x_1\cdots x_m=g}} \chi(x_1)\cdots\chi(x_m) = \left(\frac{|G|}{\chi(1)}\right)^{m-1} \chi(g).$$

Proof Let

$$e = \frac{\chi(1)}{|G|} \sum_{g\in G} \chi(g) g$$

be the central idempotent corresponding to the complex conjugate character $\bar{\chi}$ of χ. (See Theorem 1.5.) Now, we have that $e^m = e$. The lemma easily follows by comparing the coefficients of g in e^m and e. $\square$

Now, if G is a group and $m \geq 1$, then we let $\sigma \colon G^m \to G^m$ be the *shift automorphism*

$$\sigma(x_1,\ldots,x_m) = (x_m, x_1,\ldots,x_{m-1}).$$

Lemma 4.4 *If r is a divisor of m and G is a group, then $x = (x_1,\ldots,x_m) \in G^m$ is fixed by σ^r if and only if x is the concatenation of m/r identical r-tuples. In this case*

$$\prod_{i=1}^{m} x_i = h^{\frac{m}{r}}$$

for some $h \in G$.

Proof Write $x = (x_1, \ldots, x_r, x_{r+1}, \ldots, x_{2r}, \ldots, x_{m-r+1}, \ldots, x_m)$ in blocks of r elements, and notice that σ^r sends the element in the ith position to the position $i + r$ for $1 \le i \le m - r$, and the element in the $(m - r + i)$th position to the position i for $1 \le i \le r$. In particular, we see that x is fixed by σ^r if and only if

$$x = (x_1, \ldots, x_r, x_1, \ldots, x_r, \ldots, x_1, \ldots, x_r).$$

In this case, notice that

$$\prod_{i=1}^{m} x_i = h^{m/r},$$

where $h = \prod_{i=1}^{r} x_i$. $\qquad\qquad\square$

If K is a conjugacy class of G and

$$S_m(K) = \{(x_1, \ldots, x_m) \in G^m \mid x_1 \cdots x_m \in K\},$$

we observe that σ fixes $S_m(K)$ setwise since

$$x_m(x_1 \cdots x_{m-1}) = (x_1 \cdots x_{m-1} x_m)^{x_m^{-1}}.$$

In particular, the cyclic group $\langle \sigma \rangle$ of order m acts on $S_m(K)$.

Recall that if p is a prime, G is a finite group and $x \in G$, then $x = x_p x_{p'}$, where x_p is the p-part of x and $x_{p'}$ is its p'-part. Also, $x \in G$ is p-**singular** if p divides the order of x (or, in other words, if $x_p \ne 1$). Otherwise, x is p-**regular**.

Lemma 4.5 *Let G be a finite group, let p be a prime and let $n \ge 1$ be any integer. Suppose that $|G|_p = p^a$, and set $m = p^{a+n}$. Let K be a conjugacy class of G consisting of p-singular elements. If Δ is any $\langle \sigma \rangle$-orbit of $S_m(K)$, then p^n divides $|\Delta|$.*

Proof Let Δ be the $\langle \sigma \rangle$-orbit of $(x_1, \ldots, x_m) \in S_m(K)$, and write $g = x_1 \cdots x_m$, which by hypothesis is p-singular. Let H be the stabilizer of $(x_1, \ldots, x_m)$ in $\langle \sigma \rangle$. We can write $H = \langle \sigma^r \rangle$, where r divides m. Then $|\Delta| = m/|H| = r$ by the orbit-stabilizer theorem. Now, σ^r fixes $(x_1, \ldots, x_m)$ and therefore

$$g = h^{m/r}$$

for some $h \in G$ by Lemma 4.4. Thus

$$g = (h_p)^{m/r}(h_{p'})^{m/r}.$$

Since g is p-singular and m/r is a power of p, it follows that $m/r < p^a$. Thus p^n divides r. $\qquad\square$

Theorem 4.6 (Brauer–Nesbitt) *Let G be a finite group, let p be a prime, and suppose that $\chi \in \mathrm{Irr}(G)$ has p-defect zero. If $g \in G$ has order divisible by p, then $\chi(g) = 0$.*

Proof (Leitz) Write $|G|_p = p^a$, let $n \geq 1$, and set $m = p^{a+n}$. Let K be the class of g. By Lemma 4.5, we have that every orbit Δ of the action of $\langle\sigma\rangle$ on $S_m(K)$ has length divisible by p^n. By Lemma 4.3, we have that

$$\sum_{(x_1,\ldots,x_m)\in S_m(K)} \chi(x_1)\cdots\chi(x_m) = \left(\frac{|G|}{\chi(1)}\right)^{m-1} \chi(g)|K|.$$

Since the value $\chi(x_1)\cdots\chi(x_m)$ is constant on σ-orbits, we deduce that

$$p^n\alpha_n = \left(\frac{|G|}{\chi(1)}\right)^{m-1} \chi(g)|K|$$

for some algebraic integer α_n. Since p^n and $(\frac{|G|}{\chi(1)})^{m-1}$ are coprime, there are integers u, v such that

$$up^n + v\left(\frac{|G|}{\chi(1)}\right)^{m-1} = 1.$$

Multiplying this equality by $\chi(g)|K|$, we deduce that p^n divides $\chi(g)|K|$ in the ring of algebraic integers, for every n. We claim that $\chi(g) = 0$. If $\beta = \chi(g)|K|$ and $f(x) \in \mathbb{Z}[x]$ is the minimal polynomial of β, we have that p^n divides $f(0)$ in the ring of algebraic integers because p^n divides every power of β. Thus p^n divides $f(0)$ in $\mathbb{Z}$ for all n, and thus $f(0) = 0$. Hence $f(x) = x$, and the claim is proven. $\qquad\square$

Corollary 4.7 *Let G be a finite group, let p be a prime and let $P \in \mathrm{Syl}_p(G)$. Let $\chi \in \mathrm{Irr}(G)$. Then the following conditions are equivalent.*

(a) $\chi(1)_p = |G|_p$.
(b) $\chi(g) = 0$ whenever p divides $o(g)$.
(c) $\chi(g) = 0$ whenever $1 \neq g \in P$.

Proof We only have to prove that (c) implies (a). By hypothesis, we have that the characters

$$\rho_P = \sum_{\gamma\in\mathrm{Irr}(P)} \gamma(1)\gamma$$

and χ_P coincide on every nonidentity element (by using the second orthogonality relation in P). Thus

$$\chi_P = \frac{\chi(1)}{|P|}\rho_P \,.$$

Then

$$\frac{\chi(1)}{|P|} = [\chi_P, 1_P] \in \mathbb{Z} \,,$$

as desired. □

R. Knörr proved that, in fact, $\chi \in \mathrm{Irr}(G)$ has p-defect zero if and only if $\chi(x) = 0$ for all $x \in G$ of order p. The following proof of a more general result is due to J. Murray. As in Chapter 2, we choose again a maximal ideal M of the ring $\mathbf{R}$ of algebraic integers in $\mathbb{C}$ containing p. Then $F = \mathbf{R}/M$ is a field, and we denote by $*\colon \mathbf{R} \to F$ the canonical ring homomorphism. Recall from the discussion before Lemma 2.7 that, if $\chi \in \mathrm{Irr}(G)$, then $\lambda_\chi \colon \mathbf{Z}(FG) \to F$ defined by

$$\lambda_\chi(\hat{K}) = \omega_\chi(\hat{K})^* = \left(\frac{|K|\chi(x)}{\chi(1)}\right)^* ,$$

where K is a conjugacy class of G and $x \in K$, defines an algebra homomorphism.

Theorem 4.8 (Knörr) *Suppose that $\chi \in \mathrm{Irr}(G)$ is such that $\sum_{o(x)=p} \chi(x) = 0$. Then $\chi(1)_p = |G|_p$.*

Proof (Murray) We work in the group algebra FG. If $X \subseteq G$, then $\hat{X} = \sum_{x \in X} x \in FG$. Let $\Omega = \{x \in G \mid x^p = 1\}$. Hence Ω is a union of conjugacy classes of G, and therefore $\hat{\Omega} \in \mathbf{Z}(FG)$. Write

$$\hat{\Omega}^2 = \sum_{K \in \mathrm{Cl}(G)} a_K \hat{K} \,,$$

where $a_K \in F$. We claim that if $a_K \neq 0$, then $|K|_p = |G|_p$. Let $g \in K$ and let $D \in \mathrm{Syl}_p(\mathbf{C}_G(g))$. We show that $D = 1$. Assume that $D > 1$. Let

$$\Lambda = \{(x, y) \in \Omega \times \Omega \mid xy = g\}$$

and let

$$\Lambda_0 = \{(x, y) \in (\Omega \cap \mathbf{C}_G(D)) \times (\Omega \cap \mathbf{C}_G(D)) \mid xy = g\} \,.$$

We have that D acts on Λ by conjugation with fixed points Λ_0, so

$$a_K = |\Lambda| \equiv |\Lambda_0| \bmod p \,.$$

Now, we observe that the nontrivial group $Z = \{z \in \mathbf{Z}(D) \mid z^p = 1\}$ acts on Λ_0 via

$$(x, y) \cdot z = (xz, z^{-1}y).$$

Since $(x, y) \cdot z = (x, y)$ only if $z = 1$, we have that the size of the orbits of the action of Z on Λ_0 is divisible by p. The claim follows.

By hypothesis, notice that $\lambda_\chi(\hat{\Omega}) = 1$. Thus

$$\lambda_\chi(\hat{\Omega}^2) = \lambda_\chi(\hat{\Omega})^2 = 1.$$

Therefore

$$1 = \lambda_\chi(\hat{\Omega}^2) = \sum_{K \in \mathrm{Cl}(G)} a_K \lambda_\chi(\hat{K}).$$

Hence, there is a class K such that $a_K \lambda_\chi(\hat{K}) \neq 0$. By the claim, $|K| = n|G|_p$, where n is not divisible by p. Write $\chi(1) = m\chi(1)_p$, where m is not divisible by p, and let $g \in K$. Since

$$\chi(g)\frac{|G|_p}{\chi(1)_p}$$

is an algebraic integer, we have that

$$\left(\frac{\chi(g)|G|_p}{\chi(1)_p}\right)^* n^* = \lambda_\chi(\hat{K})m^* \neq 0,$$

and we conclude that

$$\chi(g)^* \left(\frac{|G|_p}{\chi(1)_p}\right)^* = \left(\frac{\chi(g)|G|_p}{\chi(1)_p}\right)^* \neq 0.$$

Thus $|G|_p = \chi(1)_p$, as desired. $\qquad\square$

It immediately follows now from Theorem 4.8 that if $\chi \in \mathrm{Irr}(G)$ is such that $\chi(x) = 0$ for all $x \in G$ of order p, then $\chi(1)_p = |G|_p$.

4.3 Existence of Defect Zero Characters

In Problem 19 of his famous list of problems, Richard Brauer asked to characterize the number of characters of defect zero of a finite group by group theoretical properties. This problem was solved by G. R. Robinson using modular representation theory: this number is the rank of a certain matrix which is defined group theoretically. W. Feit, in another well-known list of open problems, asks what are some necessary and sufficient group theoretical conditions for the existence of characters of p-defect zero. In this section, we prove a theorem of S. P. Strunkov.

We start with a remarkable result.

Theorem 4.9 (Frobenius) *If G is a finite group and $g \in G$, let*

$$\gamma(g) = |\{(a, b) \in G \times G \mid a^{-1}b^{-1}ab = g\}|.$$

Then

$$\gamma = \sum_{\chi \in \mathrm{Irr}(G)} \frac{|G|}{\chi(1)} \chi.$$

Proof First of all, notice that γ is a class function of G. Also, $\gamma(g) = \gamma(g^{-1})$. We claim that the following equality

$$\sum_{g \in G} \gamma(g)g = \sum_{K \in \mathrm{Cl}(G)} |\mathbf{C}_G(x_K)| \widehat{K^{-1}} \widehat{K},$$

holds in $\mathbb{C}G$, where x_K is a fixed element in K, $\hat{X} = \sum_{x \in X} x$ whenever $X \subseteq G$, and $K^{-1} = \{k^{-1} \mid k \in K\}$. If $a, g \in G$, let $\mu(a, g) = 1$ if there is some $b \in G$ such that $a^{-1}a^b = g$, and let $\mu(a, g) = 0$ otherwise. If we fix a and g, then notice that the number of elements $b \in G$ such that $a^{-1}a^b = g$ is $|\mathbf{C}_G(a)b| = |\mathbf{C}_G(a)|$, or zero. Therefore

$$\gamma(g) = \sum_{a \in G} |\mathbf{C}_G(a)| \mu(a, g).$$

Now, the coefficient of g in $\widehat{K^{-1}} \widehat{K}$ is

$$|\{(x, y) \in K^{-1} \times K \mid xy = g\}| = \sum_{x \in K^{-1}} \mu(x^{-1}, g) = \sum_{x \in K} \mu(x, g).$$

Thus the coefficient of g in $\sum_{K \in \mathrm{Cl}(G)} |\mathbf{C}_G(x_K)| \widehat{K^{-1}} \widehat{K}$ is

$$\sum_{K \in \mathrm{Cl}(G)} |\mathbf{C}_G(x_K)| \sum_{x \in K} \mu(x, g) = \sum_{K \in \mathrm{Cl}(G)} \sum_{x \in K} |\mathbf{C}_G(x)| \mu(x, g) = \gamma(g),$$

and this proves the claim.

Let $z = \sum_{g \in G} \gamma(g)g$. Notice that

$$\omega_\chi(z) = \omega_\chi \Big(\sum_{K \in \mathrm{Cl}(G)} |\mathbf{C}_G(x_K)| \widehat{K^{-1}} \widehat{K} \Big) = \sum_{K \in \mathrm{Cl}(G)} |\mathbf{C}_G(x_K)| \omega_\chi(\widehat{K^{-1}}) \omega_\chi(\widehat{K})$$

$$= \frac{|G|}{\chi(1)^2} \sum_{K \in \mathrm{Cl}(G)} |K| \overline{\chi(x_K)} \chi(x_K)$$

$$= \left(\frac{|G|}{\chi(1)} \right)^2,$$

using that ω_χ is an algebra homomorphism, and the first orthogonality relation. Now, we use the central primitive idempotents e_χ, and the fact that

$e_\chi e_\psi = \delta_{\chi,\psi} e_\psi$ (Theorem 1.5). We have that $\{e_\chi \mid \chi \in \mathrm{Irr}(G)\}$ is a basis of $\mathbf{Z}(\mathbb{C}G)$ and furthermore

$$y = \sum_{\chi \in \mathrm{Irr}(G)} \omega_\chi(y) e_\chi$$

for every $y \in \mathbf{Z}(\mathbb{C}G)$. (This is the content of Problem 1.5, which immediately follows by using that $\omega_\chi(e_\psi) = \delta_{\chi\psi}$.) Then

$$z = \sum_{g \in G} \gamma(g) g = \sum_{\chi \in \mathrm{Irr}(G)} \left(\frac{|G|}{\chi(1)}\right)^2 e_\chi = \sum_{\chi \in \mathrm{Irr}(G)} \left(\frac{|G|}{\chi(1)}\right)^2 \frac{\chi(1)}{|G|} \sum_{g \in G} \chi(g^{-1}) g$$

$$= \sum_{g \in G} \left(\sum_{\chi \in \mathrm{Irr}(G)} \frac{|G|}{\chi(1)} \chi(g^{-1}) \right) g .$$

Since the two coefficients of $g \in G$ in the expression above should be the same, using that $\gamma(g^{-1}) = \gamma(g)$, the proof of the theorem is complete. $\square$

From Theorem 4.9, we can derive the following well-known fact: we can detect in the character table if an element of the group (a column in the character table) is a commutator. This has been widely used in the resolution of the famous Ore's conjecture, which asserts that every element in a finite non-abelian simple group is a commutator.

Corollary 4.10 *Let G be a finite group, and let $g \in G$. Then $g = [a, b]$ for some $a, b \in G$ if and only if*

$$\sum_{\chi \in \mathrm{Irr}(G)} \frac{\chi(g)}{\chi(1)} \neq 0 .$$

Proof It follows from Theorem 4.9. $\square$

If G is a finite group, p is a prime, and ψ is a character of G such that $[\psi, \chi]$ is divisible by p for all $\chi \in \mathrm{Irr}(G)$, then it is clear that the values of

$$\psi = \sum_{\chi \in \mathrm{Irr}(G)} [\psi, \chi] \chi$$

are divisible by p in the ring $\mathbf{R}$ of algebraic integers. By considering the regular character of any nontrivial p-group, for instance, we see that the converse is not in general true.

Lemma 4.11 *Let G be a finite group, let p be a prime, and suppose that ψ is a character of G such that $\psi(g) \equiv 0 \bmod p\mathbf{R}$ for every $g \in G$. Then $[\psi, \chi]$ is divisible by p for every $\chi \in \mathrm{Irr}(G)$ of p-defect zero.*

Proof Let $\chi \in \mathrm{Irr}(G)$. Notice that

$$\frac{|G|}{\chi(1)}[\chi, \psi] = \frac{1}{\chi(1)} \sum_{K \in \mathrm{Cl}(G)} |K| \chi(x_K) \overline{\psi(x_K)} = \sum_{K \in \mathrm{Cl}(G)} \omega_\chi(\hat{K}) \overline{\psi(x_K)},$$

where $x_K \in K$. Therefore, using that $\omega_\chi(\hat{K}) \in \mathbf{R}$ and the hypothesis, we have that

$$\frac{|G|}{\chi(1)}[\chi, \psi] \equiv 0 \bmod p\mathbf{R}.$$

Since $\mathbf{R} \cap \mathbb{Q} = \mathbb{Z}$, we deduce that p divides $\frac{|G|}{\chi(1)}[\chi, \psi]$. Now, if χ has p-defect zero, then p does not divide $\frac{|G|}{\chi(1)}$, and we deduce that p divides $[\chi, \psi]$. $\qquad\square$

Theorem 4.12 (Strunkov) *Let G be a finite group, and let p be a prime. Then G has a p-defect zero character if and only if there exists $g \in G$ such that p does not divide $|\{(a, b) \in G \times G \mid a^{-1}b^{-1}ab = g\}|$.*

Proof If $\gamma(g) = |\{(a, b) \in G \times G \mid a^{-1}b^{-1}ab = g\}|$ for $g \in G$, then we know by Theorem 4.9 that

$$\gamma = \sum_{\chi \in \mathrm{Irr}(G)} \frac{|G|}{\chi(1)} \chi.$$

If G has no p-defect zero character, we have that p divides $[\gamma, \chi]$ for every $\chi \in \mathrm{Irr}(G)$, and therefore $\gamma(g) \equiv 0 \bmod p\mathbf{R}$ for every $g \in G$. Since $\gamma(g)$ is an integer, we deduce that p divides $\gamma(g)$ for every $g \in G$.

Conversely, if p divides $\gamma(g)$ for all $g \in G$, then $\gamma(g) \equiv 0 \bmod p\mathbf{R}$ for all $g \in G$. Thus, by Lemma 4.11 we have that

$$\frac{|G|}{\chi(1)} = [\gamma, \chi]$$

is divisible by p for every $\chi \in \mathrm{Irr}(G)$. $\qquad\square$

4.4 Products of Conjugacy Classes

There are several open problems of similar nature to Ore's conjecture on which there is a great deal of activity lately. For instance, J. Thompson has conjectured that for every finite non-abelian simple group, there is always a conjugacy class C of G such that $C^2 = G$. (As can easily be seen, this generalizes the Ore conjecture.) Also, Z. Arad and M. Herzog have conjectured that the product of two nontrivial conjugacy classes of a simple group is never a conjugacy class. For these problems and others, an elementary character identity holds the key.

We let $\mathrm{Cl}(G) = \{K_1, \ldots, K_k\}$ be the set of conjugacy classes of G, and let $x_i \in K_i$.

Theorem 4.13 *Suppose that $C_1, \ldots, C_n$ are conjugacy classes of G, and let $c_i \in C_i$. In $\mathbf{Z}(\mathbb{C}G)$, write*

$$\widehat{C_1} \cdots \widehat{C_n} = \sum_{j=1}^{k} a_j \widehat{K_j}.$$

Then

$$a_m = \frac{|C_1| \cdots |C_n|}{|G|} \left(\sum_{\chi \in \mathrm{Irr}(G)} \frac{\chi(c_1) \cdots \chi(c_n) \overline{\chi(x_m)}}{\chi(1)^{n-1}} \right).$$

Proof Since $\widehat{C_1}, \ldots, \widehat{C_n} \in \mathbf{Z}(\mathbb{C}G)$ and $\{\widehat{K_1}, \ldots, \widehat{K_k}\}$ is a basis of $\mathbf{Z}(\mathbb{C}G)$, we certainly can write

$$\widehat{C_1} \cdots \widehat{C_n} = \sum_{j=1}^{k} a_j \widehat{K_j}$$

for some $a_j \in \mathbb{C}$. Let $\chi \in \mathrm{Irr}(G)$ and apply ω_χ to the previous identity. Using that ω_χ is multiplicative, we obtain

$$\frac{|C_1| \cdots |C_n| \chi(c_1) \cdots \chi(c_n)}{\chi(1)^{n-1}} = \sum_{j=1}^{k} a_j |K_j| \chi(x_j).$$

Now it is enough to multiply both sides by $\overline{\chi(x_m)}$, add over all $\chi \in \mathrm{Irr}(G)$ and apply the second orthogonality relation. $\qquad\square$

Corollary 4.14 *Suppose that $C_1, \ldots, C_n$ are conjugacy classes of G. Let $c_i \in C_i$, and $g \in G$. Then*

$$|\{(g_1, \ldots, g_n) \in C_1 \times \cdots \times C_n \mid g_1 \cdots g_n = g\}|$$

$$= \frac{|C_1| \cdots |C_n|}{|G|} \left(\sum_{\chi \in \mathrm{Irr}(G)} \frac{\chi(c_1) \cdots \chi(c_n) \overline{\chi(g)}}{\chi(1)^{n-1}} \right).$$

Proof Compute the coefficient of $g \in G$ in both sides of the equation in Theorem 4.13. $\qquad\square$

Corollary 4.15 (Harada) *Suppose that $K_1, \ldots, K_k$ are all the conjugacy classes of G, and let $x_i \in K_i$. Let $g = x_1 \cdots x_k$. Then*

$$\widehat{K_1} \cdots \widehat{K_k} = c\widehat{G'g},$$

where

$$c = \frac{|K_1| \cdots |K_k|}{|G'|}$$

is an integer.

Proof Let a_m be the coefficient of x_m in $\widehat{K_1} \cdots \widehat{K_k}$. By Theorem 4.13 and Burnside's zeros theorem, we have that

$$a_m = \frac{|K_1| \cdots |K_k|}{|G|} \left(\sum_{\chi \in \mathrm{Lin}(G)} \chi(x_1) \cdots \chi(x_k)\overline{\chi(x_m)} \right).$$

If $\chi \in \mathrm{Lin}(G)$, then

$$\chi(x_1) \cdots \chi(x_k) = \chi(x_1 \cdots x_k) = \chi(g).$$

Thus

$$a_m = \frac{|K_1| \cdots |K_k|}{|G|} \left(\sum_{\chi \in \mathrm{Lin}(G)} \chi(g)\overline{\chi(x_m)} \right).$$

Since

$$\sum_{\chi \in \mathrm{Lin}(G)} \chi(g)\overline{\chi(x_m)}$$

is zero if $G'x_m \neq G'g$ and it is $|G/G'|$ otherwise (by the second orthogonality relation), the theorem follows. $\qquad\square$

The following is nearly trivial. Recall that G is **perfect** if $G' = G$.

Lemma 4.16 *Suppose that $C_1, \ldots, C_n$ are conjugacy classes of G. Let $x_i \in C_i$, and let $g = x_1 \cdots x_n$.*

(a) If $g_i \in C_i$, then $g_1 \cdots g_n G' = gG'$.
(b) Suppose that $G = C_1 \cdots C_n$. Then G is perfect.

Proof Notice that if $g_i \in C_i$ then $g_i G' = x_i G'$ since G/G' is abelian. This proves (a). Now, suppose that $G = C_1 \cdots C_n$. Let $x \in G$. Then $x = g_1 \cdots g_n$ for some $g_i \in C_i$. Therefore $xG' = gG'$, by (a). Since this is for all $x \in G$, we have that $G \subseteq gG'$. In particular $|G| \leq |G'|$, and (b) follows. $\qquad\square$

Corollary 4.17 (Brauer–Wielandt) *Suppose that $K_1, \ldots, K_k$ are all the conjugacy classes of G. Then G is perfect if and only if*

$$\widehat{K_1} \cdots \widehat{K_k} = c\widehat{G}$$

for some $c \in \mathbb{Q}$.

Proof If $\widehat{K_1} \cdots \widehat{K_k} = c\widehat{G}$, since $c \neq 0$, it follows that $G = K_1 \cdots K_k$, and therefore $G' = G$ by Lemma 4.16(b). If $G = G'$, then apply Corollary 4.15. ∎

4.5 Nonvanishing Elements

A main idea in character theory is the existence of a duality between characters and classes of a finite group. This duality is not always perfect. For instance, Burnside's theorem on zeros asserts that, in the character table, every row that corresponds to a nonlinear character contains a zero. No irreducible character has a zero on a central element, but, in general, however, there might exist noncentral elements on which no irreducible character takes the value 0. The general idea is that if $\chi(x) \neq 0$ for all $\chi \in \mathrm{Irr}(G)$ (that is, if x is **nonvanishing**), then, perhaps x is not central, but at least x should lie in a nilpotent normal subgroup of G. As frequently happens in group theory, this is true with some exceptions. For instance, the reader can check that the alternating group A_7 has nonvanishing elements of orders 2 and 3.

Our next objective is to give a criterion for an element of a finite group to be nonvanishing. This is a result of J. Brough and M. Miyamoto.

Theorem 4.18 *Let N be a normal p-subgroup of G. If $x \in N$ is such that p does not divide $|G : \mathbf{C}_G(x)|$, then $\chi(x) \neq 0$ for all $\chi \in \mathrm{Irr}(G)$.*

Proof Again, we choose a maximal ideal M of the ring of algebraic integers containing p, and we let $F = \mathbf{R}/M$ and $*: \mathbf{R} \to F$ be the canonical homomorphism.

Since p does not divide $|G : \mathbf{C}_G(x)|$, we have that $N \subseteq \mathbf{C}_G(x)$. Therefore $x \in \mathbf{Z}(N)$, and, in particular, $[x, x^g] = 1$ for all $g \in G$. Now, if K is the conjugacy class of x in G, and $\hat{K} = \sum_{k \in K} k \in FG$, using that the elements of K commute we have that

$$(\hat{K})^{p^n} = |K|^* 1,$$

where $p^n = o(x)$ and 1 is the identity element of G. (In a ring of characteristic p, using the binomial theorem, we have that $(r + s)^p = r^p + s^p$, if r and s commute.) Now, recall that for all $\chi \in \mathrm{Irr}(G)$, we have that

$$\lambda_\chi(\hat{L}) = \left(\frac{|L|\chi(y)}{\chi(1)}\right)^*$$

defines an algebra homomorphism $\lambda_\chi \colon \mathbf{Z}(FG) \rightarrow F$, where L is the conjugacy class of $y \in G$. Then

$$\lambda_\chi(\hat{K})^{p^n} = \lambda_\chi((\hat{K})^{p^n}) = \lambda_\chi(|K|^*) = |K|^*.$$

Since by hypothesis $|K|^* \neq 0$, we deduce that $\lambda_\chi(\hat{K}) \neq 0$. Thus $\chi(x) \neq 0$, as desired.   □

4.6 Roots of Unity Values

We have so far written about zeros of characters but not about roots of unity values, except in Gallagher's theorem (Theorem 4.1). Our aim in this final section is to present a theorem of G. R. Robinson which uses several of the previous results and that constitutes a good example of the power of character theory.

The following general lemma is useful. In its proof, we shall use that if $\alpha_i \in \mathbf{R}$, then $(\alpha_1 + \cdots + \alpha_n)^p \equiv \alpha_1^p + \cdots + \alpha_n^p \bmod p\mathbf{R}$. (This follows by using induction on n, and the binomial theorem.) Only part (b) of Lemma 4.19 below is used in this chapter.

Lemma 4.19   *Let G be a finite group, p a prime, $g \in G$ and $\chi \in \mathrm{Char}(G)$.*

(a) We have that $\chi(g)^{o(g)_p} \equiv \chi(g_{p'})^{o(g)_p} \bmod p\mathbf{R}$.

(b) Let M be a maximal ideal of $\mathbf{R}$ containing p. Then $\chi(g) \equiv \chi(g_{p'}) \bmod M$.

Proof   Let $m = o(g)_p$. Write $g = yz$, where $y = g_p$ and $z = g_{p'}$. Let $\mathcal{X}$ be a representation affording χ. By elementary linear algebra, we can assume that

$$\mathcal{X}(g) = \mathrm{diag}(\epsilon_1, \dots, \epsilon_n)$$

is a diagonal matrix. Since y and z are powers of g, we have that

$$\mathcal{X}(y) = \mathrm{diag}(\alpha_1, \dots, \alpha_n),$$

where $\alpha_i^m = 1$, and

$$\mathcal{X}(z) = \mathrm{diag}(\beta_1, \dots, \beta_n),$$

with $\epsilon_i = \alpha_i \beta_i$. Now

$$\chi(g)^m = (\epsilon_1 + \cdots + \epsilon_n)^m \equiv \beta_1^m + \cdots + \beta_n^m \equiv (\beta_1 + \cdots + \beta_n)^m = \chi(z)^m \bmod p\mathbf{R}.$$

This shows part (a).

Now, let $F = \mathbf{R}/M$, a field of characteristic p, and let $* : \mathbf{R} \to F$ be again the associated ring epimorphism. If $\xi \in \mathbf{R}$ is a p-power root of unity, then $f = \xi^* \in F$ satisfies that $f^{p^a} = 1$ for some a. Then $0 = f^{p^a} - 1 = (f - 1)^{p^a}$ (by using the binomial theorem), and therefore $f = 1$. Hence $\xi \equiv 1 \bmod M$. Now $\alpha_i \beta_i \equiv \beta_i \bmod M$, and part (b) easily follows. $\qquad\square$

A word of caution is necessary in Lemma 4.19, since the congruence in (b) only holds modulo maximal ideals: it is false that if ξ is a pth root of unity then $\xi \equiv 1 \bmod p\mathbf{R}$.

Corollary 4.20 *Let G be a finite group and let $\chi \in \mathrm{Char}(G)$ of degree not divisible by p. If $g \in G$ is a p-element, then $\chi(g) \neq 0$.*

Proof If $\chi(g) = 0$, then $0 \equiv \chi(1) \bmod M$ by Lemma 4.19(b). Now, there are integers u and v such that $up + v\chi(1) = 1$, and we conclude that $1 \in M$. This is a contradiction. $\qquad\square$

Lemma 4.21 *Suppose that $\chi \in \mathrm{Irr}(G)$ and let $g, h \in G$. Then*

$$\chi(g)\chi(h) = \frac{\chi(1)}{|G|} \sum_{z \in G} \chi(gh^z).$$

Proof Let K and L be the G-conjugacy classes of g and h, respectively, and let $\mathcal{X}$ be a representation of G affording χ, which we extend $\mathbb{C}$-linearly to a representation $\mathcal{X} : \mathbb{C}G \to \mathrm{Mat}_n(\mathbb{C})$ of the group algebra. If $\hat{X} \in \mathbb{C}G$ is the class sum of the elements of the conjugacy class X of $x \in G$, recall that

$$\mathcal{X}(\hat{X}) = \omega_\chi(\hat{X})I_n = \left(\frac{|X|\chi(x)}{\chi(1)} \right) I_n.$$

Hence

$$\mathcal{X}(\hat{K}\hat{L}) = \mathcal{X}(\hat{K})\mathcal{X}(\hat{L}) = \left(\frac{|K||L|\chi(g)\chi(h)}{\chi(1)^2} \right) I_n.$$

On the other hand,

$$\mathcal{X}(\hat{K}\hat{L}) = \mathcal{X}(\sum_{x \in K} x\hat{L}) = \sum_{x \in K} \mathcal{X}(x\hat{L}).$$

Since $x\hat{L}$ and $x^y\hat{L}$ are G-conjugate for $y \in G$, taking traces we obtain

$$|K| \sum_{l \in L} \chi(gl) = \frac{|K||L|\chi(g)\chi(h)}{\chi(1)}.$$

Finally, if T is a set of representatives of right cosets of $\mathbf{C}_G(h)$ in G, then $L = \{h^t \mid t \in T\}$ and

$$\sum_{z \in G} \chi(gh^z) = \sum_{y \in \mathbf{C}_G(h), t \in T} \chi(gh^{yt}) = |\mathbf{C}_G(h)| \sum_{l \in L} \chi(gl),$$

and the assertion of the lemma easily follows. $\qquad\qquad\qquad\square$

We remind the reader that an element $x \in G$ is **strongly real** in G if there is an involution $t \in G$ such that $x^t = x^{-1}$. The following is standard group theory.

Lemma 4.22 *Let G be a finite group and let $u, v \in G$ be nonconjugate involutions of G. Then uv is a strongly real 2-singular element of G.*

Proof Let $H = \langle u, v \rangle = \langle uv, v \rangle$. Notice that $(uv)^v = vu = v^{-1}u^{-1} = (uv)^{-1}$, and therefore H is a dihedral group of order $2n$, where n is the order of uv. If n is odd, then $\langle u \rangle$, $\langle v \rangle$ are Sylow 2-subgroups of H, so they are H-conjugate. But this implies that u and v are conjugate in G, against the hypothesis. $\qquad\qquad\qquad\square$

Theorem 4.23 (Robinson) *Let G be a finite group of even order. Suppose that $\chi \in \mathrm{Irr}(G)$ is nonlinear and such that $\chi(g)$ is a root of unity for every 2-singular $g \in G$. Then either the Sylow 2-subgroups of G are dihedral or else G has a unique class of involutions.*

Proof Let S be a Sylow 2-subgroup of G. If S is cyclic or generalized quaternion, then S has a unique involution, and therefore G has a unique class of involutions. We may assume that S is not cyclic, generalized quaternion or dihedral, and prove that G has a unique class of involutions. Write $|S| = 2^a$, where $a \geq 3$.

Step 1 We have that $\chi(1) \equiv \epsilon \bmod 2^{a-1}$ for a uniquely defined sign ϵ.

Since $1 = [\chi, \chi] = [\chi\bar{\chi}, 1_G]$, then we can write

$$\chi\bar{\chi} = 1 + \Psi,$$

where Ψ is a character of G. By hypothesis,

$$\chi(x)\bar{\chi}(x) = |\chi(x)|^2 = 1$$

for every 2-singular $x \in G$, and thus $\Psi(x) = 0$. We have then that Ψ_S is a multiple of the regular character ρ_S of S, so $\Psi(1)$ is divisible by 2^a. Thus

$$\chi(1)^2 \equiv 1 \bmod 2^a,$$

and by elementary number theory $\chi(1) \equiv \pm 1 \bmod 2^a$ or $\chi(1) \equiv 2^{a-1} \pm 1$ mod 2^a. In any case

$$\chi(1) \equiv \epsilon \bmod 2^{a-1}$$

for a uniquely defined sign ϵ.

Step 2 We may assume that $\det(\chi)(x) = 1$ for every $x \in S$.

Since $\chi(1)$ is odd, there is an integer b such that $b\chi(1) \equiv 1 \bmod 2^a$. If $\lambda = \det(\chi)^{-b}$, then notice that

$$\det(\lambda\chi) = \lambda^{\chi(1)}\det(\chi) = \det(\chi)^{1-b\chi(1)}$$

has $2'$-order in the group of linear characters of G, and in particular, has the value 1 on 2-elements. By replacing χ by $\lambda\chi$, we may therefore assume that $\det(\chi)$ has the value 1 on the elements of S.

Step 3 If $y \in G$ is a real 2-singular element of G, then $\chi(y) = \pm 1$.

This is clear, since in this case $\chi(y)$ is a real root of unity by hypothesis.

Step 4 If $1 \neq x$ is any real 2-element of G, and δ_x is the nontrivial character of the cyclic group $\langle x \rangle / \langle x^2 \rangle$ of order 2, then $[\chi_{\langle x \rangle}, \delta_x]$ is even.

We have that all the elements in $\langle x \rangle$ are real in G, and thus $\chi_{\langle x \rangle}$ is real valued. Thus $[\chi_{\langle x \rangle}, \mu] = [\chi_{\langle x \rangle}, \bar{\mu}]$ for all $\mu \in \mathrm{Irr}(\langle x \rangle)$. By Problem 1.2(b), we have that

$$1_{\langle x \rangle} = \det(\chi)_{\langle x \rangle} = \det(\delta_x)^{[\chi_{\langle x \rangle}, \delta_x]}$$

and the step follows.

Step 5 We have that $\chi(x) = \epsilon$ for every nontrivial strongly real 2-element x of G.

We do this by induction on $o(x)$. Suppose first that x is an involution. Write

$$\chi_{\langle x \rangle} = c 1_{\langle x \rangle} + d\delta_x,$$

where we know that d is even by Step 4. Then $\chi(x) = c - d$ and $\chi(1) = c + d$. By Step 3, $\chi(x) = \pm 1$. If $\chi(x) = -\epsilon$, then $\chi(1) + \epsilon = \chi(1) - \chi(x) = 2d$ is divisible by 4, and this is not possible since $\chi(1) \equiv \epsilon \bmod 4$ by Step 1. Therefore $\chi(x) = \epsilon$, as wanted. Suppose now that x is a strongly real 2-element of order 2^n, where $n > 1$. Write $x^t = x^{-1}$, where $t \in G$ is some involution, and notice that every power of x is strongly real. Also $\langle x, t \rangle$ is a dihedral group of order 2^{n+1}. By induction we know that $\chi(y) = \epsilon$ for every $1 \neq y \in \langle x^2 \rangle$. Suppose that $\chi(x) = -\epsilon$. Then $\chi(z) = -\epsilon$ for every $z \in \langle x \rangle - \langle x^2 \rangle$, using Lemma 3.6(b). Now,

$$[\chi_{\langle x\rangle}, \delta_x] = \frac{1}{2^n}\left(\chi(1) + \sum_{1\neq y\in\langle x^2\rangle}\chi(y) - \sum_{z\in\langle x\rangle-\langle x^2\rangle}\chi(z)\right)$$

$$= \frac{1}{2^n}\left(\chi(1) + (2^{n-1}-1)\epsilon + 2^{n-1}\epsilon\right) = \epsilon + \frac{\chi(1)-\epsilon}{2^n}.$$

If $|\langle x,t\rangle| = 2^a$, then $\langle x,t\rangle \in \mathrm{Syl}_2(G)$ and we deduce that S is dihedral. Therefore $2^a > 2^{n+1}$ and $a-1 \geq n+1$. Since $\chi(1) \equiv \epsilon \bmod 2^{a-1}$ by Step 1, we have that $\chi(1) \equiv \epsilon \bmod 2^{n+1}$, and therefore that

$$\frac{\chi(1)-\epsilon}{2^n}$$

is even. But then $[\chi_{\langle x\rangle}, \delta_x]$ is odd by the previous equation, and we know that this is not possible by Step 4.

Step 6 We have that $\chi(x) = \epsilon$ for every strongly real 2-singular element x of G.

Again, we do this by induction on $o(x)$. If x is a 2-element, then we know that this is true by Step 5. So we may assume that there is an odd prime p dividing $o(x)$. Write $x = x_{p'}x_p$. Since $x^t = x^{-1}$, we have that $(x_{p'})^t = (x_{p'})^{-1}$. Therefore $x_{p'}$ is a 2-singular strongly real element of G, and by induction $\chi(x_{p'}) = \epsilon$. Now, we know that $\chi(x) = \pm 1$ by Step 3. By Lemma 4.19(b), we have that $\chi(x) \equiv \epsilon$ modulo M, for some maximal ideal M of $\mathbf{R}$ containing p. If $\chi(x) = -\epsilon$, then $2 \in M$ and since p is odd, we conclude that $1 \in M$, a contradiction. This shows that $\chi(x) = \epsilon$.

Step 7 The theorem is true.

Assume that u and v are involutions which are not G-conjugate. By Lemma 4.22, we have that uv^g is a 2-singular strongly real element of G for every $g \in G$. By Step 6, we have that $\chi(uv^g) = \epsilon$ for every $g \in G$. By Step 5, $\chi(u) = \chi(v) = \epsilon$. By Lemma 4.21, we have that

$$\chi(u)\chi(v) = \frac{\chi(1)}{|G|}\sum_{g\in G}\chi(uv^g),$$

and we deduce that $\chi(1) = 1$, the final contradiction. $\qquad\square$

A_5, A_6, A_7 and J_1, for instance, are examples of groups having nonlinear irreducible characters whose values on 2-singular elements are roots of unity. As shown by the group S_4, it may happen that a group G has dihedral Sylow 2-subgroups and more than one class of involutions. In this case, it is well known that $\mathbf{O}^2(G) \neq G$. (See Problem 4.13 for an elementary proof of this fact.)

4.7 Notes

An important and frequently used result in character theory is that, with some exceptions for the primes $p = 2$ and $p = 3$, the non-abelian simple groups always possess p-defect zero characters for every prime p. This result was proven by G. Michler and W. Willems for groups of Lie type (see [Mi86] and [Wil88]) and was later completed by A. Granville and K. Ono for alternating groups (see [GO96]).

If $\chi \in \mathrm{Irr}(G)$ does not have p-defect zero, it is my observation that the p-part of the (nonzero) integer $\sum_{o(x)=p} \chi(x)$ is $\chi(1)_p$, but the proof of this fact seems to require some elementary modular representation theory.

The proof of the Ore conjecture is a major accomplishment by M. Liebeck, E. O'Brien, A. Shalev and Pham Huu Tiep [LOST10]. Corollary 4.10 is an ingredient in the proof. We recall that not every element in the commutator subgroup of a group is a commutator, and the two smallest examples are SmallGroup(96,3) and SmallGroup(96, 203). These groups were discovered by R. M. Guralnick, when the database of small groups was not yet available. Also, if G is perfect, not every element of G is a commutator: the group $3.A_6$ is an example. The Ore conjecture asserts that non-abelian simple groups cannot be examples.

As we have said, there are significant related problems on products of classes of finite simple groups, on which there has been much recent activity. Of particular relevance is Thompson's conjecture that asserts, as we have already mentioned, that every simple group possesses a conjugacy class C such that $C^2 = G$. Essentially, the character theory tools for these and related problems are Theorem 4.13. The reader is invited to read the survey [Ma12].

We remark that we are not aware of any group theoretical proof showing that $|G'|$ divides the product of the sizes of the conjugacy classes. Slightly related, in [Ha17] K. Harada has conjectured that

$$\prod_{\chi \in \mathrm{Irr}(G)} \chi(1) \quad \text{divides} \quad \prod_{K \in \mathrm{Cl}(G)} |K|.$$

Nonvanishing elements were first studied in [INW99]. It was conjectured that every nonvanishing element x of a solvable group G lies in $\mathbf{F}(G)$, the Fitting subgroup of G, and this was proven for elements of odd order. If the order of x is not divisible by 6, then this is true for every finite group [DNPST10]. Also, it was proven in [DPSS09] that if all the p-elements are nonvanishing then the Sylow p-subgroup is normal. (A characterization of when the Sylow

p-subgroup is normal by nonvanishing elements is given in [MN12].) There is a conjecture of T. Wilde [Wi06] that asserts that if $g \in G$ and $o(g)$ does not divide $|G|/\chi(1)$, then $\chi(g) = 0$.

Characters that have roots of unity values on p-singular elements have been studied recently because of their connection with the so called *endo-trivial* modules (see [NR12]).

Problems

(4.1) Let $\alpha \in \mathbb{Q}_n$ be an algebraic integer with $|\alpha| = 1$. Show that α is a root of unity.

(*Hint*: By using the Galois conjugates of α, show that there is a finite number of minimal polynomials of algebraic integers α in $\mathbb{Q}_n$ with $|\alpha| = 1$. Hence $\{\alpha^m \mid m \geq 0\}$ is a finite set.)

(4.2) (*Burnside*) Suppose that $\chi \in \mathrm{Irr}(G)$, $g \in G$ and $\chi(g)/\chi(1)$ is an algebraic integer with $|\chi(g)| < \chi(1)$. Prove that $\chi(g) = 0$. Deduce that if $\gcd(|\mathrm{cl}_G(g)|, \chi(1)) = 1$, then either $|\chi(g)| = \chi(1)$ or $\chi(g) = 0$.

(*Hint*: If $\mathcal{G} = \mathrm{Gal}(\mathbb{Q}_{|G|}/\mathbb{Q})$ and $\sigma \in \mathcal{G}$, then show that $\chi^\sigma(g)/\chi(1)$ satisfies the same condition, and take the product over all σ.)

(4.3) Suppose that G is a finite group, $Z \subseteq \mathbf{Z}(G)$, and $\chi \in \mathrm{Irr}(G)$ is faithful. If $x \in G$ and $\mathbf{C}_{G/Z}(Zx) \neq \mathbf{C}_G(x)/Z$, then prove that $\chi(x) = 0$.

(4.4) Let G be a finite group, and suppose that $\chi \in \mathrm{Irr}(G)$ is nonlinear of degree a power of a prime p. Show that there exists $x \in G$ of order a power of p such that $\chi(x) = 0$.

(*Hint*: Use induction to assume that χ is faithful. The two previous problems are relevant.)

(4.5) Suppose that $\alpha, \beta \in \mathrm{Irr}(G)$, and let p be a prime not dividing $|G|$. If $\alpha^p = \beta^p$, then prove that $\alpha = \beta$.

(4.6) (*Gagola*) Let G be a finite group which has an irreducible character $\chi \in \mathrm{Irr}(G)$ such that χ does not vanish on exactly two conjugacy classes of G.

   (a) If $|G| > 2$, then prove that χ is unique and is the unique faithful irreducible character of G.

   (b) Prove that G contains a unique minimal normal subgroup N which is necessarily an elementary abelian p-group for some prime p.

   (c) Prove that χ vanishes on $G - N$ and is nonzero on N.

   (d) Prove that the action of G by conjugation on N is transitive on $N - \{1\}$.

(4.7) *(Arad–Herzog)* Let G be a finite simple group and let C be a nontrivial conjugacy class of G. Show that there is some integer m such that $G = C^m$.

(4.8) Suppose that $Z \leq \mathbf{Z}(G)$, where G is a finite group. Let $\chi \in \mathrm{Irr}(G)$. Let $m \geq 1$ and let

$$\Delta = \{(z_1, \ldots, z_m) \in Z^m \mid \prod_{i=1}^{m} z_i = 1\} \subseteq \mathbf{Z}(G^m).$$

(a) If $x = (x_1, \ldots, x_m)\Delta \in G^m/\Delta$, show that the element $x_1 \ldots x_m \in G$ and the value $\chi(x_1) \cdots \chi(x_m)$ only depend on x.

(b) Suppose that $\chi \in \mathrm{Irr}(G)$ and $g \in G$. Show that

$$\sum_{\substack{(x_1,\ldots,x_m)\Delta \in G^m/\Delta \\ x_1 \cdots x_m = g}} \chi(x_1) \cdots \chi(x_m) = \left(\frac{|G:Z|}{\chi(1)}\right)^{m-1} \chi(g).$$

(c) If $\chi(1)_p = |G : Z|_p$ and p divides the order of gZ show that $\chi(g) = 0$.

(*Hint:* For part (c), mimic the proof of Theorem 4.6.)

(4.9) *(Robinson)* If θ is a character of an abelian group A such that $\theta(a)$ is a root of unity for every $1 \neq a \in A$, then prove that $\theta = f\rho_A + \epsilon\lambda$ for some sign ϵ, $f \geq 0$ and integer, and some $\lambda \in \mathrm{Irr}(A)$, where ρ_A is the regular character of A.

(*Hint:* Write $\theta = \sum_{i=1}^{n} m_i \lambda_i$, where $m_i \leq m_j$ if $i \leq j$. By replacing θ by $\theta - m_1 \rho_A$, we may assume that $m_1 = 0$. Use that $n \sum_{i=1}^{n}(m_i)^2 - \left(\sum_{i=1}^{n} m_i\right)^2 = \sum_{1 \leq i < j \leq n}(m_i - m_j)^2$.)

(4.10) *(L. Solomon)* Let G be a finite group. Prove that the sum of the entries of the character table of G is $|G|$ if and only if $G/\mathbf{Z}(G)$ is abelian.

(*Hint:* Use Problem 1.12. If $G/\mathbf{Z}(G)$ is abelian, also use Lemma 4.21.)

(4.11) *(Ferguson–Isaacs)* Let G be a finite group and let $\chi \in \mathrm{Char}(G)$. Let H be a subgroup of G. Prove that $\chi(g) = 0$ whenever g is not conjugate to an element of H if and only if there is a positive integer a and a generalized character ν of H such that $a\chi = \nu^G$.

(*Hint:* Let $\pi = (1_H)^G$, and define $\alpha(g) = 1/\pi(g)$ if $\pi(g) \neq 0$ and $\alpha(g) = 0$, otherwise. Use that $\pi\alpha\chi = \chi$ and apply Problem 1.3 to α_H.)

(4.12) If $\chi \in \mathrm{Irr}(G)$, prove that

$$\sum_{x,y \in G} \chi([x, y]) = \frac{|G|^2}{\chi(1)}.$$

(4.13) *(Isaacs)* Suppose that C is a subgroup of G of index $2m$, where m is odd. Let t be an involution of G which is no G-conjugate to any element in C. Show that G has a subgroup of index 2.

(*Hint*: Consider the action of G on the right cosets of C in G.)

(4.14) *(Brauer)* Let G be a finite group and let p be a prime. Let $\mathbf{R}$ be the ring of algebraic integers in $\mathbb{C}$, let M be a maximal ideal of $\mathbf{R}$ containing p, let $F = R/M$ and let $* : \mathbf{R} \to F$ be the canonical epimorphism. If χ is a character of G, denote by χ^* the function $\chi : G \to F$ defined by $\chi^*(g) = \chi(g)^*$ for $g \in G$.

   (a) If $\chi_1, \ldots, \chi_t \in \mathrm{Irr}(G)$ are distinct and have p-defect zero, then show that the functions $\chi_1^*, \ldots, \chi_t^*$ are F-linearly independent.

   (b) Suppose that $\chi \in \mathrm{Irr}(G)$ has p-defect zero. If $x \in G$ and $\chi(x)^* \neq 0$, then show that $|G|_p = |K|_p$, where $K = \mathrm{cl}_G(x)$.

   (c) Show that the number of distinct irreducible characters of G of p-defect zero is less than or equal the number of different conjugacy classes K of G such that $|K|_p = |G|_p$.

5

Characters over a Normal Subgroup

Suppose that N is a normal subgroup of a finite group G and that $\theta \in \mathrm{Irr}(N)$ is a G-invariant character. In this case, we say that (G, N, θ) is a **character triple**. In this chapter we study the *character theory of G over θ*, and develop a technique that is frequently used with many important consequences. If we wish to study (G, N, θ) this technique often allows us to assume that N is central in G. Sometimes, it is even possible to assume that θ extends to G. We will see how to apply this tool in this chapter and in other parts of this book.

5.1 Cyclic Factors

We start with an extension theorem. Suppose that N is a normal subgroup of a finite group G and let $\theta \in \mathrm{Irr}(N)$. If there is some $\chi \in \mathrm{Irr}(G)$ such that $\chi_N = \theta$, then we have that $\theta^g = \theta$ for every $g \in G$, since χ is a class function of G. In other words, (G, N, θ) is a character triple. Of course, this is not a sufficient condition for θ to extend to G. Although we will seek sufficient conditions in the next chapter, right now we need the following one. Surprisingly, there is no purely character-theoretic proof of it, and we shall need to use representations.

Theorem 5.1 *Suppose that $N \trianglelefteq G$ and let $\theta \in \mathrm{Irr}(N)$ be G-invariant. Assume that G/N is cyclic. If $\mathcal{Y}$ is a representation of N affording θ, then there exists a representation $\mathcal{X}$ of G such $\mathcal{X}_N = \mathcal{Y}$. Hence, $\chi_N = \theta$ for every $\chi \in \mathrm{Irr}(G)$ that lies over θ.*

Proof Suppose that $G/N = \langle Ng \rangle$ for some $g \in G$. The representation $\mathcal{Y}^{g^{-1}}$ defined by $\mathcal{Y}^{g^{-1}}(n) = \mathcal{Y}(n^g)$ for $n \in N$ affords the character $\theta^{g^{-1}} = \theta$. Hence, $\mathcal{Y}^{g^{-1}}$ and $\mathcal{Y}$ are similar, and therefore there is an invertible matrix M such that

$$\mathcal{Y}(n^g) = M^{-1}\mathcal{Y}(n)M$$

for all $n \in N$. Then

$$\mathcal{Y}(n^{g^j}) = M^{-j}\mathcal{Y}(n)M^j$$

for $j \geq 1$. Now, if $m = |G : N| = o(gN)$, then $g^m \in N$ and

$$M^{-m}\mathcal{Y}(n)M^m = \mathcal{Y}(n^{g^m}) = \mathcal{Y}(g^m)^{-1}\mathcal{Y}(n)\mathcal{Y}(g^m)$$

for all $n \in N$. Therefore $\mathcal{Y}(g^m)M^{-m}$ commutes with every matrix $\mathcal{Y}(n)$ for $n \in N$. By Schur's lemma, there is $x \in \mathbb{C}^\times$ such that

$$\mathcal{Y}(g^m)M^{-m} = x I_{\theta(1)}.$$

Let $z \in \mathbb{C}^\times$ be such that $z^m = x$ and put $A = zM$. Notice that

$$\mathcal{Y}(g^m) = xM^m = A^m$$

and

$$\mathcal{Y}(n^{g^j}) = A^{-j}\mathcal{Y}(n)A^j$$

for every $j \geq 1$. Next, we check that the map

$$\mathcal{X}(g^i n) = A^i \mathcal{Y}(n),$$

where $0 \leq i < m$ and $n \in N$, defines a representation of G extending $\mathcal{Y}$. First of all, we claim that if j is any integer, then

$$\mathcal{X}(g^j n) = A^j \mathcal{Y}(n).$$

To prove the claim, write $j = md + r$, where d is some integer and $0 \leq r < m$. Then

$$\mathcal{X}(g^j n) = \mathcal{X}(g^r(g^m)^d n) = A^r \mathcal{Y}((g^m)^d n)$$
$$= A^r \mathcal{Y}(g^m)^d \mathcal{Y}(n) = A^r A^{md} \mathcal{Y}(n) = A^j \mathcal{Y}(n)$$

and the claim is proven. Finally,

$$\mathcal{X}(g^i n g^j n') = \mathcal{X}(g^{i+j} n^{g^j} n') = A^{i+j} \mathcal{Y}(n^{g^j} n') = A^{i+j} \mathcal{Y}(n^{g^j})\mathcal{Y}(n')$$
$$= A^{i+j} A^{-j} \mathcal{Y}(n) A^j \mathcal{Y}(n') = \mathcal{X}(g^i n)\mathcal{X}(g^j n')$$

for all $0 \leq i, j < m$ and $n, n' \in N$. Hence $\mathcal{X}$ is a representation. Therefore, if τ is the character afforded by $\mathcal{X}$, it follows that $\tau_N = \theta$. In particular, $\tau \in \mathrm{Irr}(G)$ because it cannot be the sum of two characters. Finally, if $\chi \in \mathrm{Irr}(G|\theta)$, then $\chi = \lambda\tau$ for some linear $\lambda \in \mathrm{Irr}(G/N)$, by the Gallagher correspondence Theorem 1.23. Thus $\chi_N = \tau_N = \theta$. $\square$

Of course, the hypothesis that G/N is cyclic cannot be removed in Theorem 5.1. The smallest counterexample is the dihedral group $G = D_8$ with $N = \mathbf{Z}(G)$, and $\theta \in \mathrm{Irr}(N)$ the nontrivial character. Hence, a representation $\mathcal{Y}$ of a normal subgroup affording a G-invariant character cannot be extended to a representation of G, in general. However it can be extended to a *projective representation*.

5.2 Projective Representations

If G is a finite group, then a **complex projective representation** of G is a function

$$\mathcal{P}: G \to \mathrm{GL}_n(\mathbb{C})$$

such that for every $x, y \in G$ there is some $\alpha(x, y) \in \mathbb{C}^\times$ satisfying

$$\mathcal{P}(x)\mathcal{P}(y) = \alpha(x, y)\mathcal{P}(xy).$$

Notice that the scalar $\alpha(x, y)$ is uniquely determined by x and y. Using that

$$(\mathcal{P}(x)\mathcal{P}(y))\mathcal{P}(z) = \mathcal{P}(x)(\mathcal{P}(y)\mathcal{P}(z)),$$

we easily check that

$$\alpha(xy, z)\alpha(x, y) = \alpha(x, yz)\alpha(y, z)$$

for all $x, y, z \in G$. The function $\alpha: G \times G \to \mathbb{C}^\times$ is called the **factor set** of $\mathcal{P}$, and the set of all functions $\alpha: G \times G \to \mathbb{C}^\times$ satisfying the above equation (which are also called **cocycles**) is $\mathbf{Z}^2(G, \mathbb{C}^\times)$.

Our interest in projective representations is that they become indispensable in order to obtain a deeper understanding of the ordinary representations. As we are about to see, projective representations naturally appear in the presence of character triples.

Definition 5.2 *Suppose that (G, N, θ) is a character triple. We say that a projective representation*

$$\mathcal{P}: G \to \mathrm{GL}_{\theta(1)}(\mathbb{C})$$

*is **associated** with θ if*

(a) $\mathcal{P}_N$ is a representation of N affording θ, and
(b) $\mathcal{P}(ng) = \mathcal{P}(n)\mathcal{P}(g)$ and $\mathcal{P}(gn) = \mathcal{P}(g)\mathcal{P}(n)$ for all $g \in G$ and $n \in N$.

We collect a few elementary properties of the factor sets of this type of projective representations.

Lemma 5.3 *Suppose that (G, N, θ) is a character triple, and let $\mathcal{P}$ be a projective representation of G associated with θ with factor set α. Then*

(a) $\alpha(1, 1) = \alpha(g, n) = \alpha(n, g) = 1$ for $n \in N$, $g \in G$;
(b) $\alpha(x, x^{-1}) = \alpha(x^{-1}, x)$ for $x \in G$;
(c) $\alpha(xn, ym) = \alpha(x, y)$ for $x, y \in G$, $n, m \in N$.

Proof Part (a) is trivial. For part (b), we have that

$$\alpha(xx^{-1}, x)\alpha(x, x^{-1}) = \alpha(x, x^{-1}x)\alpha(x^{-1}, x),$$

and then we use that $\alpha(1, x) = \alpha(x, 1) = 1$ from part (a). If $x, y \in G$, $n, m \in N$, then

$$\begin{aligned} \alpha(xn, ym)\mathcal{P}(xnym) = \mathcal{P}(xn)\mathcal{P}(ym) &= \mathcal{P}(x)\mathcal{P}(n)\mathcal{P}(ym) = \mathcal{P}(x)\mathcal{P}(nym) \\ &= \mathcal{P}(x)\mathcal{P}(yn^y m) = \mathcal{P}(x)\mathcal{P}(y)\mathcal{P}(n^y m) \\ &= \alpha(x, y)\mathcal{P}(xy)\mathcal{P}(n^y m) = \alpha(x, y)\mathcal{P}(xnym) \end{aligned}$$

and we deduce that $\alpha(xn, ym) = \alpha(x, y)$. $\qquad\qquad\square$

From Lemma 5.3(c), notice that if $\mathcal{P}$ is a projective representation of G associated with θ with factor set α, then α uniquely defines

$$\bar{\alpha} \in \mathbf{Z}^2(G/N, \mathbb{C}^\times)$$

by setting $\bar{\alpha}(Nx, Ny) = \alpha(x, y)$ for $x, y \in G$. This will become important in Chapter 10.

Our next goal is to prove that if (G, N, θ) is a character triple, then there exists a projective representation of G associated with θ. If θ is linear, this is quite easy: write

$$G = \bigcup_{j=1}^{t} Nx_j$$

as a disjoint union, with $x_1 = 1$, and define $\mathcal{P}(nx_j) = \theta(n)$ for $n \in N$. We easily check that $\mathcal{P}(gn) = \mathcal{P}(g)\mathcal{P}(n)$ and $\mathcal{P}(ng) = \mathcal{P}(n)\mathcal{P}(g)$ for $n \in N$ and $g \in G$. The general case when θ is not linear can also be easily proved with a variation of this argument, using Schur's lemma. However, we wish to obtain some control of the values of the factor set α, and for this we have to do a bit more work. We need the following lemma.

Lemma 5.4 *Suppose that $N \trianglelefteq G$ and let $\mathcal{R}$ be an irreducible representation of N affording a G-invariant character. Let $x, y \in G$ and let $\mathcal{X}$, $\mathcal{Y}$ and $\mathcal{Z}$ be* *extensions of* $\mathcal{R}$ *to* $\langle N, x \rangle$, $\langle N, y \rangle$ *and* $\langle N, xy \rangle$, *respectively. Then there is a scalar* $\alpha(x, y) \in \mathbb{C}^\times$ *such that*

$$\mathcal{X}(x)\mathcal{Y}(y) = \alpha(x, y)\mathcal{Z}(xy).$$

Proof Notice that $\mathcal{X}$, $\mathcal{Y}$ and $\mathcal{Z}$ exist by Theorem 5.1. For $n \in N$, we have

$$\mathcal{R}(yny^{-1}) = \mathcal{Y}(yny^{-1}) = \mathcal{Y}(y)\mathcal{Y}(n)\mathcal{Y}(y^{-1}) = \mathcal{Y}(y)\mathcal{R}(n)\mathcal{Y}(y^{-1}).$$

Notice that $\mathcal{Y}(y^{-1}) = \mathcal{Y}(y)^{-1}$. Conjugating by $\mathcal{X}(x)^{-1}$, we obtain

$$\mathcal{X}(x)\mathcal{Y}(y)\mathcal{R}(n)\mathcal{Y}(y)^{-1}\mathcal{X}(x)^{-1} = \mathcal{X}(x)\mathcal{R}(yny^{-1})\mathcal{X}(x^{-1})$$
$$= \mathcal{X}(x(yny^{-1})x^{-1}) = \mathcal{R}(xyny^{-1}x^{-1}).$$

Also,

$$\mathcal{R}(xyny^{-1}x^{-1}) = \mathcal{Z}(xy)\mathcal{R}(n)\mathcal{Z}(xy)^{-1}.$$

Now, the matrix $\mathcal{Z}(xy)^{-1}\mathcal{X}(x)\mathcal{Y}(y)$ commutes with the matrices $\mathcal{R}(n)$ for $n \in N$ and thus it is scalar by Schur's lemma. $\qquad\square$

Theorem 5.5 *Suppose that* (G, N, θ) *is a character triple, and let* $\mathcal{Y}$ *be a* $\mathbb{C}N$-*representation affording* θ. *Then there exists a projective representation* $\mathcal{P}$ *associated with* θ, *with factor set* α, *such that*

$$\alpha(x, y)^{|G|\theta(1)} = 1$$

for all $x, y \in G$, *and* $\mathcal{P}_N = \mathcal{Y}$.

Proof For each $\bar{g} = Ng \in G/N$, we choose a representation $\mathcal{Y}_{\bar{g}}$ of $\langle N, g \rangle$ extending $\mathcal{Y}$, using Theorem 5.1. We define

$$\mathcal{P}(g) = \mathcal{Y}_{\bar{g}}(g).$$

Note that $\mathcal{Y}_{\bar{n}} = \mathcal{Y}$ for $n \in N$. By Lemma 5.4, we have that $\mathcal{P}$ is a projective representation of G. Also, $\mathcal{P}(gn) = \mathcal{P}(g)\mathcal{Y}(n)$, $\mathcal{P}(ng) = \mathcal{Y}(n)\mathcal{P}(g)$ and $\mathcal{P}(n) = \mathcal{Y}(n)$ for $n \in N$ and $g \in G$. Therefore, $\mathcal{P}$ is a projective representation associated with θ. Furthermore, if m is any integer with $g^m \in N$, then

$$\mathcal{P}(g^m) = \mathcal{Y}(g^m) = \mathcal{Y}_{\bar{g}}(g^m) = \mathcal{Y}_{\bar{g}}(g)^m = \mathcal{P}(g)^m.$$

In particular, $\mathcal{P}(g)^{|G|} = I_{\theta(1)}$ and thus $\det(\mathcal{P}(g))^{|G|} = 1$ for all $g \in G$. Since

$$\mathcal{P}(g)\mathcal{P}(h) = \alpha(g, h)\mathcal{P}(gh),$$

we deduce that

$$\alpha(g, h)^{|G|\theta(1)} = 1$$

for all $g, h \in G$. $\qquad\square$

By Problem 5.1, it is possible to choose α in Theorem 5.5 also satisfying $\alpha(x, y) = 1$ whenever $N\langle x \rangle = N\langle y \rangle$. Sometimes, this can be useful.

5.3 A Group Associated with a Character Triple

As we shall prove below, Theorem 5.5 allows us to associate with every character triple (G, N, θ) a new finite group.

Theorem 5.6 *Suppose that (G, N, θ) is a character triple and let $\mathcal{P}$ be a projective representation of G associated with θ such that the factor set α of $\mathcal{P}$ takes roots of unity values. Let $Z \subseteq \mathbb{C}^{\times}$ be any finite subgroup containing the values of α. Let $\widehat{G} = \{(g, z) \mid g \in G, z \in Z\}$, where we define the multiplication*

$$(g_1, z_1)(g_2, z_2) = (g_1 g_2, \alpha(g_1, g_2) z_1 z_2).$$

Identify N and Z with $N \times 1$ and $1 \times Z$, respectively. Then the following hold.

(a) *$\widehat{G}$ is a finite group.*
(b) *The map $\pi : \widehat{G} \to G$ given by $\pi(g, z) = g$ is an onto homomorphism with kernel Z.*
(c) *$N \trianglelefteq \widehat{G}$, $Z \subseteq \mathbf{Z}(\widehat{G})$, and $\widehat{N} = N \times Z \trianglelefteq \widehat{G}$. Also, $\widehat{N}/N \subseteq \mathbf{Z}(\widehat{G}/N)$.*
(d) *The map defined by $\widehat{\mathcal{P}}(g, z) = z \mathcal{P}(g)$ is an irreducible representation of $\widehat{G}$ whose character $\tau \in \mathrm{Irr}(\widehat{G})$ extends θ.*
(e) *Let $\hat{\theta} = \theta \times 1_Z \in \mathrm{Irr}(\widehat{N})$ and $\hat{\lambda} \in \mathrm{Irr}(\widehat{N})$ be defined by $\hat{\lambda}(n, z) = z^{-1}$. Then $(\widehat{G}, \widehat{N}, \hat{\theta})$ is a character triple, $\hat{\lambda}$ is a linear $\widehat{G}$-invariant character with kernel N, and $\hat{\lambda}^{-1} \hat{\theta}$ extends to $\tau \in \mathrm{Irr}(\widehat{G})$.*

Proof The fact that α is a factor set implies that the multiplication in $\widehat{G}$ is associative. Since α is associated with a character triple, we know that $\alpha(g, 1) = \alpha(1, g) = \alpha(1, 1) = 1$ and $\alpha(g, g^{-1}) = \alpha(g^{-1}, g)$ for $g \in G$, by Lemma 5.3. It easily then follows that $(1, 1)$ is the identity of $\widehat{G}$ and that $(g^{-1}, \alpha(g, g^{-1})^{-1} z^{-1})$ is the inverse of (g, z) for $g \in G$ and $z \in Z$. (In fact, by Problem 5.1, we could have chosen α in Theorem 5.5 such that the inverse of (g, z) is (g^{-1}, z^{-1}).) Thus $\widehat{G}$ is a finite group, and it is clear that π is an onto homomorphism with kernel Z. This proves (a) and (b). Also, notice that $Z \subseteq \mathbf{Z}(\widehat{G})$.

Let $\widehat{N} = \{(n, z) \mid n \in N, z \in Z\}$. Note that

$$(n_1, z_1)(n_2, z_2) = (n_1 n_2, z_1 z_2)$$

since $\alpha(n_1, n_2) = 1$ for $n_1, n_2 \in N$. Hence, $\widehat{N} = N \times Z$ is a subgroup of $\widehat{G}$. Also, $\hat{\theta} = \theta \times 1_Z \in \text{Irr}(\widehat{N})$ by Theorem 1.14. If $n \in N$, $g \in G$ and $z, z' \in Z$, then, by using Lemma 5.3(c) we check that

$$(g, z)(n, z')(g, z)^{-1} = (gng^{-1}, z').$$

We deduce that N and $\widehat{N}$ are normal subgroups of $\widehat{G}$. Since $(n, z')N = (1, z')N$ in $\widehat{G}/N$, we have $\widehat{N}/N \subseteq \mathbf{Z}(\widehat{G}/N)$. This completes the proof of (c). Note also that

$$\hat{\theta}^{(g,z)}(n, z') = \hat{\theta}(gng^{-1}, z') = \theta(gng^{-1}) = \theta(n) = \hat{\theta}(n, z'),$$

and thus $\hat{\theta}$ is $\widehat{G}$-invariant.

    Now define

$$\widehat{\mathcal{P}}(g, z) = z\mathcal{P}(g)$$

for $g \in G$ and $z \in Z$. It is straightforward to check that $\widehat{\mathcal{P}}$ is an ordinary $\mathbb{C}$-representation of $\widehat{G}$. Suppose that $\widehat{\mathcal{P}}$ affords $\tau \in \text{Char}(\widehat{G})$. For $n \in N$ and $z \in Z$, note that

$$\tau(n, z) = z\theta(n)$$

because $\mathcal{P}_N$ affords θ.

    Also,

$$\hat{\theta}_N = \theta = \tau_N \in \text{Irr}(N).$$

In particular, τ cannot be a sum of two characters of $\widehat{G}$ and thus $\tau \in \text{Irr}(\widehat{G})$. If $\hat{\lambda} \in \text{Irr}(\widehat{N})$ is defined by $\hat{\lambda}(n, z) = \bar{z}$, then $\hat{\lambda}$ is a $\widehat{G}$-invariant linear character of $\widehat{N}$ with $\ker(\hat{\lambda}) = N$. Also,

$$\tau_{\widehat{N}} = \hat{\lambda}^{-1}\hat{\theta},$$

and this finishes the proof.   □

    The group that we have constructed is quite useful in character theory. If we set Z to be the group generated by the values of the factor set, then we have that $\widehat{G}$ is uniquely determined by $\mathcal{P}$. We call the group $\widehat{G}$ a **representation group for** (G, N, θ) **afforded by** $\mathcal{P}$. As we shall see, the relevance of $\widehat{G}$ is that we can construct a bijection $\Omega_{\mathcal{P}} \colon \text{Irr}(G|\theta) \to \text{Irr}(\widehat{G}|\hat{\lambda})$, depending only on $\mathcal{P}$, with many interesting properties.

5.4 Isomorphisms of Character Triples

It is time to define precisely what we mean by *replacing* a triple (G, N, θ) by another. Recall that $\text{Char}(G|\theta)$ is the set of characters of G all of whose irreducible constituents lie in $\text{Irr}(G|\theta)$. Also, write $\mathbb{C}[\text{Irr}(G|\theta)] = \text{cf}(G|\theta)$ for the complex space with basis $\text{Irr}(G|\theta)$.

Definition 5.7 (Isaacs) *Suppose that (G, N, θ) and (G^*, N^*, θ^*) are character triples and that $*: G/N \rightarrow G^*/N^*$ is an isomorphism of groups. If $N \leq U \leq G$, then we denote by U^* the unique subgroup $N^* \leq U^* \leq G^*$ such that $(U/N)^* = U^*/N^*$. Also, if β is a character of U/N, then β^* denotes the corresponding character of U^*/N^* via the isomorphism $*$. (Hence, β^* is the unique character of U^*/N^* satisfying $\beta^*(x^*) = \beta(x)$ for $x \in U/N$ by Theorem 2.1.) Assume now that for every such subgroup $N \leq U \leq G$ there is a bijection $*: \mathrm{Irr}(U \mid \theta) \rightarrow \mathrm{Irr}(U^* \mid \theta^*)$ (which we extend linearly to $*: \mathrm{Char}(U \mid \theta) \rightarrow \mathrm{Char}(U^* \mid \theta^*))$. It is said that $*$ is a **character triple isomorphism** if, for every $N \leq V \leq U \leq G$, $\chi \in \mathrm{Irr}(U \mid \theta)$ and $\beta \in \mathrm{Irr}(U/N)$, the following conditions hold:*

(a) $(\chi_V)^* = (\chi^*)_{V^*}$, and
(b) $(\chi\beta)^* = \chi^*\beta^*$.

Notice that if (G, N, θ) and (G^*, N^*, θ^*) are isomorphic, then $|\mathrm{Irr}(U|\theta)| = |\mathrm{Irr}(U^*|\theta^*)|$, by definition. Also, by using that $(\chi_N)^* = (\chi^*)_{N^*}$, we have that

$$\frac{\chi(1)}{\theta(1)} = \frac{\chi^*(1)}{\theta^*(1)}$$

for $\chi \in \mathrm{Irr}(U|\theta)$. Observe too that (a) is equivalent to

$$[\chi_V, \mu] = [(\chi^*)_{V^*}, \mu^*]$$

for $\mu \in \mathrm{Irr}(V|\theta)$. Furthermore,

$$
\begin{aligned}
(\chi^G)^* &= \left(\sum_{\tau \in \mathrm{Irr}(G|\theta)} [\chi^G, \tau]\tau \right)^* = \sum_{\tau \in \mathrm{Irr}(G|\theta)} [\chi^G, \tau]\tau^* \\
&= \sum_{\tau \in \mathrm{Irr}(G|\theta)} [\chi, \tau_U]\tau^* = \sum_{\tau \in \mathrm{Irr}(G|\theta)} [\chi^*, (\tau^*)_{U^*}]\tau^* \\
&= \sum_{\tau^* \in \mathrm{Irr}(G^*|\theta^*)} [(\chi^*)^{G^*}, \tau^*]\tau^* = (\chi^*)^{G^*}.
\end{aligned}
$$

We leave it to the reader to check that isomorphism defines an equivalence relation on character triples.

The following gives examples of character triple isomorphisms.

Lemma 5.8 *Suppose that (G, N, θ) is a character triple.*

(a) If $\alpha: G \rightarrow H$ is an isomorphism of groups, then (G, N, θ) is isomorphic to $(H, \alpha(N), \theta^\alpha)$.
(b) If $M \trianglelefteq G$, and $M \subseteq \ker(\theta)$, then (G, N, θ) and $(G/M, N/M, \bar{\theta})$ are isomorphic, where $\bar{\theta} \in \mathrm{Irr}(N/M)$ is the character defined by $\bar{\theta}(Mn) = \theta(n)$.

(c) *Suppose that* $\pi: G \to H$ *is an onto homomorphism with kernel* Z. *Suppose that* $Z \subseteq \ker(\theta)$. *Then* (G, N, θ) *is isomorphic to* (H, M, φ), *where* $M = \pi(N)$ *and* $\varphi \in \operatorname{Irr}(M)$ *is the unique character of* M *with* $\varphi(\pi(n)) = \theta(n)$ *for* $n \in N$.

(d) *Suppose that* (G, N, φ) *is a character triple. Let* $\chi \in \operatorname{Irr}(G)$ *be such that* $\varphi \chi_N = \theta \in \operatorname{Irr}(N)$. *Then* (G, N, φ) *and* (G, N, θ) *are isomorphic via an isomorphism that sends* $\beta \in \operatorname{Irr}(U|\varphi)$ *to* $\beta \chi_U \in \operatorname{Irr}(U|\theta)$ *for* $N \leq U \leq G$.

(e) *Suppose that* $\lambda \in \operatorname{Irr}(N)$ *is linear and* G-*invariant. If* $\lambda^{-1}\theta$ *extends to some* $\chi \in \operatorname{Irr}(G)$, *then* (G, N, λ) *and* (G, N, θ) *are isomorphic, via an isomorphism that sends* $\beta \in \operatorname{Irr}(U|\lambda)$ *to* $\beta \chi_U \in \operatorname{Irr}(U|\theta)$ *for* $N \leq U \leq G$.

Proof The first two parts are straightforward (using Theorem 2.1 for (a) and Theorem 1.11 for (b)). Part (c) is a direct consequence of (a) and (b), again using Theorem 2.1. Part (d) follows directly from Theorem 1.22. For part (e), notice that since $\chi_N = \lambda^{-1}\theta$, then $\lambda\chi_N = \theta$, and we apply part (d) with $\varphi = \lambda$. $\qquad\square$

Corollary 5.9 *Every character triple* (G, N, θ) *is isomorphic to some* (G^*, N^*, θ^*), *where* $N^* \subseteq \mathbf{Z}(G^*)$, *and* θ^* *is linear and faithful.*

Proof By Theorem 5.5, let $\mathcal{P}$ be a projective representation of G associated with θ such that the factor set α of $\mathcal{P}$ takes roots of unity values. Let $Z \subseteq \mathbb{C}^\times$ be any finite subgroup containing the values of α, and let $\widehat{G} = \{(g, z) \mid g \in G, z \in Z\}$ a representation group of (G, N, θ) afforded by $\mathcal{P}$. Again, identify N with $N \times 1$ and Z with $1 \times Z$, and let $\widehat{N} = N \times Z$. We have that $\pi: \widehat{G} \to G$ given by $\pi(g, z) = g$ is a group homomorphism with kernel Z. If $\hat{\theta} = \theta \times 1_Z \in \operatorname{Irr}(\widehat{N})$, and $\hat{\lambda} \in \operatorname{Irr}(\widehat{N})$ is defined by $\hat{\lambda}(n, z) = z^{-1}$, then we know by Theorem 5.6 that $(\widehat{G}, \widehat{N}, \hat{\theta})$ is a character triple, $\hat{\lambda}$ is a linear $\widehat{G}$-invariant character with $N = \ker(\hat{\lambda})$, and $\hat{\lambda}^{-1}\hat{\theta}$ extends to some $\tau \in \operatorname{Irr}(\widehat{G})$. Also, $\widehat{N}/N \subseteq \mathbf{Z}(\widehat{G}/N)$. By Lemma 5.8(e), we have that $(\widehat{G}, \widehat{N}, \hat{\lambda})$ is isomorphic to $(\widehat{G}, \widehat{N}, \hat{\theta})$. By Lemma 5.8(b), we have that $(\widehat{G}, \widehat{N}, \hat{\lambda})$ is isomorphic to $(\widehat{G}/N, \widehat{N}/N, \hat{\lambda})$, where we view $\hat{\lambda} \in \operatorname{Irr}(\widehat{N}/N)$, and notice that $\hat{\lambda}$ is now faithful. By Lemma 5.8(c), $(\widehat{G}, \widehat{N}, \hat{\theta})$ is isomorphic to (G, N, θ), and we conclude that (G, N, θ) is isomorphic to (G^*, N^*, θ^*), where $(G^*, N^*, \theta^*) = (\widehat{G}/N, \widehat{N}/N, \hat{\lambda})$. $\qquad\square$

Sometimes it is convenient to make explicit the isomorphism that we have constructed in Corollary 5.9. If $\chi \in \operatorname{Irr}(G|\theta)$, by Theorem 2.1 there is a unique $\hat{\chi} \in \operatorname{Irr}(\widehat{G})$ with Z in its kernel such that $\hat{\chi}(g, z) = \chi(g)$ for $g \in G$ and $z \in Z$. Now $\hat{\chi} \in \operatorname{Irr}(\widehat{G}|\theta)$, and since $\tau_N = \theta$, by the Gallagher correspondence there is a unique $\chi^* \in \mathrm{Irr}(\hat{G}/N)$ such that $\hat{\chi} = \chi^*\tau$. As the reader can check, χ^* lies over $\hat{\lambda}$, and $\chi \mapsto \chi^*$ is the bijection $\mathrm{Irr}(G|\theta) \to \mathrm{Irr}(\hat{G}/N|\hat{\lambda})$ constructed in Corollary 5.9.

Character triple isomorphisms have many applications. A typical one is the following, which we shall use later on.

Theorem 5.10 *Let $N \trianglelefteq G$, and let $\theta \in \mathrm{Irr}(N)$ be G-invariant. Then θ extends to G if and only if θ extends to P for every Sylow subgroup P/N of G/N.*

We need a lemma.

Lemma 5.11 *Let $N \trianglelefteq G$. Suppose that $\lambda \in \mathrm{Irr}(N)$ is linear, G-invariant and has order $o(\lambda)$ a power of a prime p. If λ extends to P, where $P/N \in \mathrm{Syl}_p(G/N)$, then λ extends to G.*

Proof Suppose that $\mu \in \mathrm{Irr}(P)$ extends λ. Then $\mu(1) = 1$ and therefore μ^G has degree $|G : P|$, which is not divisible by p. Hence μ^G has an irreducible constituent $\chi \in \mathrm{Irr}(G)$ of degree not divisible by p. Since χ lies over μ, it follows that χ lies over λ. Hence $\chi_N = e\lambda$ by Clifford's theorem, where $e = \chi(1)$ is not divisible by p. By taking determinants, we have $\delta_N = \lambda^e$, where $\delta = \det(\chi)$. (See Problem 1.2.) Now if n is an integer with $en \equiv 1$ mod $o(\lambda)$, then

$$(\delta^n)_N = (\delta_N)^n = \lambda^{en} = \lambda\,.$$

So δ^n is an extension of λ, as desired. $\qquad\square$

Proof of Theorem 5.10 If (G^*, N^*, θ^*) is an isomorphic character triple, notice that θ extends to some $N \le H \le G$ if and only if θ^* extends to H^*, by using Definition 5.7(a). Hence, by Corollary 5.9, we may assume that θ is linear. By elementary group theory, we can write $\theta = \prod_p \theta_p$, where $\theta_p \in \mathrm{Irr}(N)$ is a power of θ with order a power of the prime p. (See the discussion preceding Theorem 2.4.) Since θ extends to P, where $P/N \in \mathrm{Syl}_p(G/N)$, the same happens with any power of θ. Therefore θ_p extends to P. By Lemma 5.11, θ_p has an extension $\tau_p \in \mathrm{Irr}(G)$, and $\prod_p \tau_p$ extends θ.

Another standard application of character triple isomorphisms is the following.

Theorem 5.12 *Suppose that $N \trianglelefteq G$ and let $\theta \in \mathrm{Irr}(N)$. If $\chi \in \mathrm{Irr}(G|\theta)$, then $\chi(1)/\theta(1)$ divides $|G : N|$.*

Proof Let T be the stabilizer of θ in G, and let $\psi \in \mathrm{Irr}(T|\theta)$ be the Clifford correspondent of χ over θ. Since $\chi(1) = |G : T|\psi(1)$, it is no loss to assume that $T = G$. Now, since character triple isomorphisms preserve character degree ratios, by Corollary 5.9 we may assume that N is central in G. In this case, Theorem 1.10 concludes the proof.   □

Let us remark here that if G/N is solvable, then there is no need to use character triple isomorphisms to prove Theorem 5.12. Indeed, by using induction on $|G : N|$, we may assume that G/N has prime order and that θ is G-invariant. In this case, it suffices to apply Theorem 5.1.

We notice too that if (G, N, θ) is a character triple, then the integers $e_\chi = \chi(1)/\theta(1)$ for $\chi \in \mathrm{Irr}(G|\theta)$ tend to *behave* as character degrees of G/N. We showed after Theorem 1.19 that

$$\sum_{\chi \in \mathrm{Irr}(G|\theta)} e_\chi^2 = |G : N|.$$

Now, we have proved that e_χ divides $|G : N|$.

5.5 Counting $|\mathrm{Irr}(G|\theta)|$

Suppose that (G, N, θ) is a character triple. Our objective in this section is to prove a theorem of P. X. Gallagher that describes $|\mathrm{Irr}(G|\theta)|$ in a way that has proven to be very useful in character theory.

If $N = 1$, we know that $|\mathrm{Irr}(G|\theta)|$ is the number of conjugacy classes of G. What we are going to show is that $|\mathrm{Irr}(G|\theta)|$ is the number of θ-good conjugacy classes of G/N. We shall prove Gallagher's theorem by using character triples. (We propose another proof in Problem 5.3 which provides more insight but that is more complicated.)

Let $N \trianglelefteq G$ and suppose that $\theta \in \mathrm{Irr}(N)$ is G-invariant. Let $g \in G$, and write $D/N = \mathbf{C}_{G/N}(Ng)$. Thus $N\langle g\rangle/N \subseteq \mathbf{Z}(D/N)$ and $N\langle g\rangle \trianglelefteq D$. Since $N\langle g\rangle/N$ is cyclic, then we know that θ has an extension $\eta \in \mathrm{Irr}(N\langle g\rangle)$ (by Theorem 5.1). Assume that η is D-invariant. If $\tilde{\eta} \in \mathrm{Irr}(N\langle g\rangle)$ is any other extension of θ, then notice that $\tilde{\eta} = \lambda\eta$ for some $\lambda \in \mathrm{Irr}(N\langle g\rangle/N)$ by Gallagher's correspondence Corollary 1.23. Since $N\langle g\rangle/N \subseteq \mathbf{Z}(D/N)$, it then follows that $\tilde{\eta}$ is D-invariant too. Therefore we have that either all the extensions of θ to $N\langle g\rangle$ are D-invariant, or none of them is. An element $g \in G$ is θ-**good** in G if the extensions of θ to $N\langle g\rangle$ are D-invariant. Note that to show that g is θ-good in G, it suffices to check that there is one extension of θ to $N\langle g\rangle$ that is fixed by all elements $x \in D$.

Lemma 5.13 *Let $N \unlhd G$ and suppose that $\theta \in \mathrm{Irr}(N)$ is G-invariant. Let $g \in G$. Then the following hold.*

(a) g is θ-good in G if and only if θ extends to $N\langle g, x\rangle$ for every $x \in G$ such that $N\langle g, x\rangle/N$ is abelian.

(b) If g is not θ-good in G, then $\chi(g) = 0$ for every $\chi \in \mathrm{Irr}(G|\theta)$.

(c) If θ is linear and faithful, then g is θ-good in G if and only if $\mathbf{C}_{G/N}(Ng) = \mathbf{C}_G(g)/N$.

Proof Write $U = N\langle g\rangle$ and $D/N = \mathbf{C}_{G/N}(Ng)$. Thus $U/N \subseteq \mathbf{Z}(D/N)$.

(a) Notice that $N\langle g, x\rangle/N$ is abelian if and only if $x \in D$. Suppose that g is θ-good. Then θ has a D-invariant extension $\eta \in \mathrm{Irr}(U)$. Since $U\langle x\rangle/U$ is cyclic, then η extends to $U\langle x\rangle = N\langle g, x\rangle$ by Theorem 5.1. Conversely, assume now that for every $x \in D$ there is some $\theta_x \in \mathrm{Irr}(U\langle x\rangle)$ that extends θ. Let $\eta \in \mathrm{Irr}(U)$ be any extension of θ. We claim that η is D-invariant. Let $x \in D$. Then $(\theta_x)_U$ is an extension of θ to U, and by the Gallagher correspondence we have that $\lambda(\theta_x)_U = \eta$ for some $\lambda \in \mathrm{Irr}(U/N)$. Since $U/N \subseteq \mathbf{Z}(D/N)$ and $(\theta_x)_U$ is x-invariant, we conclude that η is x-invariant too.

(b) Suppose that g is not θ-good. Then there is some extension $\eta \in \mathrm{Irr}(U)$ and $x \in D$ such that $\eta^x \neq \eta$. By the Gallagher correspondence, write $\eta^x = \lambda\eta$ for some $1 \neq \lambda \in \mathrm{Irr}(U/N)$. Then $\lambda(g) \neq 1$ because Ng generates U/N. Now let $\chi \in \mathrm{Irr}(G|\theta)$. Every irreducible constituent μ of χ_U lies over θ, and therefore extends θ by Theorem 5.1. Hence, again by the Gallagher correspondence we can write

$$\chi_U = \Psi\eta$$

for some $\Psi \in \mathrm{Char}(U/N)$. Now

$$\chi(g^{x^{-1}}) = \Psi(g^{x^{-1}})\eta(g^{x^{-1}}) = \Psi(g)\eta^x(g) = \Psi(g)\lambda(g)\eta(g),$$

where we have used that $U/N \subseteq \mathbf{Z}(D/N)$. Also,

$$\chi(g^{x^{-1}}) = \chi(g) = \Psi(g)\eta(g).$$

Since $\lambda(g) \neq 1$, we conclude that $0 = \Psi(g)\eta(g) = \chi(g)$.

(c) Since θ is G-invariant and linear, we have that $\theta(n^g n^{-1}) = 1$ for $n \in N$ and $g \in G$. Since θ is faithful, we conclude that $N \subseteq \mathbf{Z}(G)$. Assume first that $D = \mathbf{C}_G(g)$. Then $D \subseteq \mathbf{C}_G(U)$, and therefore every extension of θ to U is D-invariant. Conversely, assume that g is θ-good. By definition, if $\eta \in \mathrm{Irr}(U|\theta)$ and $x \in D$, we have that $\eta^x(g) = \eta(g)$. Write $g^{x^{-1}} = ng$ for some $n \in N$, using that $(Ng)^x = Ng$. Since η is linear, we have that $\eta(g) = \eta^x(g) = \eta(ng) = \eta(n)\eta(g) = \theta(n)\eta(g)$. Thus $\theta(n) = 1$. Since θ is faithful, we have that $n = 1$ and $g^x = g$. $\qquad\square$

We remark that the converse of Lemma 5.13(b) does not hold. (A trivial example is to take $G = N$, $\theta \in \mathrm{Irr}(G)$, and $g \in G$ with $\theta(g) = 0$.) A less trivial example is the following. Consider the dihedral group $G = D_{12} = \langle g, t \rangle$, where $g^t = g^{-1}$, g has order 6, and t is an involution, and the normal subgroup $N = D_6 = \langle g^2, t \rangle$ of index 2. Now, the coset Ng is the union of three conjugacy classes of G: the class of g, the class of gt and the central element g^3. If $\chi \in \mathrm{Irr}(G)$ has degree 2, then $\chi_N = \theta \in \mathrm{Irr}(N)$. Notice that every element of G is θ-good. It can be checked that χ is not zero on the conjugacy classes of g and of g^3. These are *right* representatives for the coset Ng. However χ is zero on the conjugacy class of gt. Now, $\mathrm{Irr}(G|\theta) = \{\chi, \lambda\chi\}$, where $1 \neq \lambda \in \mathrm{Irr}(G/N)$, and thus every character in $\mathrm{Irr}(G|\theta)$ is zero on the θ-good element gt. (In Problem 5.3 it is shown why it is important to choose a representative of the coset Ng on which no extension of θ to $N\langle g \rangle$ vanishes. The fact that such an extension exists follows from Theorem 1.25, and we shall use this fact in Lemma 5.17.)

What is true is that g is not θ-good if and only if $\chi(ng) = 0$ for every $n \in N$ and every $\chi \in \mathrm{Irr}(G|\theta)$, and we prove this less obvious fact in Theorem 5.18 below.

Observe now that if g is θ-good, then so is every conjugate of g and every element $h \in G$ such that $N\langle g \rangle = N\langle h \rangle$. In particular, every member of the coset Ng is θ-good. We can thus unambiguously refer to the θ-**good conjugacy classes** of G/N as those consisting of θ-good elements. So from now on we write that x is θ-good in G or that Nx is θ-good in G/N, and we mean the same thing.

First, we count $|\mathrm{Irr}(G|\theta)|$ in the case where N is central, and do the general case later, by using isomorphisms of character triples. In the case where N is central, we obtain more information.

Theorem 5.14 *Suppose that $N \leq \mathbf{Z}(G)$, and let $\theta \in \mathrm{Irr}(N)$ be faithful. Assume that $Nx_1, \ldots, Nx_t$ are representatives of the different conjugacy classes of θ-good elements of G/N. Then $|\mathrm{Irr}(G|\theta)| = t$. Furthermore, if $\mathrm{Irr}(G|\theta) = \{\chi_1, \ldots, \chi_t\}$, then the matrix $(\chi_i(x_j))$ is invertible.*

Proof Since $N \subseteq \mathbf{Z}(G)$, notice that if $\chi \in \mathrm{Irr}(G|\theta)$, then $\chi(gn) = \chi(g)\theta(n)$ for all $g \in G$ and $n \in N$. (If $\mathcal{X}$ affords χ, then $\mathcal{X}(n)$ is a scalar matrix by Schur's lemma, and the scalar is $\theta(n)$.) Therefore, $\varphi(gn) = \varphi(g)\theta(n)$ for all $\varphi \in \mathrm{cf}(G|\theta)$, $g \in G$, and $n \in N$.

Now, let $X = \{x_1, \ldots, x_t\} \subseteq G$, and let $\mathbb{C}^X$ be the space of complex functions on X. This is a space of dimension t. We claim that the linear map

$f(\varphi) = \varphi_X$ is an isomorphism $f : \mathrm{cf}(G|\theta) \to \mathbb{C}^X$, where φ_X is the restriction of φ to X.

To prove that f is injective, suppose that $\varphi \in \mathrm{cf}(G|\theta)$ is zero on all the elements of X. We wish to prove that $\varphi(g) = 0$ for every $g \in G$. If g is not θ-good, this is clear by Lemma 5.13(b). Hence we can assume that g is θ-good, and thus Ng is conjugate in G/N to Nx_j, for some j. We can thus write $g^x = nx_j$ for some element $x \in G$. Then

$$\varphi(g) = \varphi(g^x) = \varphi(nx_j) = \theta(n)\varphi(x_j) = 0,$$

as required.

To prove surjectivity, suppose that $\alpha \in \mathbb{C}^X$ is an arbitrary complex valued function on X. We extend α to a class function φ on G as follows. If g is not θ-good, we set $\varphi(g) = 0$. If g is θ-good, we choose $x \in G$ and $x_j \in X$ such that $Ng^x = Nx_j$ and we set

$$\varphi(g) = \alpha(x_j)\theta(g^x x_j^{-1}).$$

To see that φ is well defined, suppose that also $Ng^y = Nx_j$, with $y \in G$. Then $(Ng)^x = (Ng)^y$ and $g^x = g^y$ by Lemma 5.13(c). Thus φ is well defined. To see that it is a class function of G, notice that $\varphi(g) = \varphi(g^h) = 0$ if g is not θ-good. Otherwise, $Ng^x = Nx_j$, $N(g^h)^{h^{-1}x} = Nx_j$ and again $\varphi(g) = \varphi(g^h)$. Also, notice that

$$\varphi(nx_j) = \alpha(x_j)\theta(n)$$

for $n \in N$. Next, we show that $\varphi \in \mathrm{cf}(G|\theta)$. Let $\gamma \in \mathrm{Irr}(G)$ be a character that does not lie over θ. Then $\gamma_N = \gamma(1)\tau$, for some $\theta \neq \tau \in \mathrm{Irr}(N)$. Hence $[\theta, \tau] = 0$. Notice that since φ and τ are class functions and $Nx_j^g = (Nx_j)^g$, we have that

$$\sum_{n \in N} \varphi(nx_j)\overline{\gamma(nx_j)} = \sum_{n \in N} \varphi(nx_j^g)\overline{\gamma(nx_j^g)}$$

for all $g \in G$. Now, summing over all the cosets of N in G, we have that

$$|G|[\varphi, \gamma] = \sum_{j=1}^{t} |G/N : \mathbf{C}_{G/N}(Nx_j)| \left(\sum_{n \in N} \varphi(nx_j)\overline{\gamma(nx_j)} \right)$$

$$= \sum_{j=1}^{t} |G/N : \mathbf{C}_{G/N}(Nx_j)| \left(\sum_{n \in N} \theta(n)\alpha(x_j)\overline{\tau(n)\gamma(x_j)} \right)$$

$$= \left(\sum_{j=1}^{t} |G/N : \mathbf{C}_{G/N}(Nx_j)|\alpha(x_j)\overline{\gamma(x_j)} \right) |N|[\theta, \tau] = 0.$$

This shows that $\varphi \in \mathrm{cf}(G|\theta)$. Hence, f is an isomorphism. In particular, the dimension of $\mathrm{cf}(G|\theta)$ is t.

For $x \in X$, let $\delta_x \in \mathbb{C}^X$ be the characteristic function of $x \in X$. Notice that

$$f(\varphi) = \varphi_X = \sum_{x \in X} \varphi(x)\delta_x.$$

Hence then the matrix of f with respect to the bases $\mathrm{Irr}(G|\theta)$ and $\{\delta_x\}$ is $(\chi_i(x_j))$, and the proof of the theorem is complete. □

Lemma 5.15 *Suppose that (G, N, θ) and (G^*, N^*, θ^*) are isomorphic character triples. Let $x \in G$ and let $(Nx)^* = N^*x^*$, where x^* is any element in $(Nx)^*$. Then Nx is θ-good in G/N if and only if N^*x^* is θ^*-good in G^*/N^*.*

Proof By Lemma 5.13(a), we have that Nx is θ-good if and only if θ extends to every $N \le H \le G$ such that $H/N = \langle Nx, Ny \rangle$ is abelian. Using property (a) of the definition of character triples, this happens if and only if θ^* extends to every $N^* \le H^* \le G^*$ such that $H^*/N^* = \langle (Nx)^*, (Ny)^* \rangle$ is abelian. Hence, the lemma follows. □

Theorem 5.16 (Gallagher) *Suppose that (G, N, θ) is a character triple. Then $|\mathrm{Irr}(G|\theta)|$ is the number of conjugacy classes of G/N which consist of θ-good elements.*

Proof By Corollary 5.9, let (G^*, N^*, θ^*) be an isomorphic character triple, where $N^* \subseteq \mathbf{Z}(G^*)$ and θ^* is faithful. By Lemma 5.15, the conjugacy class of Ng in G/N is θ-good if and only if the conjugacy class of $(Ng)^*$ is θ^*-good. Since $|\mathrm{Irr}(G|\theta)| = |\mathrm{Irr}(G^*|\theta^*)|$, the assertion of the theorem follows from Theorem 5.14. □

As in Theorem 5.14, there is a convenient invertible matrix associated with (G, N, θ). (See Problem 5.3.)

5.6 Some Applications

In this section, we give some applications of character triple isomorphisms which perhaps are not so well known. The following lemma is quite useful.

Lemma 5.17 *Suppose that (G, N, θ) and (G^*, N^*, θ^*) are isomorphic character triples. Let $x \in G$ and let $x^* \in G^*$ be such that $(Nx)^* = N^*x^*$, where $^* : G/N \to G^*/N^*$ is the associated group isomorphism. Let $\chi \in \mathrm{Irr}(G|\theta)$.*

(a) If θ^* is linear, then $\chi(x) = \alpha \chi^*(x^*)$, where α is an algebraic integer.
(b) We have that $\chi(nx) = 0$ for all $n \in N$ if and only if $\chi^*(n^*x^*) = 0$ for all $n^* \in N^*$.

Proof Let $X = N\langle x \rangle$ and let τ be an extension of θ to X, by Theorem 5.1. Notice that

$$X^*/N^* = (X/N)^* = (N\langle x \rangle / N)^* = \langle Nx \rangle^* = \langle (Nx)^* \rangle = N^*\langle x^* \rangle / N^*.$$

Also notice that $\tau^* \in \text{Irr}(X^*)$ is an extension of θ^*. Now, by the Gallagher correspondence Corollary 1.23, we have that $\chi_X = \tau \psi$ for some character ψ of X/N. Then

$$(\chi^*)_{X^*} = \tau^* \psi^*.$$

We also have that $\psi(x) = \psi^*(x^*)$.

(a) Suppose that θ^* is linear. Then τ^* is linear, and therefore $\tau^*(x^*)$ is a root of unity. Now

$$\chi(x) = \tau(x)\psi(x) = \tau(x)\psi^*(x^*) = \frac{\tau(x)}{\tau^*(x^*)}\chi^*(x^*).$$

Since $\tau(x)/\tau^*(x^*)$ is an algebraic integer, this proves (a).

To prove (b), assume that $\chi(nx) = 0$ for all $n \in N$. By Theorem 1.25, we have that

$$\sum_{y \in Nx} |\tau(y)|^2 = |N|.$$

Then there exists $y \in Nx$ such that $\tau(y) \neq 0$. Then $0 = \chi(y) = \tau(y)\psi(y)$, and we conclude that $\psi(y) = 0$. Since ψ is a character of X/N, we have that $\psi(ny) = 0$ for all $n \in N$. Hence $\psi(nx) = 0$ for all $n \in N$. Therefore $\psi^*(n^*x^*) = 0$ for all $n^* \in N^*$, and we conclude that $\chi^*(n^*x^*) = 0$ for all $n^* \in N^*$. $\qquad\square$

Before the applications, we prove the following characterization of θ-good elements by zeros of characters. This explains many zeros in the character table.

Theorem 5.18 *Suppose that $N \trianglelefteq G$, and that $\theta \in \text{Irr}(N)$ is G-invariant. Then $g \in G$ is not θ-good in G if and only if $\chi(ng) = 0$ for all $n \in N$ and all $\chi \in \text{Irr}(G|\theta)$.*

Proof If $g \in G$ is not θ-good in G, then ng is not θ-good in G for all $n \in N$, and then $\chi(ng) = 0$ for all $\chi \in \text{Irr}(G|\theta)$, by Lemma 5.13(b).

Assume now that $\chi(ng) = 0$ for all $n \in N$ and for all $\chi \in \mathrm{Irr}(G|\theta)$. Let (G^*, N^*, θ^*) be an isomorphic triple with $N^* \subseteq \mathbf{Z}(G^*)$ and θ^* faithful. By Lemma 5.15, we know that Ng is θ-good if and only if N^*g^* is θ^*-good, where $(Ng)^* = N^*g^*$, for some $g^* \in G^*$. Now, by Lemma 5.17(b), we have that $\chi^*(n^*g^*) = 0$ for all $n^* \in N^*$ and all $\chi^* \in \mathrm{Irr}(G^*|\theta^*)$. Hence, it is no loss to assume that $N \subseteq \mathbf{Z}(G)$ and that θ is faithful. If g is θ-good, then we have that the matrix $(\chi_i(x_j))$ in Theorem 5.14 has a column of zeros. But this is not possible, since it is an invertible matrix. $\qquad\square$

The following is the *relative version* of Theorem 1.8.

Theorem 5.19 *Suppose that $N \trianglelefteq G$, and that $\theta \in \mathrm{Irr}(N)$ is G-invariant. Let $x \in G$ and $D/N = \mathbf{C}_{G/N}(Nx)$. If $\chi \in \mathrm{Irr}(G|\theta)$, then*

$$\frac{\theta(1)|G:D|\chi(x)}{\chi(1)}$$

is an algebraic integer.

Proof If Nx is not θ-good, then $\chi(x) = 0$ by Lemma 5.13(b), and in this case the theorem is true. So we may assume that Nx is θ-good. Now, let (G^*, N^*, θ^*) be an isomorphic character triple with $N^* \subseteq \mathbf{Z}(G^*)$ and θ^* faithful. Let $x^* \in G^*$ be such that $N^*x^* = (Nx)^*$. We have that N^*x^* is θ^*-good by Lemma 5.15. Since θ^* is linear and faithful, this means that

$$D^*/N^* = (D/N)^* = \mathbf{C}_{G^*/N^*}(N^*x^*) = \mathbf{C}_{G^*}(x^*)/N^*,$$

by Lemma 5.13(c).

Now, by Theorem 1.8, we have that

$$\omega = \frac{|G^*:D^*|\chi^*(x^*)}{\chi^*(1)}$$

is an algebraic integer. By Lemma 5.17, we know that $\chi(x) = \alpha\chi^*(x^*)$ for some algebraic integer α. We have that

$$\frac{\theta(1)|G:D|\chi(x)}{\chi(1)} = \alpha\omega,$$

and this is an algebraic integer. $\qquad\square$

Next we present a generalization of Burnside's theorem on zeros of characters. Indeed, Corollary 4.2 is the case when $N = 1$.

Theorem 5.20 *Suppose that $N \trianglelefteq G$ and let $\chi \in \mathrm{Irr}(G)$. Then χ_N is not irreducible if and only if there is some $g \in G$ such that $\chi(ng) = 0$ for all $n \in N$.*

Proof If χ_N is irreducible, then for every $g \in G$ there exists some $x \in Ng$ such that $\chi(x) \neq 0$ by Theorem 1.25. Suppose now that χ_N is not irreducible. We wish to prove that χ has the value 0 on a whole right coset of N in G. Let $\theta \in \mathrm{Irr}(N)$ be under χ. Let T be inertia group of θ in G. By the Clifford correspondence, let $\psi \in \mathrm{Irr}(T|\theta)$ be such that $\psi^G = \chi$. If $T < G$, then there exists

$$g \in G - \bigcup_{x \in G} T^x$$

by elementary group theory. Now, since $N \subseteq T$, we have that $ng \in G - \bigcup_{x \in G} T^x$ for all $n \in N$, and by the induction formula, we conclude that $\chi(ng) = 0$ for all $n \in N$. Therefore, we may assume that $T = G$. Hence (G, N, θ) is a character triple. By Corollary 5.9, let (G^*, N^*, θ^*) be an isomorphic character triple with $N^* \subseteq \mathbf{Z}(G^*)$. Since χ_N is not irreducible, we have that

$$1 < \chi(1)/\theta(1) = \chi^*(1)/\theta^*(1) = \chi^*(1).$$

Hence, χ^* is not linear. By Burnside's theorem, there exists $g^* \in G^*$ such that $\chi^*(g^*) = 0$. Since $N^* \subseteq \mathbf{Z}(G^*)$, we have that $\chi^*(n^*g^*) = \theta^*(n^*)\chi^*(g^*) = 0$ for all $n^* \in N^*$. Now, if $(Ng)^* = N^*g^*$ for some $g \in G$, by Lemma 5.17(b) we conclude that $\chi(ng) = 0$ for all $n \in N$. $\qquad\square$

To end this section, we present yet another application. It is a generalization by R. Knörr of the second orthogonality relation.

Suppose that (G, N, θ) is a character triple. If $g \in G$, then we define

$$\hat{\theta}(g) = \eta(g)\overline{\eta(g)},$$

where η is any extension of θ to $N\langle g \rangle$ (using Theorem 5.1). By the Gallagher correspondence, this is a well-defined class function: if $\mu \in \mathrm{Irr}(N\langle g \rangle)$ is another extension, then $\mu = \lambda\eta$ for some linear character, and therefore $\mu(g)\overline{\mu(g)} = \hat{\theta}(g)$. (In fact, $\hat{\theta}$ is a character of G, as we shall see in the Problems.)

Theorem 5.21 (Knörr) *Suppose that $N \trianglelefteq G$ and $\theta \in \mathrm{Irr}(N)$ is G-invariant. Let $x, y \in G$. Then*

$$\sum_{\chi \in \mathrm{Irr}(G|\theta)} \chi(x)\overline{\chi(y)} = 0$$

if Nx and Ny are not G/N-conjugate. Also,

$$\sum_{\chi \in \mathrm{Irr}(G|\theta)} |\chi(x)|^2 = |\mathbf{C}_{G/N}(Nx)|\hat{\theta}(x)$$

if x is θ-good.

Proof (Isaacs) Suppose that $\{Ng_1, \ldots, Ng_t\}$ is any set of representatives of the conjugacy classes of G/N that consist of θ-good elements. Let $\eta_k \in \text{Irr}(N\langle g_k \rangle)$ be any fixed but arbitrary extension of θ (using Theorem 5.1). By Theorem 5.16, we may write $\text{Irr}(G|\theta) = \{\chi_1, \ldots, \chi_t\}$. By Lemma 5.13(b), $\chi_i(g) = 0$ if $g \in G$ is not θ-good.

By the Gallagher correspondence we can write

$$(\chi_i)_{N\langle g_k \rangle} = \psi_{i,k} \eta_k,$$

where $\psi_{i,k}$ is a character of $N\langle g_k \rangle / N$, which we view as a character of $N\langle g_k \rangle$ with N in its kernel. Now

$$\sum_{n \in N} \chi_i(ng_k)\overline{\chi_j(ng_k)} = \psi_{i,k}(g_k)\overline{\psi_{j,k}(g_k)} \sum_{n \in N} \eta_k(ng_k)\overline{\eta_k(ng_k)}$$

$$= \psi_{i,k}(g_k)\overline{\psi_{j,k}(g_k)}|N|,$$

using Theorem 1.25. Since χ_i and χ_j are class functions of G, notice that the left-hand side of the above sum does not change if we replace g_k by g_k^h for any $h \in G$. Now

$$\delta_{i,j} = [\chi_i, \chi_j] = \frac{1}{|G|} \sum_{k=1}^{t} \left(\sum_{Ng \in \text{cl}_{G/N}(Ng_k)} \sum_{n \in N} \chi_i(ng_k)\overline{\chi_j(ng_k)} \right)$$

$$= \frac{1}{|G|} \sum_{k=1}^{t} |G/N : \mathbf{C}_{G/N}(Ng_k)||N|\psi_{i,k}(g_k)\overline{\psi_{j,k}(g_k)}.$$

Let Y be the $t \times t$ matrix with (i, j)-entry

$$\frac{\delta_{i,j}}{|\mathbf{C}_{G/N}(Ng_i)|}$$

and let $X = (\psi_{i,j}(g_j))$, a $t \times t$ matrix. The above equation tells us that

$$XY\overline{X}^t = I_t.$$

Therefore

$$Y\overline{X}^t X = I_t.$$

We deduce that

$$\sum_{l=1}^{t} \overline{\psi_{l,i}(g_i)}\psi_{l,j}(g_j) = \delta_{i,j}|\mathbf{C}_{G/N}(Ng_i)|.$$

Assume now that $Nx, Ny \in G/N$ are not G/N-conjugate. If, for instance, Nx is not θ-good, then $\chi_i(x) = 0$ for all i (by Lemma 5.13(b)), and the theorem is correct in this case. Hence, we may assume that both Nx and Ny

are θ-good. Since by hypothesis, they are not G/N-conjugate, they can be chosen, say $x = g_1$ and $y = g_2$, in a set of representatives $\{Ng_1, \ldots, Ng_t\}$ of the θ-good conjugacy classes of G/N. With the previous notation, recall that

$$(\chi_l)_{N\langle g_i \rangle} = \psi_{l,i} \eta_i .$$

Now we have that

$$\sum_{l=1}^{t} \chi_l(g_1)\overline{\chi_l(g_2)} = \sum_{l=1}^{t} \psi_{l,1}(g_1)\eta_1(g_1)\overline{\psi_{l,2}(g_2)}\eta_2(g_2)$$

$$= \left(\sum_{l=1}^{t} \psi_{l,1}(g_1)\overline{\psi_{l,2}(g_2)} \right) \eta_1(g_1)\overline{\eta_2(g_2)} = 0.$$

Also, we have that

$$\sum_{l=1}^{t} \chi_l(g_1)\overline{\chi_l(g_1)} = |\mathbf{C}_{G/N}(Ng_1)|\eta_1(g_1)\overline{\eta_1(g_1)} = |\mathbf{C}_{G/N}(Ng_1)|\hat{\theta}(g_1),$$

and the proof of the theorem is complete. $\qquad\square$

5.7 Notes

There is a powerful idea in this chapter: many results in character theory admit a version relative to a character of a normal subgroup, what we usually call a *projective* or a *relative* version. Sometimes, it is possible to prove these *projective* results by using character triples which allow us to assume that the normal subgroup is central. These generalizations turn out to be essential in order to obtain a deeper insight into the *ordinary* version. This is especially true about conjectures or theorems whose proofs involve (or we think should involve) the Classification of Finite Simple Groups. As a paradigmatic example, let us come back to the McKay conjecture. This conjecture, as we know, asserts that if G is a finite group, and $\mathrm{Irr}_{p'}(G)$ is the set of irreducible characters of G of degree not divisible by p, then

$$|\mathrm{Irr}_{p'}(G)| = |\mathrm{Irr}_{p'}(\mathbf{N}_G(P))|,$$

where $P \in \mathrm{Syl}_p(G)$. The only approach that so far has been devised in order to prove this statement is to strengthen it into another statement that, only superficially, looks the same: if $L \trianglelefteq G$ and $\theta \in \mathrm{Irr}(L)$ is P-invariant, then

$$|\mathrm{Irr}_{p'}(G|\theta)| = |\mathrm{Irr}_{p'}(L\mathbf{N}_G(P)|\theta)|,$$

where now $\mathrm{Irr}_{p'}(G|\theta) = \mathrm{Irr}_{p'}(G) \cap \mathrm{Irr}(G|\theta)$. This *projective* form of the conjecture, which in the case $L = 1$ is the McKay conjecture, allows us to work by induction on $|G : L|$, and to put simple groups into the picture. In Chapter 10, we will make this precise.

There are some properties of characters that cannot be be studied using character triple isomorphisms. For instance, fields of values or determinantal orders are not usually preserved under character triple isomorphisms. Some other times, we are studying a property that does not seem to be preserved under character triple isomorphisms. But under specific circumstances, it is still possible to construct the group $\widehat{G}$ of Theorem 5.6 in a way that might be useful. For instance, if $o(\theta)\theta(1)$ is a p-power, where recall that $o(\theta)$ is the determinantal order of θ, then it is possible to construct an isomorphic triple (G^*, N^*, θ^*) with N^* central, and such that $o(\chi)$ is a p-power if and only if $o(\chi^*)$ is a p-power, for characters over θ. (See Problem 6.3.)

There is a theory by A. Turull that constructs a *universal character triple*, in a very general way. This is his theory of Brauer–Clifford groups [Tu09].

Problems

(5.1) Suppose that (G, N, θ) is a character triple. Then there exists a projective representation associated with θ with factor set α such that

$$\alpha(x, y)^{|G|\theta(1)} = 1$$

for all $x, y \in G$, and $\alpha(x, y) = 1$ if $N\langle x \rangle = N\langle y \rangle$. Furthermore, if $N \leq H \leq G$ is a fixed subgroup of G such that θ extends to H, then it is possible to choose α also satisfying that $\alpha(x, y) = 1$ for all $x, y \in H$. (*Hint:* Modify the proof of Theorem 5.5.)

(5.2) Suppose that (G, N, θ) is a character triple.
   (a) If $\lambda \in \mathrm{Irr}(G)$ is a linear character, show that multiplication by λ gives an isomorphism between the triples (G, N, θ) and $(G, N, \lambda_N\theta)$.
   (b) Assume that θ extends to P, where $P/N \in \mathrm{Syl}_p(G/N)$. Prove that there exists a triple (G^*, N^*, θ^*) isomorphic to (G, N, θ) such that N^* is a central p'-subgroup of G^*.

   (*Hint:* For part (b), we may assume that N is central. Show that θ_p extends to G, and use part (a).)

(5.3) Suppose that (G, N, θ) is a character triple. Let $X \subseteq G$ be such that $\{Nx \mid x \in X\}$ is a complete set of representatives of the θ-good conjugacy classes of G/N. For every $x \in X$, we fix an extension $\theta_x \in \mathrm{Irr}(N\langle x \rangle)$ of θ.
   (a) We let $\theta_X(g) = 0$ if $g \in G$ is not θ-good. If g is θ-good, then $Ng^y = Nx$ for some $x \in X$ and $y \in G$, and we let $\theta_X(g) = \theta_x(g^y)$. Show that θ_X is a well-defined class function of G.

(b) Let $\alpha : X \rightarrow \mathbb{C}$ be any function. Let $\tilde{\alpha} : G \rightarrow \mathbb{C}$ be defined as follows. Let $\tilde{\alpha}(g) = 0$ if g is not θ-good. If g is θ-good, then Ng is conjugate to some Nx for a unique $x \in X$, and we set $\tilde{\alpha}(g) = \alpha(x)$. Show that $\tilde{\alpha}\theta_X \in \mathrm{cf}(G|\theta)$.

(c) Show that we can choose X such that $\theta_x(x) \neq 0$ for all $x \in X$.

(d) If X is as in (c) and $\varphi \in \mathrm{cf}(G|\theta)$, show that

$$\varphi(nx) = \varphi(x) \frac{\theta_x(nx)}{\theta_x(x)}$$

for every $x \in X$, and $n \in N$.

(e) If X is as in (c), prove that restriction $f : \mathrm{cf}(G|\theta) \rightarrow \mathbb{C}^X$ defines an isomorphism of vector spaces. Conclude that $|X| = |\mathrm{Irr}(G|\theta)|$.

(f) If $X = \{x_1, \ldots, x_t\}$ is as in (c), and $\mathrm{Irr}(G|\theta) = \{\chi_1, \ldots, \chi_t\}$, then prove that the matrix $(\chi_i(x_j))$ is invertible.

(5.4) Assume that (G, N, θ) is a character triple, and suppose that $\alpha \in \mathrm{Irr}(U|\theta)$, where $N \leq U \leq G$. If $g \in G$, then we know that $\alpha^g \in \mathrm{Irr}(U^g|\theta)$, where $\alpha^g(u^g) = \alpha(u)$ for $u \in U$. If $n \in N$, then notice that

$$\alpha^g = \alpha^{ng}$$

because $N \subseteq U^g$, and characters are class functions. In particular, if $\overline{G} = G/N$, $g \in G$ and $\bar{g} = Ng \in G/N$, then we can define

$$\alpha^{\bar{g}} = \alpha^g \in \mathrm{Irr}(U^g|\theta).$$

We say that a character triple isomorphism * between (G, N, θ) and (G^*, N^*, θ^*) is **strong** if the following extra condition is also satisfied: for every $N \leq U \leq G$, $g \in G$, and $\alpha \in \mathrm{Irr}(U|\theta)$, we have that

$$(\alpha^{\bar{g}})^* = (\alpha^*)^{\bar{g}^*}.$$

In this case, we will say that (G, N, θ) and (G^*, N^*, θ^*) are **strongly isomorphic**.

(a) Show that strong isomorphism of character triples establishes an equivalence relation on character triples.

(b) If (G, N, θ) is a character triple, then show that there is a strongly isomorphic character triple (G^*, N^*, θ^*), where N^* is central and θ^* is faithful.

(5.5) (*Relative Glauberman correspondence*) Suppose that a p-group P acts by automorphisms on G, and let $N \trianglelefteq G$ be P-invariant. Assume that G/N has order not divisible by p. Let $C/N = \mathbf{C}_{G/N}(P)$. Let $\mathrm{Irr}_P(G)$

be the set of P-invariant irreducible characters of G. In this problem we show how to construct a natural bijection $': \mathrm{Irr}_P(G) \to \mathrm{Irr}_P(C)$. In fact, if $\chi \in \mathrm{Irr}_P(G)$, then

$$\chi_C = e\chi' + p\Delta + \Xi,$$

where Δ and Ξ are characters of C or zero, $e \equiv \pm 1 \bmod p$, and no irreducible constituent of Ξ lies over some P-invariant character of N.

(a) Show that C acts on $\mathrm{Irr}_P(N)$ by conjugation.

(b) Let R be a complete set of representatives of the action of C on $\mathrm{Irr}_P(N)$. Show that

$$\mathrm{Irr}_P(G) = \bigcup_{\theta \in R} \mathrm{Irr}_P(G|\theta) \quad \text{and} \quad \bigcup_{\theta \in R} \mathrm{Irr}_P(C|\theta),$$

are disjoint unions, where $\mathrm{Irr}_P(G|\theta) = \mathrm{Irr}_P(G) \cap \mathrm{Irr}(G|\theta)$. Conclude that it is enough to show that, for every $\theta \in R$, there exists a bijection

$$': \mathrm{Irr}_P(G|\theta) \to \mathrm{Irr}_P(C|\theta)$$

satisfying $\chi_C = e\chi + p\Delta + \Xi$ where Δ and Ξ are characters of C or zero, $e \equiv \pm 1 \bmod p$, and no irreducible constituent of Ξ lies over some P-invariant character of N.

(c) By using Mackey, show that it is possible to assume that $G_\theta = G$.

(d) Let $\Gamma = GP$ be the semidirect product. By using character triple isomorphisms, show that we may assume that $[N, P] = 1$ and that $N \subseteq \mathbf{Z}(G)$. Therefore $G = K \times N_p$, where N_p is a Sylow p-subgroup of G.

(e) Show that $C = \mathbf{C}_K(P) \times N_p$ and use the Glauberman correspondence Theorem 2.9.

(*Hint:* Problems 2.5 and 5.5 are useful.)

(5.6) Suppose that (G, N, θ) is a character triple. We have defined

$$\hat{\theta}(g) = \eta(g)\overline{\eta(g)},$$

where η is any extension of θ to $N\langle g \rangle$. Prove that $\hat{\theta}$ is character of G that extends $\theta\bar{\theta}$.

(5.7) Suppose that $\chi \in \mathrm{Irr}(G)$ and let $N \trianglelefteq G$. Assume that χ has p-defect zero. If $\theta \in \mathrm{Irr}(N)$, prove that θ has p-defect zero and that $(\chi(1)/\theta(1))_p = |G/N|_p$.

(5.8) (*Gallagher*) Suppose that (G, N, θ) is a character triple with $(\chi(1)/\theta(1))_p = |G : N|_p$. If p divides $o(gN)$, prove that $\chi(g) = 0$.
(*Hint:* Use character triple isomorphisms. Then use Problem 4.8(c).)

(5.9) Let $N \trianglelefteq G$, and suppose that $\theta \in \text{Irr}(N)$ is G-invariant. If $G/N = \mathbf{Q}_8$ show that θ extends to G.

(*Hint*: We may assume that $N \subseteq \mathbf{Z}(G)$ and that θ is faithful. Write $Z/N = \mathbf{Z}(G/N)$, $G/N = \langle Nx, Ny \rangle$ with $Nx^2 = Ny^2$ of order 2. Show that $Z \subseteq \mathbf{Z}(G)$ and that $G' \cap N = 1$, by proving that $[x, y]^2 = 1$.)

6

Extension of Characters

Suppose that N is a normal subgroup of a finite group G and let $\theta \in \text{Irr}(N)$. We study various criteria that guarantee that θ extends to a character of G. In the cases we study, this extension can be chosen canonically, and this allows us to control some of its properties.

6.1 Determinants

If $N \trianglelefteq G$ and $\theta \in \text{Irr}(N)$, let us write $\text{Ext}(G|\theta)$ for the (possibly empty) set of $\chi \in \text{Irr}(G)$ such that $\chi_N = \theta$. These are the **extensions** of θ to G. If θ extends to some $\chi \in \text{Irr}(G)$, then we know that θ is G-invariant. We notice that if θ extends to G, then $|\text{Ext}(G|\theta)|$ is the number of linear characters of G/N, by the Gallagher correspondence Corollary 1.23. In particular, if $|\text{Ext}(G|\theta)| \neq 0$, then

$$|\text{Ext}(G|\theta)| = |\text{Lin}(G/N)| = |G : G'N|$$

is independent of θ.

We have already proved two of the most basic extendibility criteria: characters of subgroups of abelian groups extend to the group (Corollary 1.17), and if θ is G-invariant and G/N is cyclic, then θ extends to G, by Theorem 5.1.

By Theorem 5.10, we also know that if θ is G-invariant, then θ extends to G if and only if θ extends to P, for every Sylow subgroup P/N of G/N. This result brings primes and some arithmetic into the extension problem.

Theorem 6.1 (Gallagher) *Let $N \trianglelefteq G$ and let $\theta \in \text{Irr}(N)$ be G-invariant. Assume that $\gcd(\theta(1), |G : N|) = 1$. Then θ extends to G if and only if $\det(\theta)$ extends to G, and in this case, the map $\chi \mapsto \det(\chi)$ defines a natural bijection*

$$\text{Ext}(G|\theta) \to \text{Ext}(G|\det(\theta)).$$

Proof Let $\lambda = \det(\theta)$. First we prove that if $\chi \in \mathrm{Irr}(G)$ extends θ, then $\nu = \det(\chi)$ extends λ, and that the map $\psi \mapsto \det(\psi)$ is a bijection $\mathrm{Ext}(G|\theta) \to \mathrm{Ext}(G|\lambda)$. If $\mathcal{X}$ is a representation affording χ, then it is clear that

$$\nu_N = \det(\mathcal{X})_N = \det(\mathcal{X}_N) = \det(\theta) = \lambda.$$

Suppose now that $\psi \in \mathrm{Ext}(G|\theta)$ is such that $\det(\psi) = \nu$. We know that $\psi = \tau \chi$ for some $\tau \in \mathrm{Lin}(G/N)$ by the Gallagher correspondence. Then

$$\det(\psi) = \tau^{\chi(1)}\det(\chi)$$

and therefore

$$1 = \tau^{\chi(1)} = \tau^{\theta(1)}.$$

Since τ lies in the group $\mathrm{Lin}(G/N)$ which has order coprime with $\theta(1)$, we deduce that $\tau = 1$. Since $|\mathrm{Ext}(G|\theta)| = |G : G'N| = |\mathrm{Ext}(G|\lambda)|$, we deduce that the map $\psi \mapsto \det(\psi)$ is a bijection.

Now we prove by induction on $|G : N|$, that if λ extends to $\mu \in \mathrm{Irr}(G)$, then θ extends to G. By Theorem 5.10, we may assume that G/N is a p-group for some prime p. Also, we may assume that $N < G$. Let $N \leq K \triangleleft G$ of index p. By induction, we know that θ extends to K. By the first paragraph, there is a unique extension $\rho \in \mathrm{Ext}(K|\theta)$ such that $\det(\rho) = \mu_K$. Since μ_K is G-invariant and

$$\det(\rho^g) = \det(\rho)^g = \mu_K,$$

it follows by uniqueness that $\rho = \rho^g$ for all $g \in G$. Since G/K is cyclic, we deduce that ρ extends to G by Theorem 5.1. In particular, θ extends to G. $\square$

We see that under the hypothesis that $\gcd(\theta(1), |G : N|) = 1$, in order to extend θ we may replace θ by its determinant and assume that $\theta(1) = 1$. If we further assume that $\gcd(o(\theta), |G : N|) = 1$, then we are going to conclude next that θ does extend to G, and that some extension can be picked up canonically.

We remind the reader of some useful terminology in finite group theory. If π is a set of primes, an integer n is a π**-number** if every prime dividing n is in π. Usually, π' denotes the set of primes which are not in π. A finite group G is a π**-group** if $|G|$ is a π-number, and a π**-subgroup** H of G is a subgroup that is a π-group. A **Hall** π**-subgroup** of G is a π-subgroup H of G such that $|G : H|$ is a π'-number. Of course, when $\pi = \{p\}$ consists of a single prime, then we recover the standard Sylow terminology. Finally, an element $x \in G$ can be uniquely written as $x = x_\pi x_{\pi'}$, where x_π and $x_{\pi'}$ are elements of G that commute, $o(x_\pi)$ is a π-number and $o(x_{\pi'})$ is a π'-number. In fact, x_π is the product of x_p for $p \in \pi$. If $x = x_\pi$, then we say that x is a π**-element**, or that x has π**-order**.

Corollary 6.2 *Let* $N \trianglelefteq G$ *and suppose that* $\theta \in \mathrm{Irr}(N)$ *is G-invariant with* $\gcd(o(\theta)\theta(1), |G : N|) = 1$. *Then* θ *extends to G and has a unique extension* χ *such that* $\gcd(o(\chi), |G : N|) = 1$. *In fact,* $o(\chi) = o(\theta)$.

Proof By Theorem 6.1 we may assume that θ is linear. Since θ is G-invariant, then $\ker(\theta) \trianglelefteq G$. By working in $G/\ker(\theta)$, we may assume that θ is faithful. Now $\theta^g(n) = \theta(n)$ for $g \in G$ and $n \in N$, and this implies that $N \subseteq \mathbf{Z}(G)$. By hypothesis and Lemma 1.12, we have that $\gcd(|N|, |G : N|) = 1$. To show that θ extends, we may assume that G/N is a p-group by Theorem 5.10. In this case, if $P \in \mathrm{Syl}_p(G)$, then $G = N \times P$ by elementary group theory, and $\theta \times 1_P$ extends θ.

Now, let $\nu \in \mathrm{Irr}(G)$ be an extension of θ. Let π be the set of prime divisors of $|N|$. If we write $\nu = \nu_\pi \nu_{\pi'}$ in the group $\mathrm{Lin}(G)$, then $(\nu_{\pi'})_N = 1_N$ because $(\nu_{\pi'})_N$ is a character of a π-group and has π'-order. Hence, we may assume that ν has π-order. Now

$$(\nu^{o(\theta)})_N = \theta^{o(\theta)} = 1_N,$$

and therefore $\nu^{o(\theta)} = 1_G$ because $\nu^{o(\theta)}$ is the character of the π'-group G/N and has π-order. Since $o(\theta)$ divides $o(\nu)$ because ν extends θ, we conclude that $o(\nu) = o(\theta)$. If $\mu \in \mathrm{Irr}(G)$ is any other extension of θ with π-order, then $\mu^{-1}\nu = 1_G$ because $\mu^{-1}\nu$ is a character of the π'-group G/N and has π-order. $\square$

Corollary 6.3 *Suppose that N is a normal Hall π-subgroup of G, and let* $\theta \in \mathrm{Irr}(N)$ *be G-invariant. Then* θ *extends to G. In fact, there is a unique extension* $\eta \in \mathrm{Irr}(G)$ *of* θ *such that* $o(\eta)$ *is a π-number.*

Proof Since by hypothesis $\gcd(|N|, |G : N|) = 1$, the assertion immediately follows from Corollary 6.2. $\square$

Sometimes the character χ in Corollary 6.2 is called the **canonical extension** of θ to G, and it is frequently written as $\hat{\theta}$ to emphasize that it is uniquely determined by θ. As we shall see repeatedly in this chapter, once two characters uniquely determine each other, then they have the same field of values. Let us check this fact with respect to θ and $\hat{\theta}$ in Corollary 6.2.

Corollary 6.4 *Let* $N \trianglelefteq G$ *and suppose that* $\theta \in \mathrm{Irr}(N)$ *is G-invariant with* $\gcd(o(\theta)\theta(1), |G : N|) = 1$. *Let* $\hat{\theta} \in \mathrm{Irr}(G)$ *be the unique extension of* θ *to G such that* $\gcd(o(\hat{\theta}), |G : N|) = 1$. *Then* $\mathbb{Q}(\hat{\theta}) = \mathbb{Q}(\theta)$.

Proof Write $\chi = \hat{\theta}$. Since $\chi_N = \theta$, we have that $\mathbb{Q}(\theta) \subseteq \mathbb{Q}(\chi)$. Now, let $\sigma \in \mathrm{Gal}(\mathbb{Q}(\chi)/\mathbb{Q}(\theta))$. Then $(\chi^\sigma)_N = (\chi_N)^\sigma = \theta^\sigma = \theta$. Since $o(\chi^\sigma) = o(\chi)$ (see Problem 3.8), we conclude that $\chi^\sigma = \chi$, by the uniqueness in Corollary 6.2. Thus σ is the trivial automorphism, and $\mathbb{Q}(\theta) = \mathbb{Q}(\chi)$ by elementary Galois theory. $\qquad\qquad\qquad\qquad\qquad\qquad\qquad\qquad\qquad\qquad\qquad\qquad\square$

6.2 Field of Values

In this section, we prove an extension theorem whose proof relies on Galois conjugation. Recall from Chapter 3 that if G is a finite group of order dividing n, then $\mathrm{Gal}(\mathbb{Q}_n/\mathbb{Q})$ acts on $\mathrm{Irr}(G)$, on the elements of G and on the conjugacy classes of G. In fact, if $\sigma \in \mathrm{Gal}(\mathbb{Q}_n/\mathbb{Q})$, then σ sends the nth roots of unity ξ to ξ^m, where m is an integer coprime with n, and we have defined $x^\sigma = x^m$ for $x \in G$. In particular, if $x \in G$ is an involution, then $|G|$ is even, m is odd and $x^m = x$.

Theorem 6.5 *Let $N \trianglelefteq G$ and suppose that $\theta \in \mathrm{Irr}(N)$ is G-invariant. Let $\sigma \in \mathrm{Gal}(\mathbb{Q}_{|G|}/\mathbb{Q})$. Assume that $\theta^\sigma = \theta$, and that σ fixes no nontrivial irreducible character of G/N. Then there is a unique $\chi \in \mathrm{Irr}(G|\theta)$ which is σ-invariant. In fact, $\chi_N = \theta$ and $\mathbb{Q}(\chi) = \mathbb{Q}(\theta)$.*

Proof First, we claim that if there is a σ-invariant extension $\chi \in \mathrm{Irr}(G)$ of θ, then χ is the unique σ-invariant irreducible character over θ, and that $\mathbb{Q}(\chi) = \mathbb{Q}(\theta)$. Indeed, suppose that $\psi \in \mathrm{Irr}(G|\theta)$ is σ-invariant. By the Gallagher correspondence, we have that $\psi = \mu\chi$ for some $\mu \in \mathrm{Irr}(G/N)$. By the uniqueness in the Gallagher correspondence, we deduce that $\mu^\sigma = \mu$, and then $\mu = 1$ by hypothesis. Since $\chi_N = \theta$, we certainly have that $\mathbb{Q}(\theta) \subseteq \mathbb{Q}(\chi)$. Now, if $\tau \in \mathrm{Gal}(\mathbb{Q}(\chi)/\mathbb{Q}(\theta))$, then

$$(\chi^\tau)^\sigma = (\chi^\sigma)^\tau = \chi^\tau,$$

using that τ and σ commute on $\mathbb{Q}(\chi)$. Since χ^τ also lies over $\theta^\tau = \theta$, by uniqueness we conclude that $\chi^\tau = \chi$. Therefore τ is the trivial automorphism and we deduce that $\mathbb{Q}(\chi) = \mathbb{Q}(\theta)$, by elementary Galois theory. (This is the same argument as in Corollary 6.4.)

Next, we prove that if θ has some extension $\mu \in \mathrm{Irr}(G)$, then θ has some σ-invariant extension. By hypothesis, notice that the map $\mathrm{Lin}(G/N) \to \mathrm{Lin}(G/N)$ given by $\lambda \mapsto \lambda^{-1}\lambda^\sigma$ is a bijection. Now, μ^σ extends $\theta^\sigma = \theta$, and by the Gallagher correspondence it follows that $\mu^\sigma = \nu\mu$, for some $\nu \in \mathrm{Lin}(G/N)$. Then $\nu = \lambda^{-1}\lambda^\sigma$ for some $\lambda \in \mathrm{Lin}(G/N)$, and we easily check that $\chi = \lambda^{-1}\mu \in \mathrm{Irr}(G)$ is a σ-invariant extension of θ.

Finally, we prove that θ extends to G by induction on $|G : N|$. By Theorem 3.3, we have that σ fixes no nontrivial conjugacy class of G/N. Therefore, if H/N is any subgroup of G/N, then σ fixes no nontrivial conjugacy class of H/N, and again by Theorem 3.3, we have that σ fixes no nontrivial irreducible character of H/N. By induction and Theorem 5.10, we may therefore assume that G/N is a p-group. Now, we choose any $N \leq M \triangleleft G$ of prime index. By induction, it follows that θ extends to M. By the previous paragraph, θ has a σ-invariant extension $\psi \in \mathrm{Irr}(M)$. By the first paragraph of this proof, ψ is the unique σ-invariant character of M over θ. Since $(\psi^g)^\sigma = (\psi^\sigma)^g = \psi^g$ for $g \in G$, we deduce that ψ is G-invariant by uniqueness. Therefore it extends to G because G/M is cyclic (by Theorem 5.1). $\qquad\square$

Observe that the hypotheses of Theorem 6.5 imply that G/N has odd order. Indeed, if σ fixes no nontrivial irreducible character of G/N, then σ fixes no nontrivial conjugacy classes of G/N by Theorem 3.3, and therefore G/N has no involutions.

There are two useful consequences of Theorem 6.5. If p is a prime, an irreducible character χ of a finite group G is called *p-rational* if $\mathbb{Q}(\chi) \subseteq \mathbb{Q}_m$ for some m not divisible by p.

Corollary 6.6 *Suppose that $N \trianglelefteq G$ and let $\theta \in \mathrm{Irr}(N)$.*

(a) *If θ is p-rational for some odd prime p, and G/N is a p-group, then there is a unique p-rational $\chi \in \mathrm{Irr}(G|\theta)$. If θ is G-invariant, then $\chi_N = \theta$, and $\mathbb{Q}(\chi) = \mathbb{Q}(\theta)$.*

(b) *If G/N has odd order and θ is real-valued, then there is a unique real-valued $\chi \in \mathrm{Irr}(G|\theta)$. If θ is G-invariant, then $\chi_N = \theta$, and $\mathbb{Q}(\chi) = \mathbb{Q}(\theta)$.*

Proof If ψ is any character of a subgroup of G and $\sigma \in \mathrm{Gal}(\mathbb{Q}_{|G|}/\mathbb{Q})$, notice that

$$(\psi^\sigma)^G = (\psi^G)^\sigma$$

by the induction formula. Let $T = G_\theta$ be the stabilizer of θ in G.

(a) Write $|G| = n = p^a m$, where m is not divisible by p. By elementary Galois theory, we know that

$$\mathrm{Gal}(\mathbb{Q}_n/\mathbb{Q}_m) = \langle \sigma \rangle$$

is cyclic since p is odd. Also, notice that if $\mu \in \mathrm{Irr}(H)$, where H is any group of order dividing n, then μ is p-rational if and only if μ is σ-invariant. Now, since $\theta^\sigma = \theta$, induction of characters commutes with σ, and the uniqueness in the Clifford correspondence, we have that $\psi \in \mathrm{Irr}(T|\theta)$ is p-rational if and only if ψ^G is p-rational. Hence, we may assume that θ is G-invariant. In order to apply Theorem 6.5, it is enough to check that in a p-group the only p-rational irreducible character is the trivial one, for p odd. But this is clear: if P is a p-group and $\psi \in \mathrm{Irr}(P)$ is p-rational, then $\mathbb{Q}(\psi) \subseteq \mathbb{Q}_m$ for some m not divisible by p. Then $\mathbb{Q}(\psi) \subseteq \mathbb{Q}_m \cap \mathbb{Q}_{|P|} = \mathbb{Q}$, and ψ is real-valued. Thus $\psi = 1_P$ by Corollary 3.5.

To prove part (b) we use the same argument as in part (a) with $\sigma \in \mathrm{Gal}(\mathbb{Q}_{|G|}/\mathbb{Q})$ being complex-conjugation. Since G/N has odd order, then σ fixes no character of G/N again by Corollary 3.5. $\square$

Let us see an application of Corollary 6.6(b). This the first theorem in this book whose proof is going to use the CFSG. (See Theorem A.1 in Appendix A.)

In order to prove some theorems in this book (such as the Itô–Michler theorem, the Howlett–Isaacs theorem, or Theorem 6.7), we are going to assume some specific facts on non-abelian simple groups, whose proof is omitted. For instance, the simple groups of Lie type $G(q)$, where q is a power of p, possess an irreducible character St, called the **Steinberg character**, of fundamental importance. This character has p-defect zero, and therefore has the value 0 on p-singular elements, by the Brauer–Nesbitt theorem. If $x \in G(q)$ is p-regular, then $\mathrm{St}(x) = \pm|\mathbf{C}_{G(q)}(x)|$, and therefore St is a rational-valued character. The ideal reader should learn the techniques of the representation theory of alternating groups and of groups of Lie type necessary in order to complete these proofs. It is often the case that the sporadic groups can be handled with computers using the information that has been gathered for decades by the specialists, but not always.

Let us mention that Theorem 6.7 is one of those examples where a statement is true because it holds for simple groups. The McKay conjecture, however, is known to hold for simple groups, but still it is not known if it holds for finite groups in general.

Theorem 6.7 *A finite group G has even order if and only if there is a nontrivial rational-valued $\chi \in \mathrm{Irr}(G)$.*

Proof If G has odd order, then any real-valued irreducible character of G is the trivial one by Corollary 3.5.

Next we show by induction on $|G|$ that if G has even order, then there is $1 \neq \chi \in \mathrm{Irr}(G)$ rational-valued. Let N be a minimal normal subgroup of G. If G/N has even order, then there is $1 \neq \chi \in \mathrm{Irr}(G/N)$ rational-valued by induction, and the theorem is proven in this case. We can thus assume that G/N has odd order. If $N < G$, then by induction there is $1 \neq \theta \in \mathrm{Irr}(N)$ rational-valued. Let T be the stabilizer of θ in G. By Corollary 6.6(b), there exists a unique real-valued $\psi \in \mathrm{Irr}(T|\theta)$. Also, $\mathbb{Q}(\psi) = \mathbb{Q}(\theta) = \mathbb{Q}$. By the Clifford correspondence, we have that $\chi = \psi^G \in \mathrm{Irr}(G)$. Also, χ is not principal because $\theta \neq 1_N$, and χ is rational-valued by the induction formula. Hence, we may assume that $N = G$. In other words, G is simple. We check in the [ATLAS] (or using [GAP]) that the theorem is true for the sporadic simple groups. (As a matter of fact, each sporadic finite simple group has at least six rational-valued irreducible characters.) By using the Steinberg character for simple groups of Lie type, we are left with the alternating groups A_n for $n \geq 5$. Now, we know that A_n acts 2-transitively on $\Omega = \{1, \ldots, n\}$. By Problem 1.11(d), we conclude that A_n has a rational-valued irreducible character of degree $n - 1$. $\qquad\square$

6.3 Multiplicities

In character theory, we often have the following configuration: G is a finite group, $N \trianglelefteq G$, $H \leq G$ with $NH = G$, and we have characters $\theta \in \mathrm{Irr}(N)$ and $\varphi \in \mathrm{Irr}(M)$, where $M = H \cap N$. We want to relate the character theory of G over θ with that of H over φ. For instance, we may have that G/N is a p-group (which as we know now is the key situation when studying extendability of characters) and $H = P \in \mathrm{Syl}_p(G)$, or we may have that $H = \mathbf{N}_G(Q)$, for some $Q \in \mathrm{Syl}_p(N)$ (as in a typical McKay situation). It is frequently useful to draw diagrams, and we urge the reader to do so. In the diamond below, for instance, the lower node corresponds to the intersection of the subgroups, and the upper node to the product (which is only written if the product of the corresponding subgroups is a group).

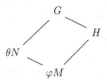

The following lemma, which only uses Frobenius reciprocity, is elementary but useful.

Lemma 6.8 (Isaacs) *Suppose that $N \trianglelefteq G$, $NH = G$, $M = N \cap H$. Suppose that $\theta \in \mathrm{Irr}(N)$ is G-invariant, $\varphi \in \mathrm{Irr}(M)$ is H-invariant and $[\theta_M, \varphi] = 1$.*

(a) *If* $\chi \in \mathrm{Irr}(G|\theta)$, *then*

$$\chi_H = \chi' + \Delta,$$

where $\chi' \in \mathrm{Irr}(H|\varphi)$, *and* Δ *is a character of* H *(or zero) such that none of its irreducible constituents lie over* φ.

(b) *If* $\xi \in \mathrm{Irr}(H|\varphi)$, *then*

$$\xi^G = \xi' + \Xi,$$

where $\xi' \in \mathrm{Irr}(G|\theta)$, *and* Ξ *is a character of* G *(or zero) such that none of its irreducible constituents lie over* θ.

(c) *The map* $\chi \mapsto \chi'$ *is a bijection* $\mathrm{Irr}(G|\theta) \to \mathrm{Irr}(H|\varphi)$, *with inverse* $\xi \mapsto \xi'$. *Also,* $\chi(1)/\theta(1) = \chi'(1)/\varphi(1)$, *and therefore* θ *extends to* G *if and only if* ξ *extends to* H. *In fact, this map defines a character triple isomorphism* $(G, N, \theta) \to (H, M, \varphi)$.

(d) *If* $\theta_M = \varphi$, *then restriction defines a bijection* $\mathrm{Irr}(G|\theta) \to \mathrm{Irr}(H|\varphi)$.

Proof Let $\chi \in \mathrm{Irr}(G|\theta)$ and write $\chi_N = e\theta$. Then $[\chi_M, \varphi] = e$. Hence, there is at least one irreducible constituent χ' of χ_H which lies over φ. Let $u = [\chi_H, \chi']$ and write $\chi'_M = f\varphi$. Also, write $\chi_H = u\chi' + \Delta$, where Δ is a character of H or zero such that $[\Delta, \chi'] = 0$. Now,

$$\chi_M = uf\varphi + \Delta_M,$$

and thus $uf \leq e$.

By Frobenius reciprocity we can write $(\chi')^G = u\chi + \Xi$, where Ξ is a character of G or zero such that $[\Xi, \chi] = 0$. By Mackey (Theorem 1.16), we have that

$$((\chi')^G)_N = (\chi'_M)^N,$$

and therefore

$$ue\theta + \Xi_N = f\varphi^N.$$

Since $[f\varphi^N, \theta] = f[\varphi, \theta_M] = f$, we conclude that $ue \leq f$. Combining this with $uf \leq e$, we have that $u = 1$, $e = f$ and no irreducible constituent of Δ lies over φ. Also we have that no irreducible constituent of Ξ lies over θ. This proves (a), (b), and (c), except for the last part. If $\theta_M = \varphi$, then χ_M is a multiple of φ, and necessarily $\Delta = 0$, which proves (d).

Recall from Theorem 1.18 that restriction defines a bijection $\mathrm{Irr}(G/N) \to \mathrm{Irr}(H/M)$. (We could also obtain this result by applying (d) with $\theta = 1_N$.) Using this, it is straightforward to check that $'$ defines a character triple isomorphism $(G, N, \theta) \to (H, M, \varphi)$. We leave the details of this for the reader to check. $\square$

The following is a similar result. The deeper part is that we can construct choice-free bijections between certain extensions.

Theorem 6.9 (Thompson)   *Suppose that G/N is a p-group, and let $H \leq G$ such that $G = NH$. Write $M = N \cap H$. Let $\theta \in \mathrm{Irr}(N)$ be G-invariant and let $\varphi \in \mathrm{Irr}(M)$ be H-invariant such that $[\theta_M, \varphi]$ is not divisible by p.*

(a) Suppose that $\xi \in \mathrm{Irr}(H)$ extends φ. Then there is a unique extension $\chi \in \mathrm{Irr}(G)$ of θ such that

$$\chi_H = \Psi\xi + \Delta,$$

where Δ is a character of H (or zero) with $[\Delta_M, \varphi] = 0$, and Ψ is a character of H/M with $\det(\Psi) = 1$.

(b) Suppose that $\chi \in \mathrm{Irr}(G)$ extends θ. Then there is a unique extension $\xi \in \mathrm{Irr}(H)$ of φ such that

$$\xi^G = \Psi\chi + \Delta,$$

where Δ is a character of G (or zero) with $[\Delta_N, \theta] = 0$, and Ψ is a character of G/N with $\det(\Psi) = 1$.

Proof   By Mackey (Theorem 1.16) and Frobenius reciprocity, we have that

$$[(\xi^G)_N, \theta] = [(\xi_M)^N, \theta] = [\varphi^N, \theta] = [\varphi, \theta_M]$$

is not divisible by p. Now

$$[(\xi^G)_N, \theta] = \sum_{\chi \in \mathrm{Irr}(G|\theta)} [\xi^G, \chi][\chi_N, \theta],$$

and therefore there exists $\chi \in \mathrm{Irr}(G|\theta)$ such that $[\chi_N, \theta][\xi^G, \chi]$ is not divisible by p. Since θ is G-invariant, notice that $[\chi_N, \theta] = \chi(1)/\theta(1)$. By Theorem 5.12, we have that $\chi_N = \theta$, since G/N is a p-group. Thus θ extends to G.

Now, if $\chi \in \mathrm{Irr}(G)$ is an arbitrary extension of θ, we can write

$$\chi_H = \alpha + \Delta,$$

where all the irreducible constituents of the character α lie over φ, and Δ is a character (or zero) such that none of the irreducible constituents of Δ lie over φ. By the Gallagher correspondence, we can write $\alpha = \Psi_\chi\xi$, for a unique character Ψ_χ of H/M. Thus

$$\chi_H = \Psi_\chi\xi + \Delta.$$

Set

$$\delta_\chi = \det(\Psi_\chi) \in \mathrm{Lin}(H/M).$$

Our aim is to show that there exists a unique χ such that $\delta_\chi = 1$. Write $m = \Psi_\chi(1)$, and notice that

$$[\theta_M, \varphi] = [\chi_M, \varphi] = [\alpha_M, \varphi] = [(\Psi_\chi \xi)_M, \varphi] = [\Psi_\chi(1)\varphi, \varphi] = m$$

is not divisible by p by hypothesis. If $\lambda \in \mathrm{Irr}(G/N)$ is linear, then $\lambda\chi$ is another extension of θ, and notice that

$$\delta_{\lambda\chi} = (\lambda_H)^m \delta_\chi$$

because $\Psi_{\lambda\chi} = \lambda_H \Psi_\chi$.

Now fix an extension $\chi \in \mathrm{Irr}(G)$ of θ and write $\mu = \delta_\chi$. We claim that there is a unique linear $\lambda \in \mathrm{Irr}(G/N)$ such that $\delta_{\lambda\chi} = 1_H$. This will prove what we want. Since $hM \mapsto hN$ is a group isomorphism $H/M \to G/N$, it easily follows that restriction of linear characters is a bijection $\mathrm{Lin}(G/N) \to \mathrm{Lin}(H/M)$. Furthermore, since m is coprime with p and H/M is a p-group, we have that the map $\epsilon \mapsto \epsilon^m$ permutes $\mathrm{Lin}(H/M)$. Now let $\lambda \in \mathrm{Irr}(G/N)$ be the unique linear character such that $(\lambda_H)^m = \bar{\mu}$, the complex conjugate of μ. Then

$$\delta_{\lambda\chi} = (\lambda_H)^m \mu = 1_H.$$

Hence, we have shown that $\rho = \lambda\chi \in \mathrm{Irr}(G)$ satisfies the conclusions of the theorem. We show next that it is unique. Let $\tau \in \mathrm{Irr}(G)$ be an extension of θ and assume that $\delta_\tau = \det(\Psi_\tau) = 1$. By the Gallagher correspondence, we can write $\tau = \epsilon\rho$ for some linear character $\epsilon \in \mathrm{Irr}(G/N)$. Now we have that

$$1_H = \delta_{\epsilon\rho} = (\epsilon_H)^m \delta_\rho = (\epsilon_H)^m.$$

We conclude that $\epsilon_H = 1_H$, $\epsilon = 1_G$ and $\tau = \rho$, as desired.

Part (b) is proven similarly. Again notice that to show that φ extends to H is easier. Simply note that

$$[\theta_M, \varphi] = [\chi_M, \varphi] = \sum_{\xi \in \mathrm{Irr}(H|\varphi)} [\chi_H, \xi][\xi_M, \varphi]$$

is not divisible by p and therefore there is $\xi \in \mathrm{Irr}(H|\varphi)$ such that $[\xi_M, \varphi]$ is not divisible by p. Since H/M is a p-group and φ is H-invariant, we apply Theorem 5.12 to conclude that $\xi_M = \varphi$. $\qquad\square$

We remark that the corresponding characters χ and ξ in Theorem 6.9 need not necessarily satisfy that $[\chi_H, \xi] \neq 0$. For instance, if $G = \mathrm{SL}_2(3)$, $N = \mathbf{Q}_8$, $H \in \mathrm{Syl}_3(G)$, $\theta \in \mathrm{Irr}(N)$ has degree 2, $\varphi = 1_1$ and $p = 3$, then the extension of θ that corresponds to $\xi = 1_H$, is the unique rational-valued irreducible character $\hat{\theta}$ of G. We have that $\hat{\theta}_H$ is the sum of the two nontrivial characters of H.

Corollary 6.10 *Suppose that G/N is a p-group. Let $P \in \mathrm{Syl}_p(G)$ and $Q = P \cap N$. Let $\theta \in \mathrm{Irr}(N)$ be G-invariant such that $[\theta_Q, 1_Q]$ is not divisible by p. Then θ has a unique extension $\hat{\theta} \in \mathrm{Irr}(G)$ such that, if we write*

$$\hat{\theta}_P = \Psi + \Delta$$

where $\Psi \in \mathrm{Char}(P/Q)$ and Δ is a character of P (or zero) with $[\Delta_Q, 1_Q] = 0$, then $\det(\Psi) = 1_P$. Furthermore, $\mathbb{Q}(\theta) = \mathbb{Q}(\hat{\theta})$.

Proof We simply apply Theorem 6.9 with $H = P \in \mathrm{Syl}_p(G)$ and $\xi = 1_P$. Let $\sigma \in \mathrm{Gal}(\mathbb{Q}(\hat{\theta})/\mathbb{Q}(\theta))$, and notice that $\hat{\theta}^\sigma \in \mathrm{Irr}(G)$ extends θ. Since $\det(\Psi^\sigma) = \det(\Psi)^\sigma = 1_P$ by using Problem 3.8, we necessarily have that $\hat{\theta}^\sigma = \hat{\theta}$ by uniqueness. Thus $\mathbb{Q}(\hat{\theta}) = \mathbb{Q}(\theta)$. $\square$

Corollary 6.10 can be applied to p-defect zero characters.

Corollary 6.11 *Suppose that G/N is a p-group. Let $\theta \in \mathrm{Irr}(N)$ be G-invariant of p-defect zero. Then θ has a unique extension $\hat{\theta} \in \mathrm{Irr}(G)$ such that, if $P \in \mathrm{Syl}_p(G)$, $Q = P \cap N$, and we write*

$$\hat{\theta}_P = \Psi + \Delta$$

where $\Psi \in \mathrm{Char}(P/Q)$ and Δ is a character of P (or zero) with $[\Delta_Q, 1_Q] = 0$, then $\det(\Psi) = 1_P$. Furthermore, $\mathbb{Q}(\theta) = \mathbb{Q}(\hat{\theta})$.

Proof Let $P \in \mathrm{Syl}_p(G)$ and let $Q = P \cap N$. Since θ has p-defect zero, by the Brauer–Nesbitt theorem (Theorem 4.6) we have that θ_Q vanishes on the nontrivial elements of Q. Hence ρ_Q is a multiple of the regular character ρ_Q of Q. By degrees,

$$\theta_Q = \frac{\theta(1)}{|Q|}\rho_Q.$$

Since $\theta(1)_p = |Q|$, we have that $[\theta_Q, 1_Q]$ is not divisible by p, and we can apply Corollary 6.10. $\square$

Observe that a p-defect zero character θ vanishes on p-singular elements, and therefore it is p-rational. If p is odd, then Corollary 6.6(a) tells us that there is a unique p-rational extension $\hat{\theta} \in \mathrm{Irr}(G)$. (By the uniqueness, we conclude that this has also to be the extension obtained in Corollary 6.11.) What we did not know before Corollary 6.11 is that for $p = 2$, θ also possesses a canonical extension to G, which necessarily is 2-rational.

From what we have accomplished so far, we can derive some group theoretical consequences. If G is a finite group, recall that $\mathbf{O}^p(G)$ is the smallest normal subgroup $N \trianglelefteq G$ such that G/N is a p-group.

Theorem 6.12 *Let G be a finite group, let $K = \mathbf{O}^p(G)$, $P \in \mathrm{Syl}_p(G)$ and $Q = K \cap P$. Then $[Q, P] = P' \cap Q$.*

Proof We have that $G = KP$, and $Q \in \mathrm{Syl}_p(K)$. It is clear that $Q' = [Q, Q] \subseteq [Q, P] \subseteq Q$, because $Q \trianglelefteq P$. Write $\mathrm{Irr}_P(K)$ for the P-invariant characters of K.

Let $\varphi \in \mathrm{Irr}(Q/[Q, P])$. Hence, we have that $\varphi \in \mathrm{Irr}(Q)$ is linear. Also $\varphi^x(y) = \varphi(y)$ for $x \in P$, $y \in Q$, using that φ is linear and $[Q, P] \subseteq \ker(\varphi)$. We can write φ^K as a sum

$$\sum_{\substack{\theta \in \mathrm{Irr}_P(K) \\ [\varphi^K, \theta]\theta(1) \not\equiv 0 \bmod p}} [\varphi^K, \theta]\theta + \sum_{\substack{\theta \in \mathrm{Irr}_P(K) \\ [\varphi^K, \theta]\theta(1) \equiv 0 \bmod p}} [\varphi^K, \theta]\theta + \sum_{\theta \in \mathrm{Irr}(K) - \mathrm{Irr}_P(K)} [\varphi^K, \theta]\theta.$$

Note that the characters in $\mathrm{Irr}(K)$ which are not P-invariant lie in P-orbits of nontrivial p-power size, and that each character of this P-orbit occurs with the same multiplicity in the P-invariant character φ^K. Since φ^K has degree not divisible by p, we deduce that there exists $\theta \in \mathrm{Irr}(K)$ of degree not divisible by p, P-invariant, such that $[\theta, \varphi^K] = [\theta_Q, \varphi]$ is not divisible by p. Now, since $\mathbf{O}^p(K) = K$, we have that $o(\theta)$ is not divisible by p. By Corollary 6.2, we have that θ extends to G. Then, by Theorem 6.9(b) we have that φ extends to some linear $\lambda \in \mathrm{Irr}(P)$. Therefore

$$P' \cap Q \subseteq \ker(\lambda) \cap Q = \ker(\lambda_Q) = \ker(\varphi),$$

and we deduce that

$$P' \cap Q \subseteq \bigcap_{\varphi \in \mathrm{Irr}(Q/[Q,P])} \ker(\varphi) = [Q, P],$$

as required. $\qquad\square$

Recall that a **normal p-complement** K of a group G is a normal subgroup $K \trianglelefteq G$ with order not divisible by p and with index a power of p. (In other words, K is a normal Hall p'-subgroup of G.)

Corollary 6.13 *Let G be a finite group, and let $P \in \mathrm{Syl}_p(G)$. Then G has a normal p-complement if and only if every linear character of P extends to G.*

Proof Let $K = \mathbf{O}^p(G)$ and $Q = P \cap K$. Assume that every linear $\lambda \in \mathrm{Irr}(P)$ has an extension $\nu \in \mathrm{Irr}(G)$. By writing $\nu = \nu_p\nu_{p'}$ as the product of its p-part and its p'-part, notice that $(\nu_{p'})_P = 1$ (because the order of $(\nu_{p'})_P$ is not divisible by p). Therefore we may assume that $o(\nu)$ is a power of p. Therefore every p-regular element of G lies in the kernel of ν. Since K is the subgroup generated by the p-regular elements of G, we have that $K \subseteq \ker(\nu)$ and $K \cap P \subseteq \ker(\nu_P) = \ker(\lambda)$. Then

$$Q = K \cap P \subseteq \bigcap_{\lambda \in \mathrm{Irr}(P/P')} \ker(\lambda) = P'.$$

By Theorem 6.12, we deduce that $Q = Q \cap P' = [Q, P]$. But now, since $Q \trianglelefteq P$ is a p-group, we have that $[Q, P] < Q$ unless $Q = 1$, by elementary group theory. Hence $Q = P \cap K = 1$ and K is a normal p-complement of G. The other direction is trivial. If K is a normal p-complement of G, then $KP = G$ and $K \cap P = 1$. Thus restriction of characters defines a bijection $\mathrm{Irr}(G/K) \to \mathrm{Irr}(P)$ by Theorem 1.18. $\qquad\square$

Corollary 6.14 (Tate) *Suppose that $N \trianglelefteq G$ and $P \in \mathrm{Syl}_p(G)$. If $P \cap N \subseteq P'$, then N has a normal p-complement.*

Proof We may assume that $NP = G$. Let $K = \mathbf{O}^p(G) \subseteq N$ and let $Q = K \cap P$. By hypothesis, $Q \subseteq N \cap P \subseteq P'$. By Theorem 6.12, we have that $[Q, P] = P' \cap Q = Q$. Since P is a p-group and $Q \trianglelefteq P$ with $[Q, P] = Q$, it follows that $Q = 1$, by elementary group theory. Thus K is a p'-group. Hence G has a normal p-complement and so N does too. $\qquad\square$

6.4 An Induction Theorem

To finish this chapter, we leave irreducible restriction of characters and prove a theorem on irreducible induction.

As we shall realize when we introduce wreath products, it is quite easy to construct groups that have irreducible characters that are induced from characters of normal abelian subgroups. What is surprising is that there is a nontrivial partial converse. Recall that a subgroup A of G is **subnormal** in G if there is a chain $A = A_0 \trianglelefteq A_1 \trianglelefteq \ldots \trianglelefteq A_k \trianglelefteq G$. In this case, we write $A \trianglelefteq \trianglelefteq G$.

Theorem 6.15 (Riese) *Let $A \le G$, where A is abelian, and assume that λ^G is irreducible for some $\lambda \in \mathrm{Irr}(A)$. Then $A \trianglelefteq \trianglelefteq G$.*

Proof (Isaacs) We prove the theorem by induction on $|G|$. Write $\chi = \lambda^G \in \mathrm{Irr}(G)$, and let V be the (normal) subgroup of G generated by the elements $g \in G$ with $\chi(g) \ne 0$. Since A is abelian, each irreducible constituent of χ_A has degree $1 = \lambda(1)$. Write

$$\chi_A = e_1 \lambda_1 + \cdots + e_t \lambda_t,$$

where the $\lambda_i \in \mathrm{Irr}(A)$ are the different irreducible constituents of χ_A. Then

$$|G : A| = \chi(1) = e_1 + \cdots + e_t \le e_1^2 + \cdots + e_t^2 = [\chi_A, \chi_A].$$

Hence

$$|G| \le |A|[\chi_A, \chi_A] = \sum_{a \in A} |\chi(a)|^2 \le \sum_{g \in G} |\chi(g)|^2 = |G|$$

and we deduce that $\chi(g) = 0$ for all $g \in G - A$. Therefore $V \subseteq A$. Also, if $Z = \mathbf{Z}(G)$, we have $Z \subseteq V$.

If $A \subseteq H < G$, then since λ^H is irreducible, the inductive hypothesis yields $A \lhd \lhd H$. Assuming that A is not subnormal in G, then Wielandt's Zipper Lemma (Theorem 2.4 of [Is06]) guarantees that there is a unique maximal subgroup M of G with $A \subseteq M$. If the normal closure $A^G < G$, then $A \lhd \lhd A^G \lhd G$, and we are done. We can thus suppose that $A^G = G$, and so $A^g \not\subseteq M$ for some element $g \in G$. By the uniqueness of M, therefore, we have $\langle A, A^g \rangle = G$. In particular, since A and A^g are abelian, we have that $A \cap A^g \subseteq Z$. But $V \lhd G$ and $V \subseteq A$, and thus $V \subseteq A \cap A^g \subseteq Z$, and we have $V = Z = A \cap A^g$. Thus χ vanishes off Z. If $\chi_Z = \chi(1)\mu$ for some $\mu \in \mathrm{Irr}(Z)$, we have that

$$\mu^G = \frac{|G : Z|}{\chi(1)} \chi$$

because both characters coincide on Z and on $G - Z$ (using the induction formula for μ^G). Since $[\chi_Z, \mu] = [\chi, \mu^G]$, we have that

$$|G : Z| = \chi(1)^2 = |G : A|^2$$

and we deduce that $|G : A| = |A : Z|$. Hence $|G : A| = |A^g : A^g \cap A|$, and it follows that $|AA^g| = |G|$, by using the formula on the size of the product of two subgroups. This implies that $AA^g = G$. Write $g = ab$, where $a \in A$ and $b \in A^g$. Then

$$A^g = (A^g)^{b^{-1}} = A^a = A,$$

and thus $A = G$. This is a contradiction since the subgroup A was assumed to be not subnormal. □

6.5 Notes

All the criteria to extend characters that we have discussed in this chapter have produced *canonical extensions*. By this we mean that from a G-invariant $\theta \in \mathrm{Irr}(N)$ we have constructed an extension $\hat{\theta} \in \mathrm{Irr}(G)$ which is independent of any choices made to produce that extension. This uniqueness has allowed us to work by induction, under the hypothesis of G/N being solvable. (This hypothesis is reasonable since θ extends to G if it extends to P for every Sylow subgroup P/N of G/N.) Of course, several of these criteria can be applied simultaneously to the same character, with respect to different Sylow subgroups. For instance, if $\theta \in \mathrm{Irr}(N)$ is a G-invariant real-valued character of odd degree and $o(\theta) = 1$ (or of 2-defect zero), then we have that θ extends to G.

Notice that we have not been able to use character triple isomorphisms in this chapter since character triple isomorphisms might not respect some of our hypotheses. If we are only trying to show that $\theta \in \mathrm{Irr}(N)$ extends to G, then we may assume that $N \subseteq \mathbf{Z}(G)$, and we are led to study central extensions of G/N (in fact, of their Sylow subgroups). This is, of course, related with Corollary 5.9 and the *Schur multiplier* $\mathbf{M}(G/N)$ of G/N, which will appear later in this book in another context.

A remarkable extension theorem which we would like to mention now uses some modular representation theory, and therefore we are not able to prove it here with our techniques. If G is a finite group, we say that $\chi \in \mathrm{Irr}(G)$ is in the **principal p-block** of G if

$$\sum_{x \in G_{p'}} \chi(x) \neq 0,$$

where $G_{p'}$ is the set of elements of G with order not divisible by p. If $N \trianglelefteq G$, $|G/N|$ is not divisible by p and $G = N\mathbf{C}_G(P)$, where $P \in \mathrm{Syl}_p(G)$, it is a fact that restriction gives a bijection between the irreducible characters of the principal blocks of G and of N. This is a result of J. L. Alperin and E. C. Dade (see [Al76], [Da77]).

Finally, about the existence of nontrivial rational-valued characters in groups of even order (Theorem 6.7), it was a conjecture of R. Gow that every group of even order has, in fact, a nontrivial rational-valued odd-degree irreducible character. This problem was solved by Pham Huu Tiep and me in [NT08].

Problems

(6.1) Suppose that G is a finite group, $N \trianglelefteq G$ and $\theta \in \mathrm{Irr}(N)$ is linear and G-invariant. Assume that there exists $H \leq G$ such that $G = NH$ and $N \cap H = 1$. Show that θ extends to G.

(6.2) Suppose that $N \trianglelefteq G$ and that $\theta \in \mathrm{Irr}(N)$ is G-invariant. Assume that $o(\theta)\theta(1)$ is a π-number for a set of primes π. (That is, every prime divisor of $o(\theta)\theta(1)$ lies in π.) If G/N is cyclic, show that there is an extension χ of θ to G such that $o(\chi)$ is a π-number.

(6.3) Suppose that (G, N, θ) is a character triple, and assume that $o(\theta)\theta(1)$ is a π-number, for some set π of primes. Show that there is an isomorphic character triple (G^*, N^*, θ^*) such that N^* is a central π-group, θ^* is faithful, and such that $o(\eta)$ is a π-number if and only if $o(\eta^*)$ is a π-number for $\eta \in \mathrm{Irr}(U|\theta)$ and $N \le U \le G$.

(*Hint:* Using Problem 6.2, modify the proofs of Theorems 5.5 and 5.6.)

(6.4) Let G be a finite group. Suppose that K is a normal Hall π-subgroup of G, and Z is a normal π'-subgroup of G. Let $\theta \in \mathrm{Irr}(K)$ and $\lambda \in \mathrm{Irr}(Z)$ be G-invariant. Show that the triples $(G, KZ, \theta \times \lambda)$ and $(G/K, KZ/K, \hat{\lambda})$ are isomorphic, where $\hat{\lambda}$ is the character of $KZ/K \cong Z$ that corresponds to λ under the natural isomorphism.

(6.5) Let G be a finite group, and assume that $P \in \mathrm{Syl}_p(G)$ is abelian. Suppose that $N \trianglelefteq G$ has p-power index. Show that every G-invariant $\theta \in \mathrm{Irr}(N)$ of degree not divisible by p extends to G.

(6.6) Let $N \trianglelefteq G$, and let $\theta \in \mathrm{Irr}(N)$ be G-invariant, real-valued, of odd degree with $o(\theta) = 1$.

(a) Show that θ extends to G.

(b) If $G'N = G$, show that θ has a unique extension to G. Show that this unique extension is real-valued and has determinantal order 1.

(c) Show by induction on $|G : N|$ that θ has a unique real-valued extension χ with $o(\chi) = 1$ in any case.

(*Hint:* By induction, we may assume that G/N is abelian. Let H/N be a 2-complement, and let P/N be a Sylow 2-subgroup. Use Lemma 6.8(d).)

(6.7) (*Riese–Schmid*) Let p be a prime. If $\chi \in \mathrm{Irr}(G)$ has q-defect zero for every prime $q \ne p$, prove that $\chi = \nu^G$ for some $\nu \in \mathrm{Irr}(P)$, where $P \in \mathrm{Syl}_p(G)$.

7

Degrees of Characters

How much information the character degrees of a finite group provide about its structure is one of the main questions in this chapter. Since we know that the degrees of the irreducible complex characters divide the order of the group, the first question is to detect the prime numbers that do not divide the degree of any irreducible character. In order to solve this problem it seems inevitable that we will have to use the CFSG. Later in the chapter, we shall study complex group algebras, which, according to Theorem 1.3, are determined by the character degrees and their multiplicities, and character tables.

7.1 The Itô–Michler Theorem

Let us start by stating the following key theorem of character theory.

Theorem 7.1 (Itô–Michler) *Let G be a finite group and let p be a prime number. Then p does not divide $\chi(1)$ for any $\chi \in \mathrm{Irr}(G)$ if and only if G has a normal and abelian Sylow p-subgroup.*

If G has a normal abelian Sylow p-subgroup P then we know that $\chi(1)$ divides $|G : P|$ for every $\chi \in \mathrm{Irr}(G)$ by Itô's Corollary 1.21, and therefore $\chi(1)$ is not divisible by p. For the converse, I. M. Isaacs proved that it was enough to check the theorem for simple groups ([Is06], Theorem 12.33). Then G. Michler showed that every finite non-abelian simple group S of order divisible by p possesses an irreducible character of degree divisible by p ([Mi86]), thus completing the proof of Theorem 7.1. This result, together with the Howlett–Isaacs theorem that we shall see in Chapter 8, was one of the first applications of the CFSG to character theory. To this day, none has been able to prove Theorem 7.1 without using the Classification, and it is conceivable that such a proof is simply impossible. This is suggesting that in the representation theory of finite groups we should accept that there are results that are true for every finite group because finite simple groups satisfy certain properties. This seems to be a feature of our theory.

Another remark is the following. The Itô–Michler theorem, like Theorem 6.7 on rational-valued characters, is another example of a theorem on finite groups that is true because it holds for simple groups. As we said, this *straight* reduction to simple groups does not always happen. In the Howlett–Isaacs theorem or in the McKay conjecture, for instance, we will have to verify more complicated statements on simple groups in order to achieve a proof or a reduction theorem.

The two conclusions on the Sylow p-subgroup in the Itô–Michler theorem (normal and abelian) have been inextricably linked for years. A famous conjecture of R. Brauer asserts that $P \in \mathrm{Syl}_p(G)$ is abelian if and only if the irreducible characters of G in the principal p-block of G have the degrees not divisible by p. (This is the celebrated **Brauer's height zero conjecture** for the principal block.) Hence the condition that P is abelian should be equivalent to the condition that some subset of the irreducible characters of G have degree not divisible by p. In the next section, we *isolate* the normality condition of P from the character degrees of G.

7.2 Normal Sylow p-Subgroups

We need the following result on finite simple groups.

Theorem 7.2 *Suppose that S is a non-abelian finite simple group. Then let p be a prime dividing $|S|$ and let $P \in \mathrm{Syl}_p(S)$. Then there exists $\chi \in \mathrm{Irr}(S)$ of degree divisible by p lying over the trivial character of P.*

Proof This is Theorem 2.1 of [MN12]. □

Let us mention that Theorem 7.2 is clear if S has a p-defect zero character χ. (As mentioned in Chapter 4, non-abelian simple groups do possess p-defect zero characters for $p > 3$.) In this case, by using the proof of Corollary 4.7, we know that χ_P is a multiple of the regular character and therefore $[\chi_P, 1_P] \neq 0$.

Recall that we are denoting by $\mathrm{Irr}(\psi)$ the set of irreducible constituents of the character ψ.

Theorem 7.3 (Malle–Navarro) *Let G be a finite group and let $P \in \mathrm{Syl}_p(G)$. Then p does not divide $\chi(1)$ for any $\chi \in \mathrm{Irr}((1_P)^G)$ if and only if $P \trianglelefteq G$.*

Proof If $P \trianglelefteq G$, then we claim that $\mathrm{Irr}((1_P)^G) = \mathrm{Irr}(G/P)$. We have that $\chi \in \mathrm{Irr}((1_P)^G)$ if and only if $[\chi_P, 1_P] \neq 0$. By Clifford's theorem 1.19, this happens if and only if $\chi_P = \chi(1)1_P$, which happens if and only if $P \subseteq \ker(\chi)$. In this case, every $\chi \in \mathrm{Irr}(G/P)$ has degree dividing $|G/P|$ by Corollary 1.9, and thus $\chi(1)$ is not divisible by p.

To prove the converse, we use induction on $|G|$. Let N be a minimal normal subgroup of G. If $\chi \in \mathrm{Irr}(G/N)$ lies over $1_{PN/N}$, then we have that χ considered as a character of G lies over 1_{PN}, by Theorem 1.11. Hence χ lies over 1_P. By hypothesis, p does not divide $\chi(1)$, and by induction applied in G/N, we have that $NP \trianglelefteq G$.

Next, we claim that $NP = G$. Let $\gamma \in \mathrm{Irr}(NP)$ be over 1_P. Let $\chi \in \mathrm{Irr}(G|\gamma)$. Since χ lies over γ we have that χ lies over 1_P, and we have that p does not divide $\chi(1)$ by hypothesis. Thus p does not divide $\gamma(1)$, by Clifford's theorem. If $NP < G$, then by induction we have that $P \trianglelefteq NP$. Since $NP \trianglelefteq G$, then $P \trianglelefteq G$ by elementary group theory. This proves the claim.

Assume first that N has order not divisible by p. We claim that every $\tau \in \mathrm{Irr}(N)$ is P-invariant. By Mackey's Theorem 1.16, we have that

$$((1_P)^G)_N = (1_1)^N = \rho_N,$$

where ρ_N is the regular character of N. Then $[((1_P)^G)_N, \tau] \neq 0$, and there exists $\gamma \in \mathrm{Irr}(G)$ over τ and over 1_P. Thus $\gamma(1)$ is not divisible by p by hypothesis. Since G/N is a p-group, we have that γ_N is irreducible by Theorem 5.12. Hence, $\gamma_N = \tau$ and the claim is proved. Now, by Theorem 2.4, we deduce that $[P, N] = 1$. In particular, $P \trianglelefteq G$.

Assume now that N has order divisible by p. If N is a p-group, then $P = G$ and we are done. Hence, we may assume that N is the direct product of the distinct G-conjugates $\{S_1, \ldots, S_t\}$ of a non-abelian simple group $S \trianglelefteq N$, by elementary group theory. (See Lemmas 9.5 and 9.6 of [Is08].) Also, S has order divisible by p. Let $P_i = P \cap S_i \in \mathrm{Syl}_p(S_i)$ (using that $S_i \trianglelefteq \trianglelefteq G$). Notice that

$$Q = P \cap N = P_1 \times \cdots \times P_t \in \mathrm{Syl}_p(N).$$

We know by Theorem 7.2 that there exists an irreducible character $\gamma_i \in \mathrm{Irr}(S_i)$ of degree divisible by p such that $[(\gamma_i)_{P_i}, 1_{P_i}] \neq 0$. Let $\gamma = \gamma_1 \times \cdots \times \gamma_t \in \mathrm{Irr}(N)$. Then

$$[\gamma_Q, 1_Q] = \prod_i [(\gamma_i)_{P_i}, 1_{P_i}] \neq 0.$$

Now, by Mackey we have that $((1_P)^G)_N = (1_Q)^N$ contains γ, and therefore there exists $\tau \in \mathrm{Irr}((1_P)^G)$ over γ. By Clifford's theorem, $\gamma(1)$ divides $\tau(1)$, and we have produced an irreducible character of G that lies over 1_P that has degree divisible by p. This is not possible by hypothesis. $\square$

From Theorem 7.3, we easily deduce the Itô–Michler theorem.

Proof of the Itô–Michler theorem If all the irreducible characters of G have degree not divisible by p, then we have that $P \trianglelefteq G$ by Theorem 7.3. Suppose that $P' > 1$, and let $\tau \in \text{Irr}(P)$ be nonlinear. If $\chi \in \text{Irr}(G)$ lies over τ, then $\tau(1)$ divides $\chi(1)$, and therefore $\chi(1)$ has degree divisible by p. The converse follows from Itô's Corollary 1.21.

7.3 Normal p-Complements

There is a dual situation to the Itô–Michler theorem that was studied by J. G. Thompson: when a prime p divides all nonlinear irreducible characters of G. In this case, we shall prove that G has a normal p-complement.

We start with a theorem of I. M. Isaacs. In this book, $\text{Irr}_{p'}(G)$ denotes the set of the irreducible characters of G of degree not divisible by p. (We also say that these characters have p'-**degree**.) Recall that if $n \geq 1$ is an integer, then n_p is the largest power of p that divides n.

Theorem 7.4 (Isaacs) *Let G be a finite group. Then G has a normal p-complement if and only if the p-part of the number*

$$\sum_{\chi \in \text{Irr}_{p'}(G)} \chi(1)^2$$

is $|G : G'|_p$.

Proof Let $N = \mathbf{O}^p(G)$ and let $P \in \text{Syl}_p(G)$. Thus $G = NP$. Let $\mathcal{A}$ be the set of P-invariant irreducible characters of N of p'-degree. Since $N = \mathbf{O}^p(N)$, then we have that $o(\theta)$ is not divisible by p for every $\theta \in \text{Irr}(N)$. By Corollary 6.2, we have that every $\eta \in \mathcal{A}$ extends to G and by the Gallagher correspondence η has exactly $|\text{Lin}(G/N)| = |G : G'N|$ extensions. We have that $|G : G'N| = |G : G'|_p$ by elementary group theory. Also, if $\chi \in \text{Irr}_{p'}(G)$, then notice that $\chi_N \in \text{Irr}(N)$ by Theorem 5.12. We deduce that restriction defines an onto map

$$\text{Irr}_{p'}(G) \to \mathcal{A}$$

and that there are exactly $|G : G'|_p$ characters of the same degree that map to a given $\eta \in \mathcal{A}$. Hence

$$\sum_{\chi \in \text{Irr}_{p'}(G)} \chi(1)^2 = |G : G'|_p \sum_{\eta \in \mathcal{A}} \eta(1)^2.$$

Now, using that characters in the same P-orbit have the same degree, we have that

$$|N| = \sum_{\theta \in \mathrm{Irr}(N)} \theta(1)^2 \equiv \sum_{\eta \in \mathcal{A}} \eta(1)^2 = \frac{1}{|G : G'|_p} \sum_{\chi \in \mathrm{Irr}_{p'}(G)} \chi(1)^2 \bmod p.$$

Now, G has a normal p-complement if and only if $|N| \not\equiv 0 \bmod p$, and the result follows. $\qquad \square$

Corollary 7.5 (Thompson) *Let G be a finite group, let p be a prime and assume that p divides $\chi(1)$ for all nonlinear $\chi \in \mathrm{Irr}(G)$. Then G has a normal p-complement.*

Proof By hypothesis, we have that

$$\sum_{\chi \in \mathrm{Irr}_{p'}(G)} \chi(1)^2 = |G : G'|$$

and the result follows from Theorem 7.4. $\qquad \square$

In fact, Thompson's situation can be characterized group theoretically. It is remarkable that this characterization has a normal and an abelian part, as in the Itô–Michler theorem.

Theorem 7.6 (Gow–Humphreys) *Let G be a finite group, let p be a prime and let $P \in \mathrm{Syl}_p(G)$. Then p divides $\chi(1)$ for all nonlinear $\chi \in \mathrm{Irr}(G)$ if and only if G has a normal p-complement K and $\mathbf{C}_{K'}(P) = 1$.*

Proof Assume that G has a normal p-complement K with $\mathbf{C}_{K'}(P) = 1$. Let $\chi \in \mathrm{Irr}(G)$ be nonlinear of degree not divisible by p. Then $\chi_K = \theta \in \mathrm{Irr}(K)$ is P-invariant by Theorem 5.12. Since θ has p'-degree and K' is P-invariant, it follows that some irreducible constituent of $\theta_{K'}$ is P-invariant (using that $[\theta_{K'}, \eta] = [\theta_{K'}, \eta^x]$ for $x \in P$). By the Glauberman correspondence, the only P-invariant irreducible character of K' is the trivial character. Therefore $\chi \in \mathrm{Irr}(G/K')$ and by Theorem 1.21, $\chi(1)$ divides $|G/K' : K/K'| = |G : K| = |P|$. This is not possible since p does not divide $\chi(1)$ and $\chi(1) > 1$.

Conversely, assume that p divides $\chi(1)$ for all nonlinear $\chi \in \mathrm{Irr}(G)$. We know by Corollary 7.5 that G has a normal p-complement K. Assume that $\mathbf{C}_{K'}(P) > 1$. By the Glauberman correspondence, let $\tau \in \mathrm{Irr}(K')$ be nontrivial and P-invariant. Since τ^K has p'-degree and is P-invariant, necessarily τ^K contains a P-invariant irreducible constituent θ. We have that $\theta(1) > 1$ since K' is not contained in its kernel (because τ is not trivial). Now, θ extends to some $\chi \in \mathrm{Irr}(G)$ by Corollary 6.3, and χ is a nonlinear p'-degree irreducible character of G. $\qquad \square$

It is perhaps worth pointing out that if G satisfies the hypothesis of Theorem 7.6, then G is necessarily solvable. This follows from a well-known group theoretical fact whose proof needs the CFSG: if A acts as automorphisms on a finite group G, $\gcd(|A|, |G|) = 1$ and $\mathbf{C}_G(A) = 1$, then G is solvable.

The following generalizes Thompson's theorem.

Theorem 7.7 (Berkovich) *Let G be a finite group, and let*

$$L = \bigcap_{\substack{\chi \in \mathrm{Irr}_{p'}(G) \\ \chi(1) > 1}} \ker(\chi).$$

Then L has a normal p-complement.

Proof By Corollary 7.5, we may assume that there exists $\psi \in \mathrm{Irr}_{p'}(G)$ with $\psi(1) > 1$. Thus $L \subseteq \ker(\psi)$. Now, if $\nu \in \mathrm{Lin}(G)$, then $L \subseteq \ker(\nu\psi)$, and we deduce that $L \subseteq \ker(\nu)$. Hence, L is contained in the kernel of every irreducible character of G of p'-degree. Now, let $P \in \mathrm{Syl}_p(G)$ and let $\lambda \in \mathrm{Irr}(P)$ be linear. Since λ^G has p'-degree, there exists $\chi \in \mathrm{Irr}_{p'}(G)$ over λ. By hypothesis, $L \subseteq \ker(\chi)$, and therefore $L \cap P \subseteq \ker(\lambda)$. Since λ is an arbitrary linear character of P, we deduce that $L \cap P \subseteq P'$. By Corollary 6.14, we have that L has a normal p-complement, as required. □

7.4 Multiplicities of Character Degrees

If G is a finite group, it is customary to denote by $\mathrm{cd}(G)$ the set of the irreducible character degrees of G (without multiplicities). What is the information that $\mathrm{cd}(G)$ has of G? For instance, the Itô–Michler theorem tells us that $\mathrm{cd}(G)$ *knows* for every prime p if G has an abelian and normal Sylow p-subgroup or not. Certainly, $\mathrm{cd}(G)$ does not know if G is nilpotent, as shown by D_8 and S_3, which have the same set of character degrees. For some time it was thought that $\mathrm{cd}(G)$ could determine the solvability of G. However, groups G and H, with G nilpotent and H perfect, have been found having the same set of character degrees.

The complex group algebra of G appears to have much more information on G, however. This is in fact the content of Brauer's Problem Number 2 of his list: "When do non-isomorphic groups have isomorphic group algebras?" If G is a finite group, then $\mathbb{C}G$ is uniquely determined by the function

$$m_G(x) = |\{\chi \in \mathrm{Irr}(G) \mid \chi(1) = x\}|,$$

by Theorem 1.3.

Notice that Theorem 7.4 has the following consequence.

Theorem 7.8 (Isaacs)   *Suppose that G is a finite group, and let p be a prime. Then $\mathbb{C}G$ determines if G has a normal p-complement. In particular, $\mathbb{C}G$ determines if G is nilpotent.*

Proof   By Theorem 7.4, it suffices to notice that $\mathbb{C}G$ determines $|G : G'| = m_G(1)$ and the number $\sum_{\chi \in \mathrm{Irr}_{p'}(G)} \chi(1)^2$. $\square$

A theorem of J. Cossey and T. Hawkes that we are going to see next proves that in fact $\mathbb{C}G$ determines $|G/\mathbf{O}^p(G)|$. Since G has a normal p-complement if and only if $|G/\mathbf{O}^p(G)| = |G|_p$, it follows that Theorem 7.10 below recovers Isaacs' Theorem 7.8. It is somewhat remarkable that Isaacs uses characters of p'-degree in Theorem 7.4, while Cossey and Hawkes use characters of p-power degree. First, we need a lemma.

Lemma 7.9   *Suppose that $N \trianglelefteq G$ with $|G : N|$ a power of p. Then*

$$\sum_{\substack{\chi \in \mathrm{Irr}(G) \\ \chi(1)_p = \chi(1)}} \chi(1)^2 = |G : N| \sum_{\substack{\chi \in \mathrm{Irr}(N) \\ \chi(1)_p = \chi(1)}} \chi(1)^2.$$

Proof   Arguing by induction on $|G : N|$ it suffices to prove the lemma when N has index p. If $\theta \in \mathrm{Irr}(N)$ and $T = G_\theta$ is the stabilizer of θ in G, then either $T = G$ or either $T = N$, because $N \subseteq T \subseteq G$. In the first case, we have that θ has p different extensions to G by Theorem 5.1 and the Gallagher correspondence. In the second, $\theta^G = \chi \in \mathrm{Irr}(G)$ (by the Clifford correspondence) and θ has p distinct G-conjugates all of them inducing χ. Hence, if $\chi \in \mathrm{Irr}(G)$ has degree a power of p, then either $\chi_N = \theta \in \mathrm{Irr}(N)$ or either $\chi = \theta^G$ for every irreducible constituent θ of χ_N. Let $\mathcal{A}$ be the set of irreducible characters of N of degree a power of p. Then G acts on $\mathcal{A}$ by conjugation with orbits of size 1 or p. Let $\psi_1, \ldots, \psi_k$ be representatives of the orbits of size p, and let $\mathcal{B}$ be the G-invariant characters in $\mathcal{A}$. We have that

$$p \sum_{\theta \in \mathcal{A}} \theta(1)^2 = p \left(\sum_{\theta \in \mathcal{B}} \theta(1)^2 + p \sum_{j=1}^{k} \psi_j(1)^2 \right)$$

$$= p \sum_{\theta \in \mathcal{B}} \theta(1)^2 + \sum_{j=1}^{k} (p\psi_j(1))^2$$

$$= \sum_{\substack{\chi \in \mathrm{Irr}(G) \\ \chi(1)_p = \chi(1)}} \chi(1)^2.$$

This proves the lemma. $\square$

Theorem 7.10 (Cossey–Hawkes) *Let G be a finite group. Then the p-part of*

$$\sum_{\substack{\chi \in \mathrm{Irr}(G) \\ \chi(1)=\chi(1)_p}} \chi(1)^2$$

is $|G/\mathbf{O}^p(G)|$.

Proof Write $N = \mathbf{O}^p(G)$. By Lemma 7.9, it suffices to show that

$$\sum_{\substack{\chi \in \mathrm{Irr}(N) \\ \chi(1)=\chi(1)_p}} \chi(1)^2$$

is not divisible by p. Since $N = \mathbf{O}^p(N)$, we have that $|N/N'|$ is not divisible by p. Now

$$\sum_{\substack{\chi \in \mathrm{Irr}(N) \\ \chi(1)=\chi(1)_p}} \chi(1)^2 = \sum_{\substack{\chi \in \mathrm{Irr}(N) \\ \chi(1)=1}} \chi(1)^2 + \sum_{\substack{\chi \in \mathrm{Irr}(N) \\ \chi(1)=\chi(1)_p>1}} \chi(1)^2 \equiv |N/N'| \bmod p,$$

and the theorem follows. $\qquad\qquad\qquad\qquad\qquad\qquad\qquad\qquad\qquad\qquad\qquad\quad\Box$

7.5 Character Tables and Brauer's Induction Theorem

For a character theorist, finding how much the character table $X(G)$ knows about a finite group G and vice versa is a central problem. Of course, $\mathrm{cd}(G)$ is only the set of entries in the first column of the character table, while $\mathbb{C}G$ is determined by (and determines) its first column. If we have the whole character table at our disposal, we should be able to know much more about the group.

The examples of $\mathbf{Q}_8$ and $\mathbf{D}_8$ already tell us that the orders of elements, the isomorphism type of the Fitting subgroup or, say, the number of subgroups, are not determined by the character table. Nor is the determinantal order of characters. Of course, by the second orthogonality relation, we know the conjugacy class sizes (with multiplicities). Hence, we know $|\mathbf{Z}(G)|$. Or we know $|G/G'|$, which is the number of linear characters of G. In fact, since G/G' is isomorphic to $\mathrm{Lin}(G)$, then we know the isomorphism type of G/G' (see Problem 2.7 of [Is06]). Also the isomorphism type of $\mathbf{Z}(G)$ is determined by the character table (because an abelian group is determined by the order of its elements). Since we can determine the kernels of characters and

$$N = \bigcap_{\substack{\chi \in \mathrm{Irr}(G) \\ N \subseteq \ker(\chi)}} \ker(\chi)$$

for every $N \trianglelefteq G$, then the lattice of normal subgroups (with their orders, which are the sum of the sizes of the conjugacy class that they contain) is available from the character table. Thus we know if a finite group G is supersolvable, solvable, or simple from its character table; as we know the size of the Fitting subgroup $\mathbf{F}(G)$, for instance.

If we fix a prime p, it is much more difficult to study how much $X(G)$ knows about the *p-local structure* of G: p-elements, Sylow p-subgroups, p-Sylow normalizers, etc. Even to decide if a column of $X(G)$ corresponds to a p-element is not a triviality at all, and the only known proof of it uses Brauer's characterization of characters. (Strictly speaking, a **local subgroup** is the normalizer of a nontrivial p-subgroup. J. L. Alperin seems to be responsible for this terminology.)

This might be a good place to talk about Brauer's characterization of characters, and Brauer's induction theorem, which we have not yet used in this book, and which undoubtedly have to be mentioned in every book on characters.

Theorem 7.11 (Brauer induction)   *Suppose that G is a finite group. If $\chi \in$ Irr(G), then we can write*

$$\chi = \sum a_\lambda \lambda^G,$$

where $a_\lambda \in \mathbb{Z}$, and λ runs over the linear characters of the nilpotent subgroups of G.

Proof   This is Theorem 8.4 of [Is06]. $\square$

Corollary 7.12 (Brauer characterization)   *Suppose that $\phi \in$ cf(G). Then ϕ is a generalized character of G if and only if ϕ_H is a generalized character of H for every nilpotent subgroup H of G.*

Proof   If $\chi \in$ Irr(G), it suffices to show that $[\phi, \chi] \in \mathbb{Z}$. To show this, use Theorem 7.11, Frobenius reciprocity and the hypothesis. $\square$

Recall again that we are using $\mathbf{R}$ for the ring of algebraic integers in $\mathbb{C}$, and $\mathbb{Q}_m$ for the mth cyclotomic field.

Theorem 7.13   *Let G be a finite group, p a prime, and write $|G| = p^a m$, where p does not divide m. Let $h \in G$ be p-regular. Define $\phi \colon G \to \mathbb{C}$ by setting $\phi(g) = m$ if $g_{p'}$ is G-conjugate to h, and zero otherwise. Then $[\phi, \chi] \in \mathbf{R} \cap \mathbb{Q}_{o(h)}$ for every $\chi \in$ Irr(G).*

Proof   Notice that ϕ is a class function of G. Suppose that $H = P \times Q$ is a nilpotent subgroup of G, where P is a p-group and $|Q|$ is not divisible by p. We show first that $[\phi_H, \gamma] \in \mathbf{R} \cap \mathbb{Q}_{o(h)}$ for $\gamma \in$ Irr(H). If no G-conjugate of h lies in H, then $[\phi_H, \gamma] = 0$. Write $\gamma = \alpha \times \beta$, where $\alpha \in$ Irr(P)

and $\beta \in \text{Irr}(Q)$. Let $h_1, \ldots, h_t$ be the distinct elements of Q which are G-conjugate to h. Then, using that the elements of P and Q commute, notice that

$$Ph_1 \cup \cdots \cup Ph_t$$

is exactly the set of elements y of H such that $y_{p'}$ is G-conjugate to h, and that these are distinct right cosets of P in H. Now

$$[\phi_H, \alpha \times \beta] = \frac{1}{|P||Q|} \sum_{j=1}^{t} \sum_{x \in P} m \overline{\alpha(x)\beta(h_j)}$$

$$= \frac{m}{|Q|}[\alpha_P, 1_P] \left(\sum_{j=1}^{t} \beta(h_j^{-1}) \right) \in \mathbf{R} \cap \mathbb{Q}_{o(h)},$$

using that $|Q|$ divides m and that $\beta(h_j^{-1}) \in \mathbf{R} \cap \mathbb{Q}_{o(h)}$ because $o(h) = o(h_j)$.

Now, by Brauer's induction theorem, write

$$\chi = \sum a_\lambda \lambda^G,$$

where $a_\lambda \in \mathbb{Z}$ and λ is a linear character of a nilpotent subgroup H_λ. Then, using Frobenius reciprocity, we have that

$$[\phi, \chi] = \sum_\lambda a_\lambda [\phi_{H_\lambda}, \lambda] \in \mathbf{R} \cap \mathbb{Q}_{o(h)},$$

as desired. □

Let us see an application for which we know no proof that does not use Brauer's characterization of characters.

Corollary 7.14 *Let G be a finite group, let p be a prime, and let G_p be the set of p-elements of G. If $\chi \in \text{Irr}(G)$, then*

$$\sum_{x \in G_p} \chi(x)$$

is an integer divisible by $|G|_p$.

Proof Write $|G| = p^a m$, where p does not divide m. Define $\phi: G \to \mathbb{C}$ by setting $\phi(g) = m$ if g is a p-element of G, and zero otherwise. Then, setting $h = 1$ in Theorem 7.13 and using that $\mathbf{R} \cap \mathbb{Q} = \mathbb{Z}$, we have that ϕ is a generalized character of G. Now,

$$[\chi, \phi] = \frac{1}{|G|} \sum_{x \in G_p} \chi(x) m = \frac{1}{|G|_p} \sum_{x \in G_p} \chi(x)$$

is an integer, as desired. □

In the case where $\chi = 1_G$, Corollary 7.14 is a consequence of a well-known theorem of Frobenius on the number of solutions in G to the equation $x^n = 1$, for any integer n dividing $|G|$.

We can now prove a theorem of G. Higman.

Theorem 7.15 (Higman) *Suppose that G is a finite group, and let $x, y \in G$. Let M be a maximal ideal of $\mathbf{R}$ containing the prime p. Then $x_{p'}$ and $y_{p'}$ are G-conjugate if and only if $\chi(x) \equiv \chi(y)$ mod M for every $\chi \in \mathrm{Irr}(G)$.*

Proof If $x_{p'}$ and $y_{p'}$ are G-conjugate, and $\chi \in \mathrm{Irr}(G)$, then we know that

$$\chi(x) \equiv \chi(x_{p'}) = \chi(y_{p'}) \equiv \chi(y) \text{ mod } M$$

by Lemma 4.19(b).

For the converse, write $|G| = p^a m$, where p does not divide m. Let $h = y_{p'}$ and let ϕ be the class function on G such that $\phi(g) = m$ if $g_{p'}$ is G-conjugate to h, and 0 otherwise. We have that $\phi = \sum_{\chi \in \mathrm{Irr}(G)} a_\chi \chi$ for some $a_\chi \in \mathbf{R}$ by Theorem 7.13. Now

$$m = \phi(y) = \sum_{\chi \in \mathrm{Irr}(G)} a_\chi \chi(y) \equiv \sum_{\chi \in \mathrm{Irr}(G)} a_\chi \chi(x) = \phi(x) \text{ mod } M.$$

If $x_{p'}$ is not G-conjugate to $h = y_{p'}$, then $\phi(x) = 0$ and so $m \in M$. Since $p \in M$ and $\gcd(p, m) = 1$, we would conclude that $1 \in M$, a contradiction. $\square$

One can argue that in Theorem 7.15 we are required to calculate congruences modulo maximal ideals of $\mathbf{R}$, and that perhaps this might not be practical. But we remedy this with the following result, in which we only have to calculate mod $p\mathbf{R}$.

Theorem 7.16 *Let $x, y \in G$, and let p be a prime. Then $x_{p'}$ and $y_{p'}$ are G-conjugate if and only if*

$$\chi(x)^{|G|_p} \equiv \chi(y)^{|G|_p} \text{ mod } p\mathbf{R}$$

for all $\chi \in \mathrm{Irr}(G)$.

Proof Suppose that $x_{p'}$ and $y_{p'}$ are G-conjugate. By Lemma 4.19(a), we have that

$$\chi(x)^{|G|_p} - \chi(y)^{|G|_p} \equiv \chi(x_{p'})^{|G|_p} - \chi(y_{p'})^{|G|_p} \equiv 0 \text{ mod } p\mathbf{R}.$$

Conversely, suppose that

$$\chi(x)^{|G|_p} \equiv \chi(y)^{|G|_p} \text{ mod } p\mathbf{R}$$

for all $\chi \in \mathrm{Irr}(G)$. Let M be a maximal ideal of $\mathbf{R}$ containing $p\mathbf{R}$, so that $F = \mathbf{R}/M$ is a field. Then we have that

$$\chi(x)^{|G|_p} \equiv \chi(y)^{|G|_p} \bmod M$$

by hypothesis. Then

$$(\chi(x) - \chi(y))^{|G|_p} \equiv \chi(x)^{|G|_p} - \chi(y)^{|G|_p} \bmod M \equiv 0 \bmod M,$$

and therefore $\chi(x) \equiv \chi(y) \bmod M$, since F is a field. We now apply Theorem 7.15. $\qquad\square$

Corollary 7.17 *We have that $x \in G$ is a p-element if and only if*

$$\chi(x)^{|G|_p} \equiv \chi(1) \bmod p\mathbf{R}$$

for all $\chi \in \mathrm{Irr}(G)$.

Proof Use Theorem 7.16, and the fact that $n^p \equiv n \bmod p$ for every integer n. $\qquad\square$

Corollary 7.18 (Higman) *Let G be a finite group, and let $x \in G$. Then $X(G)$ determines the set of primes dividing $o(x)$.*

Proof Write $\pi(m)$ for the set of primes dividing the positive integer m. Let $\{p_1, \ldots, p_n\}$ be the set of primes dividing $|G| = \sum_{\chi \in \mathrm{Irr}(G)} \chi(1)^2$. Now, the column j corresponding to the trivial element of G is the column all of whose entries are positive integers a_{ij}, by using the second orthogonality relation. By Corollary 7.17, we know the columns $\mathcal{C}_{p_i}$ of the elements x with $\pi(o(x)) = \{p_i\}$. To know the columns of the elements x with $\pi(o(x)) = \{p_i, p_j\}$ we do the following. Let i and j be different indices. From the columns corresponding to elements x with $|\pi(o(x))| > 1$, we pick up those y such that $y_{p_j'}$ is G-conjugate to some of the columns in $\mathcal{C}_{p_i}$, using Theorem 7.15. We do this for every $j \neq i$, and we shall now know the prime divisors of the columns whose order is divisible by at most two primes. We continue in this way until we have exhausted all the columns in $X(G)$. $\qquad\square$

In Corollary 7.17, it is possible to replace $|G|_p$ by $o(x)_p$. However we wanted to write down an integer that could be calculated from the character table and, as we have already mentioned, the character table does not determine the order of elements. We notice now, however, that if $P \in \mathrm{Syl}_p(G)$, then the exponent of $\mathbf{Z}(P)$ can be computed by the character table of G, by Corollary 3.12. We will come back to character tables and p-local information in Chapter 9.

7.6 Notes

In Problem 12 in [Br63], Brauer asked: "Given the character table of a finite group G and a prime p dividing $n = |G|$, how much information about the structure of the p-Sylow group P can be obtained? In particular, can it be decided whether or not P is Abelian?" In this book we have already mentioned Brauer's height zero conjecture for the principal block: G has abelian Sylow p-subgroups if and only if the irreducible characters in the principal block of G have degree not divisible by p. Since we now know that the characters in the principal block are determined by the character table (using Corollary 7.18), the height zero conjecture gives us a method to check from the character table if G has abelian Sylow p-subgroups. However, Brauer's height zero conjecture remains unproven to this date. (The *only if* direction has been proved by R. Kessar and G. Malle in [KM13], using CFSG.)

W. Kimmerle and R. Sandling did prove in [KS95], using the CFSG, that if $X(G) = X(H)$, then G has abelian Sylow p-subgroups if and only if H does, although no specific method was provided to check from $X(G)$ if G has abelian Sylow p-subgroups, except to construct all the character tables of all the groups of order $|G|$, and apply their theorem.

If a finite group G has abelian Sylow p-subgroups, then notice that the conjugacy classes of the p-elements of G have size not divisible by p. We know that we can read this from the character table, and Brauer wondered if the converse was true in [Br63]. The converse is indeed true for $p = 2$ (by work of A. Camina and M. Herzog in [CH80]), but turned out to be false for $p = 3$ and $p = 5$. A few groups like Th for $p = 5$, or Ru or J_4 for $p = 3$, are counterexamples. (See [ATLAS].) Much later, Pham Huu Tiep and I proved that Brauer was indeed right for $p > 5$ (see [NT14], using CFSG). Finally it was possible to prove in [NST15] that G has abelian Sylow p-subgroups if and only if all the p-elements have class sizes not divisible by p and the irreducible characters in the principal block have degree not divisible by p.

In Brauer's Problem 11, information on subgroups from the character table of a finite group is requested. If H is any subgroup of G, then the permutation character $(1_H)^G$ satisfies a good many remarkable properties. (See Theorem 5.18 of [Is06].) However, no characterization of the characters of this form has yet been found in the character table; not even for the characters $(1_P)^G$, where $P \in \mathrm{Syl}_p(G)$. Also, we are not aware of any example of two groups G and H for which $X(G) = X(H)$, G has a subgroup of order n and H does not. In [Betal16] a natural method to detect if G has nilpotent Hall subgroups is given.

Brauer's Problem 10 asks: "Given the character table of a group G and the set of conjugate classes of G which make up a normal subgroup N of G, can it be decided whether or not N is abelian?"

A. I. Saksonov gave a negative answer to this in [Sa67]. Much later S. Mattarei constructed two groups with the same character table, one metabelian and the other with derived length 3 (see [Mat92]). Thus the character table does not determine the derived length of a solvable group. Also, the character table does not recognize the Frattini subgroup (in non-solvable groups) and a counterexample was found in [Gar76].

An interesting problem is to decide if $X(G)$ determines $|\mathbf{N}_G(P)|$, where $P \in \mathrm{Syl}_p(G)$. Some particular cases are proven in [NRi16].

Coming back to complex group algebras, in Problem 1 Brauer asks which finite dimensional algebras are group algebras: probably a problem too wide to have a solution. A. Moretó conjectured in [Mo07] a surprising restriction on group algebras: its dimension is bounded by a function of the largest multiplicity of the character degrees. Later D. A. Craven completed the proof of this conjecture (see [Cr08]).

Recall that a group G is **supersolvable** if it has a normal series whose quotients have prime order. Using computers, it can be shown that from all the groups of order less than 500 there is only one pair of groups G and H with isomorphic complex group algebras such that G is supersolvable and H is not. These are the groups SmallGroup(360,129) and SmallGroup(360,130). Hence, the complex group algebra determines nilpotency but not supersolvability.

Also, the complex group algebra does not determine if a Sylow p-subgroup is abelian, but this is not so easy to detect. B. Sambale has found that the two groups SmallGroup(1152,154124) and SmallGroup(1152,154154) constitute a counterexample for $p = 2$.

The main open question on complex group algebras right now, perhaps, is to decide if $\mathbb{C}G$ determines the solvability of G. (As we have mentioned, the set of character degrees does not suffice, as shown in [NRi14].) It would also be interesting to know if $\mathbb{C}G$ determines if G has a normal Sylow p-subgroup.

About character degrees for simple groups, there is a conjecture by B. Huppert that asserts that if H is simple and $\mathrm{cd}(H) = \mathrm{cd}(G)$, then $G = H \times A$ for some abelian group A [Hu00]. This has been recently proved for the alternating groups in [BTZ17].

About character degrees of solvable groups, there is a vast well-developed theory of character degrees that deserves a book on its own. (Some important theorems can be found in [MW93].) There are many results to be listed here by many authors, and the techniques used to prove them are of quite a different nature than those used in this book, among them the so called *regular orbits* theorems. We do mention the Isaacs–Seitz conjecture that asserts that the derived length dl(G) of a solvable group G is bounded by $|\mathrm{cd}(G)|$, or Gluck's conjecture that proposes that $|G : \mathbf{F}(G)| \le b(G)^2$, where $b(G)$ is the largest character degree of the solvable group G (see [Gl85]). Another tool that has been used is to associate graphs with character degrees, where the vertices are the primes dividing the order of G (or the irreducible nonlinear character degrees) and where (p, q) is an edge if pq divides the degree of some $\chi \in \mathrm{Irr}(G)$ (or if $\gcd(\alpha(1), \beta(1)) > 1$).

Problems

(7.1) Fix a prime p. If G is a finite group and $e \ge 1$, define

$$s_e(G) = \sum_{\chi \in \mathrm{Irr}(G), \chi(1)_p = p^e} \chi(1)^2.$$

(a) If M is normal in G with G/M a p'-group, prove that

$$s_e(G) = |G : M| s_e(M).$$

(b) If G has a normal p'-subgroup K such that G/K has a normal Sylow p-subgroup, and $P \in \mathrm{Syl}_p(G)$, show that $s_0(G)_p = |P : P'|$.
   (*Note:* If P is abelian, it can be proved that $s_0(G) \equiv |\mathbf{N}_G(P)|$ mod $p|G|_p$, using Brauer's height zero conjecture.)

(7.2) Let G be a finite group. Show that all the nonlinear real-valued characters of G have even degree if and only if G has a normal 2-complement.

(7.3) Let G be a finite group, and let p be a prime. Let G_p be the set of p-elements of G.

(a) Let T be a complete set of representatives of the conjugacy classes of p-regular elements of G. Prove that

$$|G| = \sum_{t \in T} |G : \mathbf{C}_G(t)| |\mathbf{C}_G(t)_p|.$$

(b) Suppose that H is a subgroup of G. If $|H|_p$ divides $|H_p|$, then prove that $|G|_p$ divides $|G : H| |H_p|$.

(c) Show by induction on $|G|$, that $|G|_p$ divides $|G_p|$.
   (*Note:* This is the strategy used by I. M. Isaacs and G. Robinson in [IR92] to prove Frobenius' theorem on the number of elements of order n in a finite group.)

(7.4) If G is a finite group, and p is a prime show that $X(G)$ determines if x_p and y_p are G-conjugate, where $x, y \in G$ correspond to two columns of the character table.

(7.5) Let G be a finite group. Suppose that $P \in \mathrm{Syl}_p(G)$ has order p^a, where $a \geq 2$. Then P is cyclic if and only if there is a conjugacy class K of G of p-elements such that $\mathbb{Q}(K)$ is contained in $\mathbb{Q}_{p^a}$, but $\mathbb{Q}(K)$ is not contained in $\mathbb{Q}_{p^{a-1}}$.

(7.6) Suppose that P is a normal p-subgroup of a finite group G, and let $\theta \in \mathrm{Irr}(P)$ be G-invariant. Let $\chi \in \mathrm{Irr}(G/P)$ be of p-defect zero in G/P. Define the function $\alpha(g) = 0$ if g_p is not in P, and $\alpha(g) = \theta(g_p)\chi(g_{p'})$, otherwise.

(a) Using Brauer's characterization of characters, prove that α is a generalized character of G.

(b) If $G = PC_G(P)$, prove that $\alpha \in \mathrm{Irr}(G|\theta)$ is such that $(\alpha(1)/\theta(1))_p = |G : N|_p$.

(*Note:* It is a fact that this establishes a natural bijection between defect zero characters of G/P and *relative defect zero* characters over θ.)

8

The Howlett–Isaacs Theorem

The Howlett–Isaacs theorem is one of the first applications of the Classification of Finite Simple Groups to Character Theory. In 1964, N. Iwahori and H. Matsumoto conjectured that *groups of central type* were solvable. In 1982, this conjecture was finally proven by R. Howlett and I. M. Isaacs in [HI82]. In this chapter, we give a proof of this theorem, assuming a fact about simple groups.

8.1 Fully Ramified Characters

Suppose that $N \trianglelefteq G$ and let $\theta \in \mathrm{Irr}(N)$. Recall that $\mathrm{Irr}(G|\theta)$ is the set of irreducible constituents of θ^G. Therefore we can write

$$\theta^G = \sum_{\chi \in \mathrm{Irr}(G|\theta)} [\theta^G, \chi]\chi.$$

If $\chi \in \mathrm{Irr}(G|\theta)$ and θ is G-invariant, then we know that $\chi_N = e\theta$ by Clifford's theorem. Sometimes, the integer $e = \chi(1)/\theta(1)$ is called the **ramification index** of χ over θ.

If θ is G-invariant and $\mathrm{Irr}(G|\theta)$ consists of exactly one character – that is, if $\theta^G = e\chi$ for some $\chi \in \mathrm{Irr}(G)$ – then we say that θ is **fully ramified** in G and that χ is **fully ramified** over N. The Iwahori–Matsumoto conjecture asserts that G/N should be solvable in this situation. Fully ramified characters appear quite naturally in character theory and modular representation theory.

We start with some elementary lemmas.

Lemma 8.1 *Let H be a subgroup of G, let $\chi \in \mathrm{Irr}(G)$, and let $\theta \in \mathrm{Irr}(H)$ be an irreducible constituent of χ_H with multiplicity e. Then $e^2 \le |G : H|$ and $e^2 = |G : H|$ if and only if $\chi_H = e\theta$ and $\chi(g) = 0$ for every $g \in G - H$.*

Proof We have that $\chi_H = e\theta + \Delta$, where Δ is a character of H or zero with $[\Delta, \theta] = 0$. Hence

$$e^2 \leq e^2 + [\Delta, \Delta] = [\chi_H, \chi_H] = \frac{1}{|H|} \sum_{h \in H} |\chi(h)|^2 \leq \frac{1}{|H|} \sum_{g \in G} |\chi(g)|^2 = |G : H|,$$

using that $[\chi, \chi] = 1$. The rest of the proof easily follows. $\qquad\square$

Lemma 8.2 *Let $N \trianglelefteq G$, let $\theta \in \mathrm{Irr}(N)$ be G-invariant, and let $\chi \in \mathrm{Irr}(G|\theta)$. Then the following are equivalent:*

(a) $\theta^G = e\chi$ *for some integer e;*
(b) $\chi_N = e\theta$ *and $e^2 = |G : N|$;*
(c) $\chi(g) = 0$ *for every $g \in G - N$.*

Proof If $\theta^G = e\chi$, then by Frobenius reciprocity and Clifford's theorem we have that $\chi_N = e\theta$. Then $|G : N|\theta(1) = e\chi(1) = e^2\theta(1)$ and $e^2 = |G : N|$. This proves that (a) implies (b). By Lemma 8.1, we have that (b) implies (c).

Assume (c). By the induction formula, we have that $(\theta^G)_N = |G : N|\theta$ and that θ^G is zero on $G - N$. Since χ_N is a multiple of θ, it follows that the characters θ^G and χ coincide on $G - N$ and are a multiple of each other on N. Thus $\theta^G = e\chi$ for some rational number e. Since θ^G is a character, e is necessarily an integer. This shows that (c) implies (a). $\qquad\square$

Theorem 8.3 (DeMeyer–Janusz) *Suppose that $N \trianglelefteq G$ and $\theta \in \mathrm{Irr}(N)$ is G-invariant. Assume that $\theta^G = e\chi$ for some $\chi \in \mathrm{Irr}(G)$. If P/N is a Sylow p-subgroup of G/N, then $\theta^P = e_p\eta$ for some $\eta \in \mathrm{Irr}(P)$, where e_p is the p-part of e.*

Proof By Lemma 8.2, we have that $e^2 = |G : N|$, $\chi_N = e\theta$, and $\chi(g) = 0$ for $g \in G - N$. Now, by the induction formula, θ^P is zero for $g \in P - N$ and $(\theta^P)_N = |P : N|\theta$. It follows that $\chi_P = d\theta^P$, where

$$d = \frac{\chi(1)}{|P : N|\theta(1)} = \frac{e}{|P : N|}.$$

Let $\eta \in \mathrm{Irr}(P)$ be an irreducible constituent of θ^P. Notice that

$$v = [\theta^P, \eta] = [\theta, \eta_N] = \eta(1)/\theta(1)$$

divides $|P : N|$, by Theorem 5.12. Then

$$[\chi_P, \eta] = d[\theta^P, \eta] = dv = \frac{e}{|P : N|}v$$

is an integer. We deduce that $\frac{|P:N|}{v}$ divides e. Hence

$$\frac{|P:N|}{v} \le e_p = |P:N|^{1/2}$$

and then $|P:N|^{1/2} \le v = [\eta_N, \theta]$. By Lemma 8.1, $v = |P:N|^{1/2}$ and $\eta(x) = 0$ for all $x \in P - N$. By Lemma 8.2, we conclude that $\theta^P = v\eta$, and $v = e_p$. □

The converse of Theorem 8.3 is true (and elementary): if θ is fully ramified in P for every Sylow P/N of G/N, then θ is fully ramified in G. However, we will leave this for the Problems.

We need one more elementary result on fully ramified characters.

Lemma 8.4 *Suppose that $N \trianglelefteq G$ and let $\theta \in \mathrm{Irr}(N)$ be G-invariant. Suppose that $\theta^G = e\chi$ for some $\chi \in \mathrm{Irr}(G)$. If $N \le E \trianglelefteq G$, then $\theta^E = d(\eta_1 + \cdots + \eta_t)$, where $\{\eta_1, \ldots, \eta_t\}$ is the G-orbit of $\eta \in \mathrm{Irr}(E)$. Also, η is fully ramified in the stabilizer G_η.*

Proof If $N \le H \le G$ and $\gamma \in \mathrm{Irr}(H|\theta)$, then every irreducible constituent of γ^G lies over γ, and therefore over θ. Thus γ^G is a multiple of χ. In particular, if $\eta \in \mathrm{Irr}(E|\theta)$, then η lies under χ, and we conclude that $\mathrm{Irr}(E|\theta)$ is exactly the set of irreducible constituents of χ_E. By Clifford's theorem, $\mathrm{Irr}(E|\theta)$ is the set of distinct G-conjugates of η. Now, since $[\eta_N, \theta] = [(\eta^g)_N, \theta]$ for $g \in G$, the first part follows. Finally, if $T = G_\eta$, then we have that every irreducible constituent μ of η^T lies under χ by the first observation in this proof, and therefore $\mu^G = \chi$ by the Clifford correspondence. By the uniqueness in the Clifford correspondence, we have that $\eta^T = f\mu$ for some $\mu \in \mathrm{Irr}(T)$ and some integer f. □

8.2 Groups of Central Type

Suppose that $N \trianglelefteq G$ and $\theta \in \mathrm{Irr}(N)$ is G-invariant and fully ramified in G. If we wish to study properties of the group G/N (for instance, proving that G/N is solvable), by using character triples it is no loss to assume that $N \subseteq \mathbf{Z}(G)$.

Lemma 8.5 *Suppose that $Z \subseteq \mathbf{Z}(G)$ and $\lambda \in \mathrm{Irr}(Z)$ is fully ramified in G.*

(a) We have that $Z = \mathbf{Z}(G)$.

(b) If $K \trianglelefteq G$ and p is a prime dividing $|K|$, then p divides $|K \cap Z|$.

Proof Write $\lambda^G = e\chi$ for some $\chi \in \mathrm{Irr}(G)$. Now $\chi_{\mathbf{Z}(G)} = \chi(1)\mu$ for some linear character μ. Then $\chi(z) \ne 0$ for $z \in \mathbf{Z}(G)$, and therefore $Z = \mathbf{Z}(G)$ by Lemma 8.2. This proves (a).

Let P/Z be a Sylow p-subgroup of G/Z. By Theorem 8.3, we have that $\lambda^P = e\eta$ for some $\eta \in \mathrm{Irr}(P)$. By part (a), we have that $\mathbf{Z}(P) = Z$. Now, let P_1 be a Sylow p-subgroup of P. Then $P = P_1 Z$ and $P_1 \in \mathrm{Syl}_p(G)$. Notice that $\mathbf{Z}(P_1) \subseteq \mathbf{Z}(P) = Z$.

Suppose now that $K \trianglelefteq G$ and p divides $|K|$. Then $P_1 \cap K$ is a Sylow p-subgroup of K, and thus $1 < P_1 \cap K$ is a nontrivial normal subgroup of the p-group P_1. By elementary group theory, we have that $1 < (P_1 \cap K) \cap \mathbf{Z}(P_1) = K \cap \mathbf{Z}(P_1) \subseteq K \cap Z$, as desired. $\qquad\square$

If a finite group G possesses an irreducible character $\chi \in \mathrm{Irr}(G)$ which is fully ramified over $\mathbf{Z}(G)$, then G is said to be of **central type**. By Lemmas 8.1 and 8.2, a finite group G is of central type if G possesses an irreducible character $\chi \in \mathrm{Irr}(G)$ such that $\chi(1)^2 = |G : \mathbf{Z}(G)|$. By Lemma 8.5(b), notice that every prime divisor of $|G|$ is a divisor of $|\mathbf{Z}(G)|$.

8.3 Character Correspondences

The Glauberman correspondence lies behind many fundamental results in character theory. The paradigmatic situation when it can be used is as follows. Suppose that p is a prime, K is a normal p'-subgroup of a finite group G, and P is a p-subgroup of G such that $KP \trianglelefteq G$. If $N = \mathbf{N}_G(P)$, notice that $G = KN$ by the Frattini argument, and that $N \cap K = C = \mathbf{C}_K(P)$. Now, if $\theta \in \mathrm{Irr}(K)$ is P-invariant, we have defined $\theta' \in \mathrm{Irr}(C)$ the P-Glauberman correspondent. Since the correspondence commutes with N-action (by Lemma 2.10), we also have that if $T = G_\theta$ is the stabilizer of θ in G, then $T \cap N = N_{\theta'}$ is the stabilizer of θ' in N.

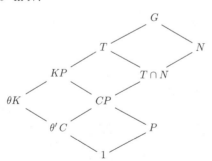

A very deep fact in character theory is that the character triples (T, K, θ) and $(T \cap N, C, \theta')$ are isomorphic. This is a theorem of E. C. Dade, also proved by L. Puig (with one more recent proof by A. Turull), that requires sophisticated machinery, as the theory of *endo-permutation modules* of p-groups.

However, T. Okuyama and W. Wajima proved a much weaker result that is good enough for many applications, and which we are going to prove next.

We shall use the Okuyama–Wajima theorem both in the proof of the Howlett–Isaacs theorem and in the reduction of the McKay conjecture. In the last step of the proof of Theorem 8.6, we use a well-known standard group-theoretical result: if an abelian group A acts faithfully on an abelian group P such that no proper subgroup of P is A-invariant, then A is cyclic. (By hypothesis, it follows that $\mathbf{C}_P(a) = 1$ for all $1 \neq a \in A$. Therefore, A is an abelian Frobenius complement, and thus A is cyclic. For other proofs, see Lemma 6.9 of [Is18] or Lemma 0.5 of [MW93].)

Theorem 8.6 (Okuyama–Wajima) *Suppose that G is a finite group, p is a prime, $K \trianglelefteq G$ has order not divisible by p, and P is a p-subgroup of G such that $KP \trianglelefteq G$. Let $N = \mathbf{N}_G(P)$ and $C = \mathbf{C}_K(P)$. Let $\theta \in \mathrm{Irr}(K)$ be P-invariant and let $\theta' \in \mathrm{Irr}(C)$ be the P-Glauberman correspondent of θ. Suppose that $C \leq A \leq N$ is such that A/C is abelian. Then θ extends to KA if and only if θ' extends to A.*

Proof We argue by induction on $|G : K||P|$. We have that $G = KN$ by the Frattini argument. Let $M = KP$. By Lemma 2.10, we have that θ is A-invariant if and only if θ' is A-invariant. By induction, we can assume that $MA = G$. Notice that θ extends to M by Corollary 6.3, and θ' extends to $CP = C \times P$. By Theorem 5.10 we may assume that A/C is a q-group for some prime $q \neq p$. Thus, notice that $P \in \mathrm{Syl}_p(G)$, A is a p-complement of N and G/M is abelian. Observe that our aim is equivalent to showing that θ extends to G if and only if θ' extends to N, again by Theorem 5.10.

We claim now that we may assume that M/K is a minimal normal subgroup of G/K. Otherwise, let $K < E < M$ be a normal subgroup of G. Let $Q = E \cap P \trianglelefteq P$, and notice that $E = KQ$. Thus, if $H = \mathbf{N}_G(Q)$, then $G = KH$ by the Frattini argument. Also $N \subseteq H$. Let $D = \mathbf{C}_K(Q)$, let η be the Q-Glauberman correspondent of θ, and recall by Theorem 2.14 that $\eta' = \theta'$, where η' is the P-Glauberman correspondent of η. (Notice too that η' is the P/Q-Glauberman correspondent.) Since $Q < P$, by induction we have that θ extends to KA if and only if η extends to DA. Now, we wish to work in $\bar{H} = H/Q$ and apply the inductive hypothesis. Notice that DA has order not divisible by p, DA normalizes Q, and therefore DA is a p-complement of $Q(DA)$. Hence, by using Theorem 1.18 (or simply Theorem 2.1), we have that the character $1_Q \times \eta$ considered as a character of DQ/Q extends to QDA/Q if and only if η extends to DA. For the same reason, θ' extends to A if and only if $1_Q \times \theta'$ considered as a character of QC/Q extends to AQ/Q. Now, since $1_Q \times \theta' \in \mathrm{Irr}(QC/Q)$ is the P/Q-Glauberman correspondent of $1_Q \times \eta \in \mathrm{Irr}(DQ/Q)$, by applying induction in $\bar{H} = H/Q$ the claim follows.

Let L/K be a normal p'-subgroup of G/K and assume that $L > K$. Since $KP \trianglelefteq G$, then P acts trivially on L/K, and thus $L = K\mathbf{C}_L(P)$ by coprime action (see Corollary 3.28 of [Is08]). Since G/M is abelian, then notice that $LP \trianglelefteq G$, and that G/LP and L/K are abelian. If $\gamma \in \mathrm{Irr}(L)$ is P-invariant, by induction we have that γ extends to G if and only if $\gamma' \in \mathrm{Irr}(\mathbf{C}_L(P))$ extends to N. Suppose, for instance, that $\chi_K = \theta$ for some $\chi \in \mathrm{Irr}(G)$. Then $\chi_L = \gamma \in \mathrm{Irr}(L)$ is P-invariant, and therefore γ' extends to N. We have that $(\gamma')_C = \theta'$ by Theorem 2.13(b). Therefore θ' extends to N. The other implication is proved the same way. Hence, we may assume that $K = L$.

Now, we have that the abelian group KA/K acts faithfully on the abelian group KP/K in such a way that no proper subgroup of KP/K is KA/K-invariant. We conclude that $KA/K \cong A/C$ is cyclic. In this case, θ extends to KA and θ' extends to C by Theorem 5.1. □

Next, we use Theorem 8.6 to prove a slightly more general version of it. First, we need a lemma.

Lemma 8.7 *Let G be a finite group. Suppose that $K \trianglelefteq G$ is a Hall subgroup of G. Suppose that $Z \trianglelefteq G$ satisfies $K \cap Z = 1$. Let $\theta \in \mathrm{Irr}(K)$ and $\lambda \in \mathrm{Irr}(Z)$ be both G-invariant. Then $\theta \times \lambda$ extends to G if and only if $\hat{\lambda}$ extends to G/K, where $\hat{\lambda} = 1_K \times \lambda \in \mathrm{Irr}(KZ/K)$.*

Proof Since $K \cap Z = 1$, notice that $KZ = K \times Z$. Let $\hat{\theta} = \theta \times 1_Z \in \mathrm{Irr}(KZ/Z)$ and $\hat{\lambda} = 1_K \times \lambda \in \mathrm{Irr}(KZ/K)$, so that $\hat{\theta}\hat{\lambda} = \theta \times \lambda \in \mathrm{Irr}(KZ)$. Now, since KZ/Z is a normal Hall subgroup of G/Z, it follows that $\hat{\theta}$ extends to some $\chi \in \mathrm{Irr}(G)$ with $Z \leq \ker(\chi)$ by Corollary 6.3. By Theorem 1.22, we have that the map $\beta \mapsto \beta\chi$ is a bijection $\mathrm{Irr}(G|\hat{\lambda}) \to \mathrm{Irr}(G|\hat{\theta}\hat{\lambda})$. By degrees, β extends $\hat{\lambda}$ if and only if $\beta\chi$ extends $\hat{\theta}\hat{\lambda}$. □

Theorem 8.8 *Suppose that G is a finite group, p is a prime, $K \trianglelefteq G$ has order not divisible by p, and P is a p-subgroup of G such that $KP \trianglelefteq G$. Let $N = \mathbf{N}_G(P)$ and $C = \mathbf{C}_K(P)$. Let $\theta \in \mathrm{Irr}(K)$ be P-invariant and let $\theta' \in \mathrm{Irr}(C)$ be the Glauberman correspondent of θ. Let $Z \trianglelefteq G$ be contained in P, and let $\lambda \in \mathrm{Irr}(Z)$ be G-invariant. Suppose that $CZ \leq A \leq N$ is such that A/CZ is abelian. Then $\theta \times \lambda$ extends to KA if and only if $\theta' \times \lambda$ extends to A.*

Proof By elementary group theory, notice that $KN = G$ and $K \cap N = C$. Since the Glauberman correspondence commutes with N-action (by Lemma 2.10), notice that the stabilizers N_θ and $N_{\theta'}$ coincide. Thus θ is A-invariant if and only if θ' is A-invariant. Hence, we may assume that both θ and θ' are

A-invariant. By Theorem 5.10, we may assume that A/CZ is a q-group for some prime q.

Suppose first that $q = p$. In this case, K is a normal p-complement of KA. By Lemma 8.7, we have that $\theta \times \lambda$ extend to KA if and only if $1_K \times \lambda$ extends to KA/K. By the same argument, we have that $\theta' \times \lambda$ extend to A if and only if $1_C \times \lambda$ extends to A/C. Since $K \cap A = C$, by Theorem 1.18 we have that restriction defines a bijection $\mathrm{Irr}(KA/K) \to \mathrm{Irr}(A/C)$. This restriction sends the characters of KA/K over $1_K \times \lambda$ onto the characters of A/C over $1_C \times \lambda$. This case easily follows.

Suppose finally that $q \neq p$. In this case, we have that Z is a normal Sylow p-subgroup of KA. By Lemma 8.7, we have that $\theta \times \lambda$ extends to KA if and only if $\theta \times 1_Z$ extends to KA/Z. By the same argument, $\theta' \times \lambda$ extends to A if and only if $\theta' \times 1_Z$ extends to A/Z. But the equivalence of these two conditions now follows from Theorem 8.6 applied in G/Z, with respect to KZ/Z and P/Z. (Notice that Z acts trivially on K, that $\mathbf{C}_{KZ/Z}(P/Z) = CZ/Z$ and that the P/Z-Glauberman correspondence on K is the same as the P-Glauberman correspondence.)   □

Corollary 8.9 *Suppose that G is a finite group, p is a prime, $K \trianglelefteq G$ has order not divisible by p, and P is a p-subgroup of G such that $KP \trianglelefteq G$. Let $N = \mathbf{N}_G(P)$ and $C = \mathbf{C}_K(P)$. Let $\theta \in \mathrm{Irr}(K)$ be P-invariant and let $\theta' \in \mathrm{Irr}(C)$ be the Glauberman correspondent of θ. Let $Z \trianglelefteq G$ be contained in P, and let $\lambda \in \mathrm{Irr}(Z)$ be G-invariant. If $CZ \le U \le N$, then*

$$|\mathrm{Irr}(KU|\theta \times \lambda)| = |\mathrm{Irr}(U|\theta' \times \lambda)|.$$

Proof Since the Glauberman correspondence commutes with the action of U by Lemma 2.10, we have that the stabilizers U_θ and $U_{\theta'}$ coincide. Since $KZ = K \times Z$ and λ is G-invariant, this also holds for the stabilizers

$$U_{\theta \times \lambda} = U_{\theta' \times \lambda}.$$

Thus

$$(KU)_{\theta \times \lambda} = KU_{\theta' \times \lambda},$$

and by the Clifford correspondence, we may assume that θ is U-invariant. (Or, equivalently, that θ' is U-invariant.)

By Theorem 5.16, we know that $|\mathrm{Irr}(KU|\theta \times \lambda)|$ is the number of conjugacy classes of KU/KZ which are $\theta \times \lambda$-good in KU. Similarly, $|\mathrm{Irr}(U|\theta' \times \lambda)|$ is the number of conjugacy classes of U/CZ which are $\theta' \times \lambda$-good in U. Since the map $CZu \mapsto KZu$ is a natural isomorphism of groups $U/CZ \to$

KU/KZ, it suffices to show that CZu is $\theta' \times \lambda$-good in U if and only if KZu is $\theta \times \lambda$-good in KU for all $u \in U$.

By Lemma 5.13(a), notice that KZu is $\theta \times \lambda$-good in KU if and only if $\theta \times \lambda$ extends to $\langle KZ, u, x \rangle$ for every element $x \in U$ such that $\langle KZ, u, x \rangle / KZ$ is abelian. By the same reason, CZu is $\theta' \times \lambda$-good in U if and only if $\theta' \times \lambda$ extends to $\langle CZ, u, x \rangle$ for every element $x \in U$ such that $\langle CZ, u, x \rangle / CZ$ is abelian. Now, if $A = \langle CZ, u, x \rangle$, then $\langle KZ, u, x \rangle = KA$, and therefore it suffices to show that $\theta \times \lambda$ extends to KA if and only if $\theta' \times \lambda$ extends to A. But this follows from Theorem 8.8.   □

The following is exactly the result on character correspondences that we shall use for proving the Howlett–Isaacs theorem.

Theorem 8.10 *Suppose that G is a finite group, p is a prime, $K \trianglelefteq G$ has order not divisible by p, and P is a p-subgroup of G such that $KP \trianglelefteq G$. Let $N = \mathbf{N}_G(P)$. Let Z be a central subgroup of G contained in KP, and let $\lambda \in \mathrm{Irr}(Z)$. If $\lambda^G = e\chi$ for some $\chi \in \mathrm{Irr}(G)$, then $\lambda^N = w\eta$ for some $\eta \in \mathrm{Irr}(N)$.*

Proof Write $Z = Z_p \times Z_{p'}$, where $Z_p = Z \cap P$ and $Z_{p'} = Z \cap K$. Write $\lambda = \lambda_p \times \lambda_{p'}$, where $\lambda_p \in \mathrm{Irr}(Z_p)$ and $\lambda_{p'} \in \mathrm{Irr}(Z_{p'})$. Let $E = KZ = K \times Z_p$. By Lemma 8.4, we have that $\lambda^E = d(\nu_1 + \cdots + \nu_t)$, where $\{\nu_1, \ldots, \nu_t\}$ is the G-orbit of $\nu \in \mathrm{Irr}(E)$. Hence $|E : Z| = dt\nu(1)$, and we deduce that t is not divisible by p. Notice that $d = [\lambda^E, \nu] = [\lambda, \nu_Z] = \nu(1)$. Therefore if $T = G_\nu$ is the stabilizer of ν in G, then $|G : T| = t$ is not divisible by p. Since KP/E is a normal p-subgroup of G/E, we conclude that $KP \subseteq T$. Also, since $KN = G$, we have that $\nu_i = \nu^{n_i}$ for some $n_i \in N$. Write $\nu = \theta \times \lambda_p$, where $\theta \in \mathrm{Irr}(K | \lambda_{p'})$. Since $\lambda_{p'}$ is G-invariant, notice that $\{\theta^{n_1}, \ldots, \theta^{n_t}\}$ is the full N-orbit of θ, of size t, and $d = \theta(1)$. Also T is the stabilizer of θ in G.

By Mackey, we have that

$$(\lambda_{p'})^K = (\lambda_{Z_{p'}})^K = (\lambda^E)_K = d(\theta^{n_1} + \cdots + \theta^{n_t}).$$

Let $C = \mathbf{C}_K(P) = N \cap K$. We know that the Glauberman correspondence defines a bijection

$$': \mathrm{Irr}_P(K) \to \mathrm{Irr}(C),$$

and since ψ' is an irreducible constituent of ψ_C for $\psi \in \mathrm{Irr}_P(K)$, it follows that

$$': \mathrm{Irr}_P(K | \lambda_{p'}) \to \mathrm{Irr}(C | \lambda_{p'})$$

is also a bijection. Now, the Glauberman correspondence commutes with the action of N by Lemma 2.10, and we conclude that $\mathrm{Irr}(C|\lambda_{p'}) = \{(\theta')^{n_1}, \ldots, (\theta')^{n_t}\}$ is the full N-orbit of θ', and has size t. Then

$$(\lambda_{p'})^C = f((\theta')^{n_1} + \cdots + (\theta')^{n_t}),$$

where $f = \theta'(1)$. Thus, using Mackey and the fact that $CZ = C \times Z_p$, we have

$$\lambda^{CZ} = f((\theta')^{n_1} \times \lambda_p + \cdots + (\theta')^{n_t} \times \lambda_p).$$

By Corollary 8.9 with $U = N$, we have that

$$|\mathrm{Irr}(G|\theta \times \lambda_p)| = |\mathrm{Irr}(N|\theta' \times \lambda_p)|.$$

Since $\theta \times \lambda_p$ lies over λ and $\mathrm{Irr}(G|\lambda) = \{\chi\}$, we conclude that

$$|\mathrm{Irr}(N|\theta' \times \lambda_p)| = 1.$$

In other words, $(\theta' \times \lambda_p)^N = u\mu$ for some $\mu \in \mathrm{Irr}(N)$. Finally

$$\lambda^N = (\lambda^{CZ})^N = uft\mu,$$

using that $(\theta')^{n_i} \times \lambda_p = (\theta' \times \lambda_p)^{n_i}$. $\square$

We momentarily digress in order to show how the Okuyama–Wajima argument (Theorem 8.6) is going to be used in the p-solvable case of the McKay conjecture. We shall use the next result in the last theorem in this book (but not for the Howlett–Isaacs theorem). It has the same flavour as Theorem 8.10, but there is a significant difference: we wish to control the degrees of the characters in question. To make our argument work, we also need to assume in this case that P is a Sylow p-subgroup of G. It is possible to remove this condition, but only by appealing to Dade's deep theorem, on which we have commented at the beginning of this section.

If $N \trianglelefteq G$ and $\theta \in \mathrm{Irr}(N)$, then recall that $\mathrm{Irr}_{p'}(G|\theta)$ is the set of irreducible characters of G of degree not divisible by p which lie over θ.

Theorem 8.11 *Let G be a finite group, let p be a prime, and let $P \in \mathrm{Syl}_p(G)$. Suppose that $K \trianglelefteq G$ has order not divisible by p, and that $KP \trianglelefteq G$. Let $Z \trianglelefteq G$ be a normal p-subgroup of G, and let $\lambda \in \mathrm{Irr}(Z)$ be G-invariant. Let $N = \mathbf{N}_G(P)$ and $C = \mathbf{C}_K(P)$. Let $\theta \in \mathrm{Irr}(K)$ be P-invariant, and let $\theta' \in \mathrm{Irr}(C)$ be the Glauberman correspondent of θ with respect to P. Then*

$$|\mathrm{Irr}_{p'}(G|\theta \times \lambda)| = |\mathrm{Irr}_{p'}(N|\theta' \times \lambda)|.$$

Proof Notice that $G = KN$ because $KP \trianglelefteq G$ and the Frattini argument. By Lemma 2.10, we have that the action of N commutes with the P-Glauberman correspondence. Therefore, $N_\theta = N_{\theta'}$. Also notice that $G_{\theta \times \lambda} = G_\theta \cap G_\lambda = G_\theta$. By the same argument, $N_{\theta \times \lambda} = N_\theta = N_{\theta'} = N_{\theta' \times \lambda}$. Since $P \subseteq N_\theta$, we have that the Clifford correspondence sends characters of degree not divisible by p onto characters of degree not divisible by p. Since $G_{\theta \times \lambda} = KN_{\theta \times \lambda}$, we may assume that θ and θ' are N-invariant. Now, let $\hat\theta \in \mathrm{Irr}(KP)$ be the canonical extension of θ (by Corollary 6.3). Since $\hat\theta_{KZ}$ and $\theta \times 1_Z$ are two extensions of θ with determinantal order coprime with p, by the uniqueness in Corollary 6.3 we deduce that $\hat\theta_{KZ} = \theta \times 1_Z$. Therefore $\hat\theta \in \mathrm{Irr}(KP/Z)$.

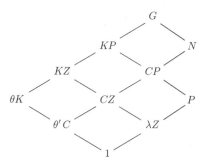

Suppose now that $\mathcal{A}$ is a complete set of representatives of the orbits of the action of N on the linear characters of KP/K that lie over $1_K \times \lambda$. (Notice that $KP/K, CP/C$, and P are naturally isomorphic, so we can view the characters of $\mathcal{A}$ as characters of any of these groups.) We claim that

$$\mathrm{Irr}_{p'}(G|\theta \times \lambda) = \bigcup_{\mu \in \mathcal{A}} \mathrm{Irr}(G|\hat\theta\mu)$$

and

$$\mathrm{Irr}_{p'}(N|\theta' \times \lambda) = \bigcup_{\mu \in \mathcal{A}} \mathrm{Irr}(N|\widehat{\theta'}\mu)$$

are disjoint unions, where in the second equality $\widehat{\theta'} = \theta' \times 1_P$, and we view μ as a character of CP/C. First of all, notice that the characters in $\mathrm{Irr}(G|\hat\theta\mu)$ have degree not divisible by p by Theorem 5.12. Also $\hat\theta\mu$ and $\hat\theta\mu_1$ are G-conjugate if and only if μ and μ_1 are G-conjugate (by the uniqueness in the Gallagher correspondence). This happens if and only if they are N-conjugate, because $G = KN$. Also, the characters in $\mathrm{Irr}_{p'}(G|\hat\theta\mu)$ lie over $\theta \times \lambda$, since μ lies over λ and $Z \subseteq \ker(\hat\theta)$. On the other hand, if $\chi \in \mathrm{Irr}_{p'}(G|\theta \times \lambda)$ and $\rho \in \mathrm{Irr}(KP)$ is under χ, then ρ has p'-degree. Thus $\rho_K = \theta$ by Theorem 5.12. Therefore $\rho = \hat\theta\mu$ for some linear $\mu \in \mathrm{Irr}(KP/K)$ by the Gallagher correspondence. Replacing ρ by some N-conjugate, we may assume that $\mu \in \mathcal{A}$. By using the same argument for $\mathrm{Irr}_{p'}(N|\theta' \times \lambda)$, the claim is proved.

We claim now that if $KP \leq W \leq G$, then

$$|\mathrm{Irr}(W|\hat{\theta})| = |\mathrm{Irr}(N \cap W|\widehat{\theta'})|.$$

To prove the claim, it is clear that we may assume that $W = G$. We use some group theory now. Since CP/C is a normal Sylow p-subgroup of N/C, then by the Schur–Zassenhaus theorem there is some subgroup U/C of N/C such that $U(CP) = N$ and $U \cap CP = C$. In particular, U is a p'-subgroup of N. (In fact, U is a p-complement of N.) Notice now that $(KP)(KU) = G$ because $(KP)(KU)$ contains $U(CP) = N$ and K. Since KP/K is a p-group and KU/K is a p'-group, then $KP \cap KU = K$. Now, by Lemma 6.8(d), we have that restriction defines a bijection

$$\mathrm{Irr}(G|\hat{\theta}) \to \mathrm{Irr}(KU|\theta).$$

By the same reason, we have that restriction defines a bijection

$$\mathrm{Irr}(N|\widehat{\theta'}) \to \mathrm{Irr}(U|\theta').$$

By Corollary 8.9 (applied with $Z = 1$), we have that $|\mathrm{Irr}(KU|\theta)| = |\mathrm{Irr}(U|\theta')|$, and the claim is proven.

Suppose now that $\mu \in \mathrm{Irr}(P)$ is any linear character. Since $K \cap P = 1$, by Theorem 1.18 we can view $\mu \in \mathrm{Irr}(KP/K)$ (or $\mu \in \mathrm{Irr}(CP/C)$). If $g \in G$, then, by the Gallagher correspondence, $\mu^g\hat{\theta} = \mu\hat{\theta}$ if and only if $\mu = \mu^g$. Therefore

$$G_{\mu\hat{\theta}} = G_\mu,$$

using that $\hat{\theta}$ is G-invariant. Notice too that μ extends to some $\nu \in \mathrm{Irr}(G_\mu)$, by Corollary 6.3. Now, by Theorem 1.22 we have that $\beta \mapsto \beta\nu$ is a bijection $\mathrm{Irr}(G_\mu|\hat{\theta}) \to \mathrm{Irr}(G_\mu|\hat{\theta}\mu)$. By the Clifford correspondence

$$|\mathrm{Irr}(G|\hat{\theta}\mu)| = |\mathrm{Irr}(G_{\hat{\theta}\mu}|\hat{\theta}\mu)| = |\mathrm{Irr}(G_\mu|\hat{\theta}\mu)| = |\mathrm{Irr}(G_\mu|\hat{\theta})|.$$

By the same argument applied in N with respect to the character $\widehat{\theta'}$ and the character $\mu \in \mathrm{Irr}(CP/C)$, we have that

$$|\mathrm{Irr}(N|\widehat{\theta'}\mu)| = |\mathrm{Irr}(N_\mu|\widehat{\theta'})|.$$

By the previous claim with $W = G_\mu$, we conclude that

$$|\mathrm{Irr}(G|\hat{\theta}\mu)| = |\mathrm{Irr}(N|\widehat{\theta'}\mu)|$$

for any linear $\mu \in \mathrm{Irr}(P)$. By the claim in the second paragraph of this proof, the proof of the theorem is now complete. $\qquad\square$

8.4 Simple Groups for the Howlett–Isaacs Theorem

In this section, we write down the property that simple groups satisfy which will allow us to prove the Howlett–Isaacs theorem.

When dealing with a general finite group, simple groups usually appear when we consider a non-abelian chief factor K/L of G. In this case, K/L is the direct product of the different G-conjugates of a non-abelian simple group U/L. In some problems – for instance, in Theorems 6.7 or 7.3 – we are able to assume that $L = 1$. In other cases, one can only assume that $L \subseteq \mathbf{Z}(K)$, and then central extensions come into play.

Theorem 8.12 *Let S be a non-abelian simple group. Then there exists a prime p dividing $|S|$ such that the following conditions hold:*

(a) There is no solvable subgroup of S having p-power index.
(b) Whenever Y is a finite group with $Y/\mathbf{Z}(Y) \cong S^n$ for some integer n, then p does not divide $|Y' \cap \mathbf{Z}(Y)|$.

Proof This is Theorem 2.1 and Corollary 7.2 of [HI82]. □

Assertion (b) in Theorem 8.12 is equivalent to the following: if $\hat{S}$ is a *universal covering group* of S, then p does not divide $|\mathbf{Z}(\hat{S})|$. (See Theorem C.1.) For the reader who is not familiar with covering groups, we shall properly introduce these and the Schur multiplier of S, which is $\mathbf{Z}(\hat{S})$, in Appendix B.

8.5 Proof of the Iwahori–Matsumoto Conjecture

Theorem 8.13 (Howlett–Isaacs) *Suppose that $Z \trianglelefteq G$. Assume that $\lambda \in \mathrm{Irr}(Z)$ is fully ramified with respect to G. Then G/Z is solvable.*

Proof We argue by induction on $|G : Z|$.
Step 1 We may assume that $Z = \mathbf{Z}(G)$ and that the Fitting subgroup F of G is the largest normal solvable subgroup of G.

By using character triple isomorphisms Corollary 5.9, it is no loss to assume that $Z \subseteq \mathbf{Z}(G)$. By Lemma 8.5(a), we have that $Z = \mathbf{Z}(G)$. Let $F = \mathbf{F}(G)$ be the Fitting subgroup of G, and let R be the largest normal solvable subgroup of G. Suppose that $F < R$. Let L/F be a chief factor of G with $L \subseteq R$. Since L/F is solvable, we have that L/F is a p-group for some prime p. Now, let $K = F_{p'}$ be the p-complement of F, let $P \in \mathrm{Syl}_p(L)$, and let $N = \mathbf{N}_G(P)$. Notice that $L = KP \trianglelefteq G$ and that $N < G$, since L is not nilpotent. Also, $G = KN$, by the Frattini argument. By Theorem 8.10, we have that λ is fully ramified in N. Since $|N : Z| < |G : Z|$, we conclude by induction that N

is solvable. Thus $G/K \cong N/(N \cap K)$ is solvable. Since K is solvable, we deduce that G is solvable in this case.

Step 2 If p divides $|F : Z|$, then every composition factor of G has a solvable subgroup of p-power index.

Let Q/Z be a Sylow p-subgroup of F/Z. Let $\theta \in \mathrm{Irr}(Q|\lambda)$ and let $I = G_\theta$ be the stabilizer of θ in G. By Lemma 8.4, we have that $\lambda^Q = e(\theta^{x_1} + \cdots + \theta^{x_t})$, where $\{\theta^{x_1}, \ldots, \theta^{x_t}\}$ is the full G-orbit of θ, $t = |G : I|$, and θ is fully ramified in I. Since $|I : Q| < |G : Z|$, by induction we have that I/Q (and therefore I) is solvable. Now, $|Q : Z| = e\theta(1)t$, and thus t is a power of p. We deduce that I is a solvable subgroup of G with p-power index. If U/V is a composition factor of G, using that $U \trianglelefteq \trianglelefteq G$, we easily prove that $U \cap I$ has p-power index in U. Hence, $V(U \cap I)/V$ is a solvable subgroup of U/V with p-power index.

Step 3 Final contradiction.

Let M/F be a chief factor of G. By Step 1, we have that M/F is a direct product of groups isomorphic to a non-abelian simple group S. By Theorem 8.12, there exists a prime p dividing $|M/F|$ such that S has no solvable subgroup of index a power of p, and whenever Y is a finite group with $Y/\mathbf{Z}(Y) \cong S^n$ for some integer n, then p does not divide $|Y' \cap \mathbf{Z}(Y)|$. By the previous step, we have that p does not divide $|F : Z|$. Let W be the p-complement of F. Then $F = WZ$ and thus $F/W \subseteq \mathbf{Z}(M/W)$. Since M/F is a non-abelian chief factor of G, we have $F/W = \mathbf{Z}(M/W)$. Now if $X = M/W$, then $X/\mathbf{Z}(X) \cong S^n$ for some n, and thus we conclude that p does not divide $|(M/W)' \cap F/W|$. Since F/W is a p-group, we have that $(M/W)' \cap F/W$ is trivial. Hence $M' \cap F \subseteq W$ and p does not divide $|M' \cap F|$. Thus p does not divide $|M' \cap Z|$. Since M/F is perfect, $M = FM'$ and p divides $|M'|$ (because p divides $|M : F|$). This contradicts Lemma 8.5(b) with $K = M'$. $\qquad\square$

8.6 Notes

In 1979, R. A. Liebler and J. E. Yellen proposed a proof of the solvability of groups of central type which unfortunately contained an error (see [LY79]).

Groups of central type are solvable but need not be nilpotent. The smallest counterexample is SmallGroup(216,39) which is $(\mathbf{C}_9 \times \mathbf{Q}_8) : \mathbf{C}_3$.

S. Gagola proved in [Ga74] that if G is any solvable group, then there exists a central type group H such that G is isomorphic to a subgroup of $H/\mathbf{Z}(H)$. Therefore, besides solvability, nothing can be said about subgroups of central type groups.

Groups of central type appear quite naturally in the theory of complex twisted algebras, since these are exactly the groups for which these algebras are simple. Also, groups of central type appear in modular representation theory in relation to blocks with a unique modular character.

Finally, J. F. Humphreys made a generalization of the Iwahori–Matsumoto conjecture which remains open. Suppose that $N \trianglelefteq G$ and that $\theta \in \mathrm{Irr}(N)$ is G-invariant. Humphreys' conjecture asserts that if all the irreducible characters in $\mathrm{Irr}(G|\theta)$ have the same degree, then G/N is solvable. (This conjecture is mentioned in [Hi11].)

Problems

(8.1) Let $N \trianglelefteq G$, and suppose that $\theta \in \mathrm{Irr}(N)$ is G-invariant. Assume that θ is fully ramified in P for every Sylow p-subgroup P/N of G/N. Prove that θ is fully ramified in G.

(8.2) Let $N \trianglelefteq G$, and suppose that $\theta \in \mathrm{Irr}(N)$ is G-invariant. Assume that θ is fully ramified in G. If H/N is a Hall π-subgroup of G/N, prove that θ is fully ramified in H.

(8.3) Let $N \trianglelefteq G$, and suppose that $\theta \in \mathrm{Irr}(N)$ is G-invariant. Assume that θ is fully ramified in G. If $N < G$, show that G/N has no cyclic self-centralizing subgroup.
(*Hint:* Use that G/N has a unique θ-good class.)

(8.4) Let $N \trianglelefteq G$, with G/N abelian, and suppose that $\theta \in \mathrm{Irr}(N)$ is G-invariant. In this problem, we show that there is a unique maximal subgroup $N \le U \le G$ such that every extension of θ to U is G-invariant and fully ramified with respect to G/U.

(a) Suppose that V and W are subgroups of G containing N, and assume that $\alpha \in \mathrm{Irr}(V)$ and $\beta \in \mathrm{Irr}(W)$ are G-invariant and extend θ. Then show that θ extends to VW. Conclude that there exists a unique largest subgroup $N \le U \le G$ such that θ has G-invariant extension to U.

(b) Suppose that $U = N$. Let M/N be cyclic, and let $\chi \in \mathrm{Irr}(G|\theta)$. By the Gallagher correspondence, write

$$\chi_M = e\left(\sum_{\lambda \in S} \lambda \phi\right)$$

where $\phi \in \mathrm{Irr}(M)$ and $S \subseteq \mathrm{Irr}(M/N)$. Using Clifford's theorem, show that S is a subgroup of $\mathrm{Irr}(M/N)$.

(c) As before, if $S = \langle \lambda \rangle$, show that $K/N = \ker(\lambda)$ is trivial, and conclude that $\chi(g) = 0$ for all $g \in G - N$.
(*Hint:* For (a), use Lemma 6.8(d).)

9

Global–Local Counting Conjectures

We give an overview of the main global–local counting conjectures in the representation theory of finite groups, some of their variations, and their relationship.

9.1 The McKay Conjecture

The McKay conjecture was the first of the global–local counting conjectures that was proposed. Recall that $\mathrm{Irr}_{p'}(G)$ is the set of irreducible characters of G of degree not divisible by the prime p.

Conjecture 9.1 (McKay) *Let G be a finite group, let p be a prime and let $P \in \mathrm{Syl}_p(G)$. Then*

$$|\mathrm{Irr}_{p'}(G)| = |\mathrm{Irr}_{p'}(\mathbf{N}_G(P))|.$$

Let us analyze the McKay conjecture in the case where G has a normal p-complement K. In global–local group theory, this is the easiest case to consider. If $\chi \in \mathrm{Irr}_{p'}(G)$, then notice that $\chi_K = \theta \in \mathrm{Irr}(K)$ by Theorem 5.12. Also, $\theta \in \mathrm{Irr}_P(K)$ (that is, θ is P-invariant) because χ is a class function. On the other hand, if $\theta \in \mathrm{Irr}_P(K)$, then θ extends to some $\chi \in \mathrm{Irr}_{p'}(G)$ by Corollary 6.3. And in fact, there is a unique extension $\hat{\theta} \in \mathrm{Irr}(G)$ whose determinantal order $o(\hat{\theta})$ is not divisible by p. (Recall that this is called the canonical extension of θ to G.) Hence, we can write

$$\mathrm{Irr}_{p'}(G) = \bigcup_{\theta \in \mathrm{Irr}_P(K)} \mathrm{Irr}_{p'}(G|\theta),$$

where $\mathrm{Irr}_{p'}(G|\theta) = \mathrm{Irr}_{p'}(G) \cap \mathrm{Irr}(G|\theta)$. Now, using the Gallagher correspondence Corollary 1.23, we have that

$$\mathrm{Irr}_{p'}(G|\theta) = \{\lambda\hat{\theta} \mid \lambda \in \mathrm{Lin}(G/K)\}.$$

Since $G/K \cong P$, we conclude that

$$|\mathrm{Irr}_{p'}(G)| = |P/P'||\mathrm{Irr}_P(K)|.$$

Now, $\mathbf{N}_K(P)$ is a normal p-complement of $\mathbf{N}_G(P)$. Since $[\mathbf{N}_K(P), P] = K \cap P = 1$, we easily see that $\mathbf{N}_G(P) = \mathbf{C}_K(P) \times P$. Therefore

$$|\mathrm{Irr}_{p'}(\mathbf{N}_G(P))| = |P/P'||\mathrm{Irr}(\mathbf{C}_K(P))|.$$

We see that the McKay conjecture is true for a group G with a normal p-complement K if and only if

$$|\mathrm{Irr}_P(K)| = |\mathrm{Irr}(\mathbf{C}_K(P))|.$$

But, of course, this equality holds by the Glauberman correspondence!

We see that the *easiest* case of the McKay conjecture follows from the equality $|\mathrm{Irr}_P(K)| = |\mathrm{Irr}(\mathbf{C}_K(P))|$, which we proved in Chapter 2, and that definitely is not a triviality. As we shall see in Chapter 10, it turns out that the Glauberman correspondence is an essential ingredient in the proof of the McKay conjecture for p-solvable groups (by Okuyama and Wajima), and in the reduction of the McKay conjecture to simple groups.

But there is more to learn from the normal p-complement case, since we really have proved the following.

Theorem 9.2 *Suppose that G has a normal p-complement. Let $P \in \mathrm{Syl}_p(G)$. Then there is a canonical bijection $': \mathrm{Irr}_{p'}(G) \rightarrow \mathrm{Irr}_{p'}(\mathbf{N}_G(P))$. Also, $\mathbb{Q}(\chi) = \mathbb{Q}(\chi')$ for $\chi \in \mathrm{Irr}(G)$.*

Proof Again, let K be the normal p-complement of G. Then we have that $\mathbf{N}_G(P) = \mathbf{C}_K(P) \times P$. Given $\chi \in \mathrm{Irr}_{p'}(G)$, using the Gallagher correspondence, there is a unique linear character $\lambda \in \mathrm{Irr}(G/K)$ such that $\chi = \lambda\hat{\theta}$, where $\hat{\theta} \in \mathrm{Irr}(G)$ is the canonical extension of $\chi_K = \theta$. By Problem 3.7, we know that the Glauberman correspondence $': \mathrm{Irr}_P(K) \rightarrow \mathrm{Irr}(\mathbf{C}_K(P))$ commutes with the action of $\mathcal{G}$, where $\mathcal{G} = \mathrm{Gal}(\mathbb{Q}_{|G|}/\mathbb{Q})$. Then the map $\chi \mapsto \chi'$, where $\chi' = \theta' \times \lambda$, is a canonical bijection that commutes with the action of $\mathcal{G}$. In particular, $\mathbb{Q}(\chi) = \mathbb{Q}(\chi')$. $\square$

In Theorem 9.2, we have used the word *canonical*, as we did before in Chapter 2 with respect to the Glauberman correspondence. We agree that it is not easy to define formally when a map is *natural* or *canonical*. Another expression that we could have used, *choice-free*, describes what we mean: we understand that a map f is *canonical* if the image of every x by f is independent of any choice made in order to define $f(x)$.

In the search for a proof of the McKay conjecture, one might speculate on how to associate with every $\chi \in \text{Irr}_{p'}(G)$ a character $\chi' \in \text{Irr}(\mathbf{N}_G(P))$, and a natural place to look is in the character restriction $\chi_{\mathbf{N}_G(P)}$. In general, however, it seems that this is not going to be the right place to search. The group $G = \text{SL}_2(3)$ has a normal p-complement for $p = 3$. If $\chi \in \text{Irr}(G)$ is the unique rational-valued character of degree 2, then the image of χ under the natural correspondence in Theorem 9.2 is $\chi' = \lambda \times 1_P$, where $\lambda \in \text{Irr}(\mathbf{Z}(G))$ is nontrivial. However, if ϵ is any nontrivial character of P, then the restriction

$$\chi_{\mathbf{N}_G(P)} = (\lambda \times \epsilon) + (\lambda \times \bar{\epsilon})$$

does not contain χ'.

Also, one soon loses the hope that there should always exist a character correspondence $\text{Irr}_{p'}(G) \rightarrow \text{Irr}_{p'}(\mathbf{N}_G(P))$ that commutes with the action of $\mathcal{G} = \text{Gal}(\mathbb{Q}_{|G|}/\mathbb{Q})$. For instance, $G = \text{GL}_2(3)$ for $p = 3$ has exactly four rational-valued irreducible characters of degree not divisible by p, while $\mathbf{N}_G(P) = C_2 \times S_3$ has six rational-valued irreducible characters of p'-degree. (Whether or not there are correspondences that commute with the action of $\text{Gal}(\mathbf{Q}_p(\xi)/\mathbf{Q}_p)$, where $\mathbf{Q}_p$ is the field of p-adics and ξ is a primitive $|G|$th root of unity, is the subject of another conjecture.) The fact that $p = 3$ in the above solvable example deserves an explanation: I. M. Isaacs proved that there is a canonical bijection $\text{Irr}_{p'}(G) \rightarrow \text{Irr}_{p'}(\mathbf{N}_G(P))$ whenever G is solvable and $p = 2$. (Other canonical bijections for $p = 2$ have been obtained for special classes of groups, as symmetric and general linear groups in odd characteristic.)

By Clifford's theorem and Itô's Corollary 1.21, we notice that $\text{Irr}_{p'}(\mathbf{N}_G(P)) = \text{Irr}(\mathbf{N}_G(P)/P')$, and therefore the section $\mathbf{N}_G(P)/P'$ and the complex group algebra $\mathbb{C}[\mathbf{N}_G(P)/P']$ might play a role in the McKay conjecture.

Two cases are quite natural to consider in the McKay conjecture: when $P' = 1$, that is, when P is abelian, or when $P = \mathbf{N}_G(P)$, that is, when P is self-normalizing. If P is cyclic, then the McKay conjecture is true (with a nontrivial proof), and it follows from Dade's deep *cyclic defect group theory*. If P is abelian, then M. Broué has proposed a structural explanation for the McKay conjecture, which has further consequences at the character theory level (the existence of the so called *perfect isometries*) and that extends the cyclic defect theory. Broué's conjecture (and the McKay conjecture) is still open for P abelian.

On the other hand, the case $\mathbf{N}_G(P) = P$ is now solved. It started with a fairly easy but surprising result: Theorem 9.4.

We shall use the next lemma frequently. It is elementary, but it gives the first connection between irreducible characters of p'-degree and Sylow normalizers.

Lemma 9.3 *Let G be a finite group, let $P \in \mathrm{Syl}_p(G)$ and let $L \trianglelefteq G$. If $\chi \in \mathrm{Irr}_{p'}(G)$, then χ_L has a P-invariant irreducible constituent and any two of them are $\mathbf{N}_G(P)$-conjugate.*

Proof Let $\theta \in \mathrm{Irr}(L)$. Since $[\chi_L, \theta] = [\chi_L, \theta^x]$ for $x \in G$, we can write $\chi_L = e(\Delta_1 + \cdots + \Delta_k)$, where Δ_i is the sum of the elements in the P-orbit $\mathcal{O}_i$ of some $\eta_i \in \mathrm{Irr}(L)$. Since $\Delta_i(1) = \eta_i(1)|\mathcal{O}_i|$ and p does not divide $\chi(1)$, it follows that there exists a P-orbit of length 1. Now, if θ and η are irreducible P-invariant constituents of χ_L, by Clifford's theorem we have that $\theta^g = \eta$ for some $g \in G$. If G_θ is the stabilizer of θ in G, we have that P and P^g are contained in the stabilizer $T = G_\eta$. By Sylow's theorem $(P^g)^x = P$ for some $x \in T$. Then

$$\theta^{gx} = (\theta^g)^x = \eta^x = \eta$$

and $gx \in \mathbf{N}_G(P)$, as desired. $\qquad\square$

Recall that a finite group G is p-**solvable** if it has a normal series whose quotients are p-groups or p'-groups. Subgroups and quotients of p-solvable groups are p-solvable.

Theorem 9.4 *Let G be a p-solvable group, let $P \in \mathrm{Syl}_p(G)$, and assume that $\mathbf{N}_G(P) = P$. If $\chi \in \mathrm{Irr}_{p'}(G)$, then there exists a linear character $\lambda \in \mathrm{Irr}(P)$ such that*

$$\chi_P = \lambda + \Delta,$$

where Δ is a character of P (or zero) such that all of its irreducible constituents have degree divisible by p. Furthermore, the map $\chi \mapsto \lambda$ is a canonical bijection $\mathrm{Irr}_{p'}(G) \to \mathrm{Lin}(P)$.

Proof If $\chi \in \mathrm{Irr}_{p'}(G)$, it is clear that χ_P has a linear irreducible constituent λ, since otherwise p would divide $\chi(1)$.

First, we argue by induction on $|G|$ that if $\chi \in \mathrm{Irr}_{p'}(G)$, then λ is unique and occurs with multiplicity 1 in the restriction χ_P. Let $K = \mathbf{O}_{p'}(G)$ be the largest normal subgroup of G of order not divisible by p. We claim that $K \subseteq \ker(\chi)$. By Lemma 9.3, let $\theta \in \mathrm{Irr}(K)$ be P-invariant and lie under χ. Since $\mathbf{N}_G(P) = P$, then notice that $\mathbf{C}_K(P) = 1$. By the Glauberman correspondence, we have that $\theta = 1_K$, and the claim follows. Hence, we may assume that $K = 1$ by induction.

Let $L = \mathbf{O}_p(G)$. Since G is p-solvable and $\mathbf{O}_{p'}(G) = 1$, then $L > 1$. Write $v = \lambda_L$, and let $T = G_v$ be the stabilizer of v in G, which contains P. By Lemma 9.3 and the hypothesis, notice that v is the only P-invariant irreducible constituent of χ_L. Let $\psi \in \mathrm{Irr}(T|v)$ be the Clifford correspondent of χ over v. By the Clifford correspondence Theorem 1.20(c), notice that

$$\chi_T = \psi + \Xi,$$

where no irreducible constituent of Ξ lies over v. Notice that if Ξ_P has a linear constituent ρ, then ρ_L is a P-invariant irreducible constituent of χ_L, and therefore ρ_L should be v (because v is the only P-invariant character under χ).

Now, by Lemma 5.11 we have that v extends to T. Let $\mu \in \mathrm{Irr}(T)$ be a (linear) extension of v to G, Now, by the Gallagher correspondence, there exists $\tau \in \mathrm{Irr}(T/L)$ such that $\psi = \tau\mu$. Since $|T/L| < |G|$ and P/L is self-normalizing in T/L, by induction we have that

$$\tau_P = \tau_1 + \Delta_1$$

where $\tau_1 \in \mathrm{Irr}(P)$ is linear, and Δ_1 is a character of P (or zero) such that its irreducible constituents have degree divisible by p. Now

$$\chi_P = \tau_P\mu_P + \Xi_P = \tau_1\mu_P + \mu_P\Delta_1 + \Xi_P,$$

and we deduce that χ_P contains a unique linear constituent $\tau_1\mu_P$ (which also occurs with multiplicity one). Notice that $\tau_1\mu_P = \lambda$, because λ was chosen as a linear constituent of χ_P.

Suppose that $\chi' \in \mathrm{Irr}_{p'}(G)$ is such that $\chi'_P = \lambda + \Delta'$. We wish to prove that $\chi = \chi'$. Again, we argue by induction on $|G|$. Notice that χ' also lies over $v = \lambda_L$. If $\psi' \in \mathrm{Irr}(T)$ is the Clifford correspondent of χ' over v, then we can write $\psi' = \tau'\mu$, where $\tau' \in \mathrm{Irr}(T/L)$. Also, τ'_P contains a unique linear constituent τ'_0. By the first part of the proof, we have that $\tau'_0\mu_P = \lambda = \tau_0\mu_P$. Since μ is linear, we have that $\tau'_0 = \tau_0$, and by induction, $\tau = \tau'$. Then $\chi' = (\tau'\mu)^G = (\tau\mu)^G = \chi$, and our map is injective.

Finally, if $\lambda \in \mathrm{Irr}(P)$ is linear, then λ^G has degree $|G : P|$, which is not divisible by p. Therefore λ^G contains some $\chi \in \mathrm{Irr}_{p'}(G)$. By the first part, $\chi_P = \lambda + \Delta$, and therefore our map is surjective.   □

This type of correspondence of characters does not happen outside p-solvable groups. For instance, if $G = \mathsf{S}_5$ then the irreducible characters of G of degree 5 have three linear irreducible constituents when restricted to any Sylow 2-subgroup of G. (In symmetric groups, for $p = 2$, the conclusion of Theorem 9.4 does happen for $G = \mathsf{S}_{2^n}$, and this is an observation of J. L. Alperin, whose proof can be found in [Gi17].)

It turns out, by using the CFSG, that if $p > 3$ and $P = \mathbf{N}_G(P)$, then G is solvable. (This is a theorem of R. Guralnick, G. Malle and me ([GMN04]).) Hence, Theorem 9.4 settles the self-normalizing case of the McKay conjecture for $p > 3$. There are non-solvable groups having self-normalizing Sylow 3-subgroups, and the paradigmatic example is $G = \mathrm{PSL}_2(27).3$, which is the extension of $\mathrm{PSL}_2(27)$ by a field automorphism of order 3. Nevertheless, the assertion of Theorem 9.4 also holds for $p = 3$ for general finite groups, although its proof is much harder. Finally, there are many non-solvable groups having self-normalizing Sylow 2-subgroups. But in this case, the McKay conjecture is known to hold by the Malle–Späth solution of the McKay conjecture for $p = 2$. Hence, the full self-normalizing case of the McKay conjecture is now complete.

9.2 Strengthening the McKay Conjecture

If there is no proof of the McKay conjecture, does it have any value to find possible generalizations? In theory, it is hardly debatable that the stronger a conjecture is, the *easier* it should be to find a proof of it.

The most significant of the generalizations of the McKay conjecture was soon made by J. L. Alperin after McKay's announcement. In this book, we are restricting ourselves to ordinary character theory, but of course, in order to fully understand the global–local phenomena, modular representation theory is essential. Let us review the pieces of block theory necessary in order to be able to state the Alperin–McKay conjecture.

We have already written about the principal block in this book, in Section 5 of Chapter 6, or in Chapter 7, when we mentioned the height zero conjecture. From the character theoretical point of view, Brauer's p-blocks can be introduced mainly in two ways. If G is a finite group and $G_{p'}$ is the set of p-regular elements of G, then we **link** $\alpha, \beta \in \mathrm{Irr}(G)$ if

$$\sum_{x \in G_{p'}} \alpha(x)\beta(x^{-1}) \neq 0.$$

The connected components in $\mathrm{Irr}(G)$ are the p-**blocks** of G. Also, p-blocks are the equivalence classes in $\mathrm{Irr}(G)$ under the relation $\alpha \equiv \beta$ if and only if $\lambda_\alpha = \lambda_\beta$, where λ_α is the algebra homomorphism introduced before Lemma 2.7. (For references of these facts, see Definition 3.1 and Theorem 3.19 of [Na98].)

Now, every p-block B of G has canonically associated a G-conjugacy class of p-subgroups of G, which are called the **defect groups** of B. Moreover, every block b of a local subgroup $\mathbf{N}_G(Q)$ of G, where Q is a p-subgroup of

G, uniquely determines a block B of G, which is called the **induced block**, and which is written $B = b^G$. Brauer's first main theorem asserts that $b \mapsto b^G$ is a bijection between the blocks of $\mathbf{N}_G(D)$ with defect group D, and the blocks of G with defect group D, in what constitutes one of the first global–local bijections in the representation theory of finite groups. The block b of $\mathbf{N}_G(D)$ is called the **Brauer correspondent** of B. (See Chapter 4 of [Na98].)

If D is a defect group of the block B, and we write $|D| = p^d$ and $|G|_p = p^a$, it turns out that the minimum of $\chi(1)_p$ for $\chi \in \mathrm{Irr}(B)$ is p^{a-d}. Therefore, if $\chi \in \mathrm{Irr}(B)$ then $\chi(1)_p = p^{a-d+h}$, for some nonnegative integer h called the **height** of χ. Now, we are ready to state the Alperin–McKay conjecture.

Conjecture 9.5 (Alperin–McKay) *Let B be a block of a finite group G with defect group D, and let b be the Brauer correspondent block of $\mathbf{N}_G(D)$. Then the number of irreducible characters of height zero in B and b coincide.*

By definition, notice that characters of degree not divisible by p necessarily lie in the p-blocks having $P \in \mathrm{Syl}_p(G)$ as a defect group. Hence, $\mathrm{Irr}_{p'}(G)$ is the union of the height zero characters in the blocks with defect group $P \in \mathrm{Syl}_p(G)$, and it easily follows, using Brauer's first main theorem, that the Alperin–McKay conjecture implies the McKay conjecture.

The reader who is not familiar with block theory might not appreciate the deep step that Alperin took when proposing the Alperin–McKay conjecture. What lies behind it is that each block uniquely defines (and is defined by) a two-sided indecomposable ideal of the group algebra FG, where F is an algebraically closed field of characteristic p. Hence, blocks are also algebras over F, and to find the exact relationship between the algebras B and b is one of the main problems in representation theory. In the case when their defect common group is abelian, this exact relationship is the content of Broué's **abelian defect group conjecture**: the module categories of B and b are *derived equivalent*.

Coming back to ordinary character theory, an apparently innocent generalization of the McKay conjecture is the key for its reduction to finite simple groups. It was first used by T. R. Wolf to prove the solvable case of the McKay conjecture. Recall, again, that if $L \trianglelefteq G$ and $\theta \in \mathrm{Irr}(L)$ then $\mathrm{Irr}_{p'}(G|\theta) = \mathrm{Irr}_{p'}(G) \cap \mathrm{Irr}(G|\theta)$.

Conjecture 9.6 (Relative McKay) *Let G be a finite group, p a prime and $P \in \mathrm{Syl}_p(G)$. Let $L \trianglelefteq G$, and let $\theta \in \mathrm{Irr}(L)$ be P-invariant. Then*

$$|\mathrm{Irr}_{p'}(G|\theta)| = |\mathrm{Irr}_{p'}(L\mathbf{N}_G(P)|\theta)| .$$

This is a stronger statement than Conjecture 9.1, which of course is recovered when $L = 1$. As we shall see in Chapter 10, the fact that Conjecture 9.6 involves normal subgroups allows an inductive argument on $|G : L|$ that eventually brings simple groups into the problem.

Thirty years after McKay's announcement, the following observation was made. What lies behind it has not been yet explained.

Conjecture 9.7 (Isaacs–Navarro) *Let G be a finite group and let p be a prime. If k is any integer, let $M_k(G) = |\{\chi \in \text{Irr}_{p'}(G) \mid \chi(1) \equiv \pm k \mod p\}|$. If $P \in \text{Syl}_p(G)$, then*

$$M_k(G) = M_k(\mathbf{N}_G(P)).$$

For instance, the group $G = A_5$ has exactly four irreducible characters of degree not divisible by $p = 5$, having degrees 1, 3, 3, and 4. If $P \in \text{Syl}_p(G)$, then $\mathbf{N}_G(P)$ is a dihedral group of order 10 which has four irreducible characters, of degrees 1, 1, 2, 2. Hence $|\text{Irr}_{p'}(G)| = |\text{Irr}_{p'}(\mathbf{N}_G(P))| = 4$, as predicted by Conjecture 9.1. Conjecture 9.7 now predicts that G must have two characters of degrees ± 1 mod 5, and two characters of degrees ± 2 mod 5, which is indeed the case.

It is a fact that many results on characters have a dual version on conjugacy classes. For instance, it is not difficult to show that if G is a finite group and $P \in \text{Syl}_p(G)$, then the map $K \mapsto K \cap \mathbf{C}_G(P)$ is a bijection between the conjugacy classes of G of p'-size and the conjugacy classes of $\mathbf{N}_G(P)$ of p'-size. Also, $|K| \equiv |K \cap \mathbf{C}_G(P)|$ mod p. (The proof is similar to the proof of Lemma 2.5.) Hence, we see that there is a McKay theorem for conjugacy classes, even with congruences.

Now, we bring Galois action into the McKay conjecture. As we explained in the first section, in general there are no bijections $' : \text{Irr}_{p'}(G) \to \text{Irr}_{p'}(\mathbf{N}_G(P))$ that commute with the action of $\mathcal{G} = \text{Gal}(\mathbb{Q}_{|G|}/\mathbb{Q})$. In some sense, such an expectation is not reasonable, since $\mathcal{G}$ is *forgetting* p. The right group to look at is not $\mathcal{G}$ but $\mathcal{G}_p$, which is defined as the subgroup of $\mathcal{G}$ consisting of those σ such that $\sigma(\mathcal{P}) = \mathcal{P}$, where $\mathcal{P}$ is any prime ideal containing p in the ring of algebraic integers in $\mathbb{Q}_{|G|}$. Using elementary algebraic number theory, it is easy to check that $\mathcal{G}_p$ exactly consists of those $\sigma \in \mathcal{G}$ that send every root of unity $\xi \in \mathbb{Q}_{|G|}$ of order not divisible by p to ξ^{p^f}, where p^f is a fixed but arbitrary power of p.

Conjecture 9.8 (McKay–Galois) *Let G be a finite group, let p be a prime, and let $P \in \text{Syl}_p(G)$. Then the actions of $\mathcal{G}_p$ on $\text{Irr}_{p'}(G)$ and $\text{Irr}_{p'}(\mathbf{N}_G(P))$ are permutation isomorphic.*

In other words, Conjecture 9.8 proposes that there exists a bijection $\mathrm{Irr}_{p'}(G) \to \mathrm{Irr}_{p'}(\mathbf{N}_G(P))$ that commutes with the $\mathcal{G}_p$-action. This implies that character values between corresponding characters are connected.

The McKay–Galois conjecture allows us to detect certain local properties in character tables, which was one of the topics in Section 5 of Chapter 7. Recall that an irreducible character χ of a finite group G is p-rational if $\mathbb{Q}(\chi) \subseteq \mathbb{Q}_m$ for some m not divisible by p. Hence, $\chi \in \mathrm{Irr}(G)$ is p-rational if and only if it is fixed by $\mathrm{Gal}(\mathbb{Q}_{|G|}/\mathbb{Q}_{|G|_{p'}})$. Since $\mathrm{Gal}(\mathbb{Q}_{|G|}/\mathbb{Q}_{|G|_{p'}}) \subseteq \mathcal{G}_p$, we have that Conjecture 9.8 implies that the number of p-rational characters of p'-degree in G and in $\mathbf{N}_G(P)$ is the same. Of course, notice that all the characters of the p'-group $\mathbf{N}_G(P)/P$ are p-rational. Hence, if p is odd, observe that $\mathbf{N}_G(P) = P$ if and only if the trivial character is the only p'-degree p-rational character of $\mathbf{N}_G(P)$ (using that groups of odd order do not have rational-valued characters by Corollary 3.5). In particular, Conjecture 9.8 implies that for p odd, $\mathbf{N}_G(P) = P$ if and only if the trivial character of G is the only p-rational irreducible character of p'-degree. This is now a theorem.

Theorem 9.9 (Navarro–Tiep–Turull)   *Let G be a finite group, and let $P \in \mathrm{Syl}_p(G)$, where p is odd. Then $\mathbf{N}_G(P) = P$ if and only if the trivial character of G is the only p-rational irreducible character of G of p'-degree.*

We see that we can detect from the character table of G if G has self-normalizing Sylow p-subgroups, for odd primes. The case $p = 2$ is more subtle. We start with an easy lemma.

Lemma 9.10   *Let G be a finite group, and assume that g^2 is conjugate to g for every $g \in G$. Then G is the trivial group.*

Proof   By hypothesis, we see that G has no elements of order 2, and therefore G has odd order. Assume that $G > 1$, and let p be the smallest prime divisor of $|G|$. Let $g \in G$ of order p. Then $g^x = g^2$ for some $x \in G$, and therefore $x \in \mathbf{N}_G(\langle g \rangle)$. Now, $\mathbf{N}_G(\langle g \rangle)/\mathbf{C}_G(g)$ has order dividing $p - 1$. Since p is the smallest prime dividing $|G|$, we have that $\mathbf{N}_G(\langle g \rangle) = \mathbf{C}_G(g)$. Thus $g = g^2$, and $g = 1$, a contradiction. $\square$

Let σ be the Galois automorphism of an arbitrary cyclotomic field that fixes 2-roots of unity and squares the roots of unity of odd order, and notice that $\sigma \in \mathcal{G}_2$ by definition. In particular, if G is a finite group, then Conjecture 9.8 implies that the number of odd-degree irreducible characters of G and of $\mathbf{N}_G(P)$ that are fixed by σ coincide. We claim that all odd-degree irreducible characters of $\mathbf{N}_G(P)$ are σ-fixed if and only if $\mathbf{N}_G(P) = P$. If $\mathbf{N}_G(P) = P$ and $\lambda \in \mathrm{Irr}(P)$ has odd degree, then λ is linear, and thus $\lambda^\sigma = \lambda$ because σ fixes 2-power roots of unity. Conversely, if all the odd-degree characters of $\mathrm{Irr}_{p'}(\mathbf{N}_G(P))$ are σ-fixed, then it follows that all characters in $\mathrm{Irr}(\mathbf{N}_G(P)/P)$ are σ-fixed. By Brauer's Lemma on character tables (Theorem 2.3), we have that every conjugacy class of the odd-order group $\mathbf{N}_G(P)/P$ is σ-fixed. In particular, g^2 is conjugate to g for every $g \in \mathbf{N}_G(P)/P$. By Lemma 9.10, we conclude that $\mathbf{N}_G(P) = P$.

This consequence of Conjecture 9.8 is now a theorem.

Theorem 9.11 (A. Schaeffer-Fry)   *Let G be a finite group, $p = 2$, and $P \in$ $\mathrm{Syl}_p(G)$. Let $\sigma \in \mathrm{Gal}(\mathbb{Q}_{|G|}/\mathbb{Q})$ be the Galois automorphism that fixes 2-roots of unity, and squares odd-order roots of unity. Then $\mathbf{N}_G(P) = P$ if and only if every $\chi \in \mathrm{Irr}_{p'}(G)$ is fixed by σ.*

Among the consequences of Conjecture 9.8, there is one more related to Brauer's Problem 12, about obtaining information on Sylow subgroups from the character table of a group.

Theorem 9.12   *Let G be a finite group, and let $P \in \mathrm{Syl}_p(G)$. If $e \geq 1$, let σ be the element of $\mathrm{Gal}(\mathbb{Q}_{|G|}/\mathbb{Q})$ that fixes roots of unity of order not divisible by p, and sends p-power roots of unity ξ to ξ^{1+p^e}. Assume that Conjecture 9.8 holds for G. Then the exponent of P/P' is less than or equal to p^e if and only if all the irreducible characters of p'-degree of G are σ-fixed.*

Proof   First notice that if λ is a linear character of p-power order of any group, then $\lambda^\sigma = \lambda^{1+p^e}$, and therefore $\lambda = \lambda^\sigma$ if and only if $\lambda^{p^e} = 1$. In particular, the exponent of P/P' is less than or equal to p^e if and only if all the linear characters of P are σ-fixed.

By elementary number theory, it is easy to check that σ has order a power of p.

Now, by using Conjecture 9.8, we may assume that $P \trianglelefteq G$. Suppose first that all the irreducible characters of p'-degree of G are σ-fixed. If $\lambda \in \mathrm{Irr}(P)$ is linear, we let $\chi \in \mathrm{Irr}(G)$ be over λ. By Theorem 5.12, we have that χ has p'-degree. Hence, $\chi^\sigma = \chi$, by hypothesis. By Clifford's theorem, we have that λ^σ and λ are G-conjugate. Therefore $\lambda^\sigma = \lambda^x$ for some $x \in G$. Since $x_p \in P$, we may assume that x is an element of order not divisible by p. Now, $(\lambda^\sigma)^x = (\lambda^x)^\sigma$ and we conclude that $\lambda^{\sigma^n} = \lambda^{x^n}$ for every integer n. In particular, if $n = o(\sigma)$, then we have that $\lambda^{x^n} = \lambda$. Since n is a p-power and x is a p'-element, we have that $\langle x \rangle = \langle x^n \rangle$, and therefore $\lambda^x = \lambda$. Thus $\lambda^\sigma = \lambda$.

Conversely, assume that all the linear characters of P are σ-fixed. Let $\chi \in \mathrm{Irr}_{p'}(G)$. We wish to prove that $\chi^{\sigma} = \chi$. If $\lambda \in \mathrm{Irr}(P)$ is under χ, then we have that λ is linear. Let $T = G_{\lambda}$ be the stabilizer of λ in G. By Corollary 6.4, we have that λ has an extension $\hat{\lambda} \in \mathrm{Irr}(T)$ which is σ-fixed. Now, by the Clifford and the Gallagher correspondences we can write $\chi = (\hat{\lambda}\tau)^{G}$, where $\tau \in \mathrm{Irr}(T/P)$. Since T/P is a p'-group, we have that τ is also σ-fixed. Hence $\hat{\lambda}\tau$ and $(\hat{\lambda}\tau)^{G} = \chi$ are also σ-fixed. $\qquad\qquad\square$

Using some number theory, it is not difficult to show that Conjecture 9.8 is equivalent to proving that there is a bijection $f : \mathrm{Irr}_{p'}(G) \to \mathrm{Irr}_{p'}(\mathbf{N}_G(P))$ such that $\mathbf{Q}_p(\chi) = \mathbf{Q}_p(f(\chi))$, where $\mathbf{Q}_p$ is the field of p-adics. A. Turull goes even further and conjectures that χ and $f(\chi)$ should have the same *Schur index* over $\mathbf{Q}_p$. All this is hinting that, somehow, the McKay conjecture has to be solved by working over $\mathbf{Q}_p$, but nobody has figured out exactly how.

There is one further stronger version of McKay which is is also crucial in its reduction to finite simple groups.

Conjecture 9.13 (McKay with automorphisms)   *Let G be a finite group, let p be a prime, and let $P \in \mathrm{Syl}_p(G)$. Let A be a finite group that acts by automorphisms on G such that $P^a = P$ for all $a \in A$. Then the actions of A on $\mathrm{Irr}_{p'}(G)$ and $\mathrm{Irr}_{p'}(\mathbf{N}_G(P))$ are permutation isomorphic.*

Recall from Chapter 2 that if a group A acts on two sets Ω and Λ, then a bijection $f : \Omega \to \Lambda$ is A-equivariant if $f(\alpha \cdot a) = f(\alpha) \cdot a$ for every $\alpha \in \Omega$ and $a \in A$. If $A = \mathrm{Aut}(G)_P$ is the subgroup of the automorphisms of G that leave P invariant (set-wise), then another way to state Conjecture 9.13 is that there should exist an A-equivariant bijection $f : \mathrm{Irr}_{p'}(G) \to \mathrm{Irr}_{p'}(\mathbf{N}_G(P))$.

To end this section, we point out that all the variations of the McKay conjecture that we have mentioned so far seem to hold if they are combined with each other. For instance, it appears that there should be a bijection $f : \mathrm{Irr}_{p'}(G) \to \mathrm{Irr}_{p'}(\mathbf{N}_G(P))$ that respects plus or minus degrees modulo p, fields of values over the p-adics, and Brauer blocks, and that commutes with the subgroup A of automorphisms of G that fix P, etc. In fact, the same type of variation seems to hold for the remaining global–local counting conjectures, which we shall discuss next.

9.3 The Alperin Weight Conjecture

Let G be a finite group, and let p be a prime. We let $l(G)$ be the number of conjugacy classes of G consisting of p-regular elements. (Equivalently, $l(G)$ is the number of non-similar irreducible representations of G over an algebraically closed field of characteristic p. See, for instance, Corollary 2.10 of [Na98].) Inspired by the modular representation theory of groups of Lie type, J. Alperin formulated in 1987 his celebrated Alperin weight conjecture, which is often referred to as AWC. If G is a finite group, we let $k_0(G)$ be the number of defect zero characters of G.

Conjecture 9.14 (Alperin's weight conjecture) *Let G be a finite group, and let p be a prime. Then*

$$l(G) = \sum_Q \frac{|\mathbf{N}_G(Q)|}{|G|} k_0(\mathbf{N}_G(Q)/Q),$$

where Q runs over all the p-subgroups of G.

If we prefer to work with a set of representatives $\mathcal{X}$ of the orbits of the action of G on the p-subgroups of G, then AWC proposes that

$$l(G) = \sum_{Q \in \mathcal{X}} k_0(\mathbf{N}_G(Q)/Q).$$

In the right-hand side of the above equation, we only need to consider a very particular class of p-subgroups of G. To show this, we need a lemma.

Lemma 9.15 *Let G be a finite group.*

(a) *Let $N \trianglelefteq G$. Let $\chi \in \mathrm{Irr}(G)$ and let $\theta \in \mathrm{Irr}(N)$ be G-invariant under χ. If $(\chi(1)/\theta(1))_p = |G : N|_p$, then $\mathbf{O}_p(G/N) = 1$. In particular, if G has a p-defect zero character, then $\mathbf{O}_p(G) = 1$.*

(b) *Suppose that Q is a p-subgroup of G. Let $\gamma \in \mathrm{Irr}(\mathbf{N}_G(Q))$ and let $\tau \in \mathrm{Irr}(Q)$ be under γ. If τ is $\mathbf{N}_G(Q)$-invariant and $(\gamma(1)/\tau(1))_p = |\mathbf{N}_G(Q) : Q|_p$, then $Q = \mathbf{O}_p(\mathbf{N}_G(Q))$. Hence, $\mathbf{O}_p(G) \subseteq Q$.*

Proof First we prove (a). Suppose that $N \leq M \trianglelefteq G$, and let $\mu \in \mathrm{Irr}(M)$ be under χ. By Theorem 5.12, we have that $\chi(1)/\mu(1)$ divides $|G : M|$ and $\mu(1)/\theta(1)$ divides $|M : N|$. Now

$$|G : M|_p |M : N|_p = |G : N|_p = (\chi(1)/\theta(1))_p = (\chi(1)/\mu(1))_p (\mu(1)/\theta(1))_p,$$

and we deduce that $(\chi(1)/\mu(1))_p = |G : M|_p$ and $(\mu(1)/\theta(1))_p = |M : N|_p$.

Let $M/N = \mathbf{O}_p(G/N)$ now, and assume that $M > N$. If $K/N \trianglelefteq M/N$ has order p, then let $\nu \in \mathrm{Irr}(K)$ be under μ. By the previous paragraph, we have that

$$\nu(1)/\theta(1) = (\nu(1)/\theta(1))_p = |K : N|_p = p.$$

Now, since θ is K-invariant and K/N is cyclic, we have that θ extends to K by Theorem 5.1. Then $v(1) = \theta(1)$, and this is a contradiction. Thus $M = N$. For the second part of (a), apply the first to $N = 1$.

To prove (b), we use part (a) to deduce that $\mathbf{O}_p(\mathbf{N}_G(Q)/Q) = 1$. Hence $Q = \mathbf{O}_p(\mathbf{N}_G(Q))$. Let $L = \mathbf{O}_p(G)$. If $Q < QL$, then $Q < \mathbf{N}_{QL}(Q) \trianglelefteq \mathbf{N}_G(Q)$ (using that normalizers *grow* in p-groups), and this is not possible. Hence, we see that $L \subseteq Q$. $\square$

Another way to prove Lemma 9.15 is by using the Brauer–Nesbitt theorem. Indeed, if $\chi \in \mathrm{Irr}(G)$ has p-defect zero and L is a normal p-subgroup of G, then χ_L is zero on the nonidentity elements of L. Thus χ_L is a multiple of the regular character ρ_L. By Clifford's theorem, ρ_L is a multiple of 1_L and necessarily $L = 1$.

By Lemma 9.15(b), we deduce that if $k_0(\mathbf{N}_G(Q)/Q) \neq 0$ then

$$Q = \mathbf{O}_p(\mathbf{N}_G(Q)).$$

In this case, Q is said to be p-**radical** in G. Therefore, one only needs to worry about the p-radical subgroups in the right-hand side of AWC.

In 1989 R. Knörr and G. Robinson proposed in [KR89] a different point of view on AWC which turned out to be fundamental. Let G be a finite group, and let p be an arbitrary but fixed prime. A **chain** C of p-subgroups of G is

$$1 = P_0 < P_1 < \cdots < P_n\,,$$

where each P_i is a p-subgroup of G. (In this chapter we shall understand that chains are chains of p-subgroups for our fixed prime p.) We write $|C| = n$ for the **length** of C. The chain C that consists of the trivial subgroup only is the **trivial chain** and has length 0. We also say that P_n is the **last term** of C. We let $\mathcal{C}(G)$ be the set of all chains of p-subgroups of G. Of course, G acts naturally on $\mathcal{C}(G)$ by conjugation. If C is the chain $1 = P_0 < P_1 < \cdots < P_n$, and $g \in G$, then C^g is the chain $1 = P_0 < P_1^g < \cdots < P_n^g$, and the stabilizer (or normalizer) of C in G is

$$G_C = \mathbf{N}_G(C) = \mathbf{N}_G(P_1) \cap \cdots \cap \mathbf{N}_G(P_n)\,.$$

When using chains, it is often the case that authors use G_C instead of $\mathbf{N}_G(C)$. We shall use also this notation, except for the chains $1 < P$ whose normalizer is $\mathbf{N}_G(P)$.

There are other types of chains of p-subgroups. For instance, we let $\mathcal{N}(G)$ be the set of **normal chains**, which consists of those chains $1 = P_0 < P_1 < \cdots < P_n$ such that $P_i \trianglelefteq P_n$ for all i. In this case notice that $P_n \subseteq G_C$ and

$P_i \trianglelefteq G_C$ for all i. Also, $\mathcal{E}(G)$ is the set of **elementary abelian chains** $1 = P_0 < P_1 < \cdots < P_n$, where P_n is elementary abelian. Of course, $\mathcal{E}(G) \subseteq \mathcal{N}(G)$.

We will see that, for the applications that we are going to present, it does not really matter which type of chain we choose.

From now on, if G acts on a set Ω, then Ω/G denotes a complete set of representatives of the orbits of G on Ω.

Suppose that G is a finite group, $\mathcal{S}$ is the set of the subgroups of G, and A is an abelian group. Let $f : \mathcal{S} \to A$ be a G-**stable function**, which means that $f(S) = f(S^g)$ for all $S \in \mathcal{S}$ and $g \in G$. In what follows, we are going to deal with expressions of the form

$$\sum_{C \in \mathcal{C}(G)/G} (-1)^{|C|} f(G_C)$$

as in Theorem 9.16. Notice that this expression does not depend on the set of representatives that we choose. Usually the abelian group A is going to be the rational numbers, or the space of class functions of a group. In these cases, we can multiply elements of A by rational numbers and have

$$\sum_{C \in \mathcal{C}(G)/G} (-1)^{|C|} f(G_C) = \sum_{C \in \mathcal{C}(G)} (-1)^{|C|} \frac{|G_C|}{|G|} f(G_C),$$

by using that $f(H) = f(H^g)$ for $g \in G$. In what follows, we shall use either of these two expressions.

Theorem 9.16 *Let G be a finite group and let $\mathcal{S}$ be the set of all subgroups of G. Suppose that $f : \mathcal{S} \to A$ is any G-stable function, where A is an abelian group. Then*

$$\sum_{C \in \mathcal{C}(G)/G} (-1)^{|C|} f(G_C) = \sum_{C \in \mathcal{N}(G)/G} (-1)^{|C|} f(G_C) = \sum_{C \in \mathcal{E}(G)/G} (-1)^{|C|} f(G_C).$$

If $\mathbf{O}_p(G) > 1$, then this sum is 0.

Proof Suppose that $C \in \mathcal{C}(G) - \mathcal{E}(G)$, where C is the chain $1 = P_0 < P_1 < \cdots < P_n$. Let $N = \Phi(P_n)$ be the Frattini subgroup of P_n. Since C is not an elementary chain, then $N > 1$. Also, $NP_{n-1} < P_n$, by elementary group theory. Now, we define $C' \in \mathcal{C}(G) - \mathcal{E}(G)$ in the following way. Let $1 \le m \le n$ be the unique integer such that $N \subseteq P_m$ and $N \nsubseteq P_{m-1}$. If $NP_{m-1} < P_m$, then we let C' be the chain in which we insert NP_{m-1} between P_{m-1} and P_m. If $NP_{m-1} = P_m$, then we construct C' by deleting P_m from C. Now, notice that C and C' have the same last term P_n (because $P_{n-1}N < P_n$),

$|C| = |C'| \pm 1$, and that $C \in \mathcal{N}(G)$ if and only if $C' \in \mathcal{N}(G)$. We ask the reader to check that $(C')' = C$ and $(C^g)' = (C')^g$ for $g \in G$. Therefore

$$' : \mathcal{C}(G) - \mathcal{E}(G) \to \mathcal{C}(G) - \mathcal{E}(G)$$

is a bijection that commutes with G-conjugation. Hence $G_C = G_{C'}$. Also, it follows that if Z is a complete set of representatives of the orbits of the action of G on $\mathcal{C}(G) - \mathcal{E}(G)$, then $Z' = \{C' \mid C \in Z\}$ also is. Hence

$$\sum_{C \in Z} (-1)^{|C|} f(G_C) = \sum_{C \in Z} (-1)^{|C'|} f(G_{C'}) = -\sum_{C \in Z} (-1)^{|C|} f(G_C),$$

and we deduce that

$$\sum_{C \in Z} (-1)^{|C|} f(G_C) = 0.$$

Now, if Y is a complete set of representatives of the orbits of the action of G on $\mathcal{E}(G)$, then $X = Y \cup Z$ is a complete set of representatives of the orbits of the action of G on $\mathcal{C}(G)$ and

$$\sum_{C \in X} (-1)^{|C|} f(G_C) = \sum_{C \in Y} (-1)^{|C|} f(G_C).$$

Since $\mathcal{E}(G) \subseteq \mathcal{N}(G)$ and $'$ is also a bijection $\mathcal{N}(G) - \mathcal{E}(G) \to \mathcal{N}(G) - \mathcal{E}(G)$, the same argument shows that

$$\sum_{C \in X_0} (-1)^{|C|} f(G_C) = \sum_{C \in Y} (-1)^{|C|} f(G_C),$$

where now $X_0 = Y \cup Z_0$, and Z_0 is a complete set of representatives of the orbits of the action of G on $\mathcal{N}(G) - \mathcal{E}(G)$. This shows the first part of the theorem.

Suppose finally that $L = \mathbf{O}_p(G) > 1$. If $C \in \mathcal{C}(G)$ is a chain $1 = P_0 < P_1 < \cdots < P_n$, then we are going to define another chain $C^* \in \mathcal{C}(G)$ such that $|C| = |C^*| \pm 1$, $(C^*)^g = (C^g)^*$ and $(C^*)^* = C$. In this case, $*$ is a bijection $\mathcal{C}(G) \to \mathcal{C}(G)$, $G_{C^*} = G_C$, and if R is a complete set of representatives of the orbits of the action of G on $\mathcal{C}(G)$, then so is R^*. Hence

$$\sum_{C \in R} (-1)^{|C|} f(G_C) = \sum_{C \in R} (-1)^{|C^*|} f(G_{C^*}) = -\sum_{C \in R} (-1)^{|C|} f(G_C),$$

and the proof of the theorem will be complete. The chain C^* is defined as follows. If L is not contained in P_n, then C^* is formed by adding $L P_n$ to the end of the chain. Notice that

$$G_C = \mathbf{N}_G(P_1) \cap \cdots \cap \mathbf{N}_G(P_n) = \mathbf{N}_G(P_1) \cap \cdots \cap \mathbf{N}_G(P_n) \cap \mathbf{N}_G(P_n L) = G_{C^*}$$

in this case. If L is contained in P_n, we let $1 \le k \le n$ be the unique integer such that L is contained in P_k but L is not contained in P_{k-1}. If $P_k = LP_{k-1}$, then C^* is the chain C except that we delete P_k. If $LP_{k-1} < P_k$, then C^* is the chain C except that we add LP_{k-1}. We leave for the reader to check $*$ satisfies all the properties that we have listed before. □

A well-known elementary fact from group theory is relevant below: if K is a normal p-subgroup of G, then $l(G) = l(G/K)$. In fact, if $x_1, \ldots, x_l$ are representatives of the conjugacy classes of p-regular elements of G, then using the Schur–Zassenhaus theorem, it is easy to check that $Kx_1, \ldots, Kx_l$ are representatives of the conjugacy classes of G/K of p-regular elements.

Recall that a **section** of a finite group G is a group X/Y, where $Y \trianglelefteq X \le G$. The section is **proper** if $|X/Y| < G$. Notice that a section of a section of G is (isomorphic to) a section of G.

Theorem 9.17 (Knörr–Robinson)   *Let G be a finite group. Assume that AWC is true for all proper sections of G. Then AWC is true for G if and only if*

$$k_0(G) = \sum_{C \in \mathcal{N}(G)} (-1)^{|C|} \frac{|G_C|}{|G|} l(G_C).$$

Proof   We argue by induction on $|G|$. Let

$$f(G) = \sum_{C \in \mathcal{N}(G)} (-1)^{|C|} |G_C| l(G_C).$$

We want to show that AWC is true for G if and only if $|G|k_0(G) = f(G)$.

Notice that the theorem is true if $|G|$ is not divisible by p. If $1 < P$ is a non-trivial p-subgroup of G, let $\mathcal{N}_P(G)$ be the set of normal chains $1 < P_1 < \cdots < P_n$ where $P_1 = P$. Let

$$f_P(G) = \sum_{C \in \mathcal{N}_P(G)} (-1)^{|C|} |G_C| l(G_C).$$

Therefore

$$f(G) = |G|l(G) + \sum_{1 < P} f_P(G),$$

where here P runs over the nontrivial p-subgroups of G. If $C \in \mathcal{N}_P(G)$ is the chain $1 < P < P_2 < \cdots < P_n$, then using that $P_i \trianglelefteq P_n$ for all i, we form the (normal) chain $\bar{C}$ of $\mathbf{N}_G(P)/P$ as

$$P/P < P_2/P < \cdots < P_n/P.$$

We notice that $C \mapsto \bar{C}$ is a bijection $\mathcal{N}_P(G) \to \mathcal{N}(\mathbf{N}_G(P)/P)$ and that $|\bar{C}| = |C| - 1$. Also,

$$(\mathbf{N}_G(P)/P)_{\bar{C}} = G_C/P.$$

Then

$$f(\mathbf{N}_G(P)/P) = \sum_{C \in \mathcal{N}_P(G)} (-1)^{|C|-1}|G_C/P|l(G_C/P) = -\frac{1}{|P|}f_P(G),$$

using that $l(G) = l(G/K)$ if K is a normal p-subgroup of G. Since P is non-trivial, then $|\mathbf{N}_G(P)/P| < |G|$ and by hypothesis, $\mathbf{N}_G(P)/P$ (and all proper sections of $\mathbf{N}_G(P)/P$) satisfies AWC. Therefore, by induction, we have that

$$|\mathbf{N}_G(P)/P|k_0(\mathbf{N}_G(P)/P) = f(\mathbf{N}_G(P)/P) = -\frac{1}{|P|}f_P(G).$$

Thus

$$f(G) - |G|k_0(G) = |G|l(G) - \sum_{1<P} |\mathbf{N}_G(P)|k_0(\mathbf{N}_G(P)/P) - |G|k_0(G)$$

$$= |G|l(G) - \sum_{P} |\mathbf{N}_G(P)|k_0(\mathbf{N}_G(P)/P).$$

Since AWC holds if and only if the right-hand side of this equation is zero, the theorem follows. $\square$

By using Theorem 9.16, notice that Theorem 9.17 remains valid if we replace $\mathcal{N}(G)$ by $\mathcal{C}(G)$ or $\mathcal{E}(G)$. (Although, notice that we did use normal chains in our proof.)

It is a remarkable fact discovered by Knörr and Robinson that the number of p-regular classes $l(G_C)$ in Theorem 9.17 can be replaced by the number of ordinary classes $k(G_C)$. To prove this, we need to go deeper into chains.

9.4 Generalized Characters Attached to Chains

Suppose that G is a finite group and again let $\mathcal{C}(G)$ be the set of chains of p-subgroups of G. Let $y \in G$. We are going to consider the set $\mathcal{C}(G)_y$ of chains of G which are fixed by y by conjugation, so $C \in \mathcal{C}(G)_y$ if and only if $C \in \mathcal{C}(G)$ and $y \in G_C$. If C is the chain $1 = P_0 < P_1 < \cdots < P_n$, then $C \in \mathcal{C}(G)_y$ if and only if y normalizes every P_i, by definition. Suppose now that y is p-singular, and let $x = y_p \neq 1$. Since x is a power of y, notice that x also normalizes P_i for every i. We are going to divide the set $\mathcal{C}(G)_y$ into two disjoint subsets $\mathcal{A}_y$ and $\mathcal{B}_y$. We let $\mathcal{A}_y$ be the set of y-invariant chains $1 = P_0 < P_1 < \cdots < P_n$ for which there is a subscript $0 \le i \le n - 1$

such that $P_{i+1} = P_i\langle x\rangle$. (In particular, $x \in P_n$, and P_{i+1} is first term of the chain that contains x.) For instance, the chain $1 < \langle x\rangle$ lies in $\mathcal{A}_y$. We let $\mathcal{B}_y = \mathcal{C}(G)_y - \mathcal{A}_y$, that is, the rest of the y-invariant chains. For instance, the trivial chain lies in $\mathcal{B}_y$.

Lemma 9.18 *Let $y \in G$ be p-singular, and let $x = y_p$. With the previous notation, there exists a bijection $f_y \colon \mathcal{B}_y \to \mathcal{A}_y$ that satisfies the following properties.*

(a) If $f_y(C) = C'$, then

$$(-1)^{|C|+1} = (-1)^{|C'|}.$$

(b) If $f_y(C) = C'$, then

$$|\mathbf{C}_{G_C}(y)| = |\mathbf{C}_{G_{C'}}(y)|.$$

Proof Suppose that $C \in \mathcal{B}_y$, and suppose that C is the chain $1 = P_0 < P_1 < \cdots < P_n$. We have that C is a y-invariant chain such that $P_{i+1} \neq P_i\langle x\rangle$ for every $i = 0, \ldots, n - 1$. If $x \in P_n$, then choose $0 \le k \le n - 1$ so that x is in P_{k+1} but not in P_k. Then $P_k < P_k\langle x\rangle < P_{k+1}$, so we construct the y-invariant chain $f_y(C)$ by inserting the group $P_k\langle x\rangle$ between P_k and P_{k+1}. (Recall that P_k is x-invariant, so $P_k\langle x\rangle$ is a subgroup, and $x^y = x$.) If x is not in P_n, then define the y-invariant chain $f_y(C)$ by adding $P_n\langle x\rangle$ at the end of the chain. Then f_y is a well-defined map from $\mathcal{B}_y$ into $\mathcal{A}_y$ and the length of $f_y(C) = C'$ is the length of C plus 1. We now define $g_y \colon \mathcal{A}_y \to \mathcal{B}_y$. If $C \in \mathcal{A}_y$, then $g_y(C)$ is the y-invariant chain in which we have removed from the chain C the first term containing x. The reader is invited to check that $f_y g_y$ and $g_y f_y$ are the identity maps.

For the second part, again write $C \in \mathcal{B}_y$ as the chain $1 = P_0 < P_1 < \cdots < P_n$. Notice that the subgroups that form $f_y(C)$ are $P_1, \ldots, P_n$ and some $P_i\langle x\rangle$. In particular, $G_{C'} \subseteq G_C$ and $G_C \cap \mathbf{C}_G(x) \subseteq G_{C'}$. Since x is a power of y, notice that $\mathbf{C}_G(y) \subseteq \mathbf{C}_G(x)$. Now

$$\mathbf{C}_G(y) \cap G_C = \mathbf{C}_G(y) \cap G_C \cap \mathbf{C}_G(x) \subseteq \mathbf{C}_G(y) \cap G_{C'} \subseteq \mathbf{C}_G(y) \cap G_C,$$

as desired. $\qquad\qquad\square$

Theorem 9.19 *Let G be a finite group. Suppose that for each subgroup H of G we have defined a generalized character $\tau(H) \in \mathbb{Z}[\mathrm{Irr}(H)]$. Assume that the following conditions are satisfied.*

(a) For $g \in G$, we have that $\tau(H^g) = \tau(H)^g$, where as usual $\tau(H)^g(h^g) = \tau(H)(h)$ for $h \in H$.

(b) With the notation of Lemma 9.18, if $y \in G$ is p-singular and $C \in \mathcal{B}_y$, then
$$\tau(G_C)(y) = \tau(G_{C'})(y), \text{ where } C' = f_y(C).$$

Then the generalized character

$$\Psi = \sum_{C \in \mathcal{C}(G)} (-1)^{|C|} \frac{|G_C|}{|G|} \tau(G_C)^G$$

vanishes on the p-singular elements of G.

Proof Suppose that $C \in \mathcal{C}(G)$, and let C^G be the G-orbit of C. Notice that the size of the G-orbit of C is $|G : G_C|$. In fact, $C^G = \{C^{z_1}, \ldots, C^{z_t}\}$, where $G = \bigcup_{j=1}^{t} G_C z_j$ is a disjoint union. Also, for $g \in G$, notice that $|C^g| = |C|$, and

$$\tau(G_{C^g})^G = \tau((G_C)^g)^G = (\tau(G_C)^g)^G = \tau(G_C)^G$$

using the hypothesis. Hence, if R is a complete set of representatives of the G-orbits on $\mathcal{C}(G)$, then we can write

$$\Psi = \sum_{C \in R} (-1)^{|C|} \tau(G_C)^G .$$

Let $y \in G$ and let $\mathcal{C}(G)_y$ be the set of y-invariant chains in $\mathcal{C}(G)$. If $C \in \mathcal{C}(G)$, we claim that

$$\tau(G_C)^G(y) = \sum_{D \in C^G \cap \mathcal{C}(G)_y} \tau(G_D)(y) .$$

By the definition of the induced character, we have that

$$\tau(G_C)^G(y) = \frac{1}{|G_C|} \sum_{\substack{z \in G \\ zyz^{-1} \in G_C}} \tau(G_C)(zyz^{-1})$$

$$= \frac{1}{|G_C|} \sum_{\substack{z \in G \\ y \in G_{C^z}}} \tau(G_{C^z})(y) ,$$

using that $\tau(G_C)^z = \tau(G_{C^z})$, by hypothesis. If we write $C^G = \{C^{z_1}, \ldots, C^{z_t}\}$, where $G = \bigcup_{j=1}^{t} G_C z_j$, then the claim easily follows.

Suppose finally that y is p-singular. Then, using that the length of a chain is constant on its G-orbit and the claim, we can write

$$\Psi(y) = \sum_{C \in R} (-1)^{|C|} \tau(G_C)^G(y) = \sum_{C \in R} \left(\sum_{D \in C^G \cap \mathcal{C}(G)_y} (-1)^{|D|} \tau(G_D)(y) \right)$$

$$= \sum_{D \in \mathcal{C}(G)_y} (-1)^{|D|} \tau(G_D)(y) .$$

Now, we use Lemma 9.18. We know that we can partition $\mathcal{C}(G)_y = \mathcal{A}_y \cup \mathcal{B}_y$, and that there is a bijection $f_y \colon \mathcal{B}_y \to \mathcal{A}_y$ such that if $f_y(D) = D'$, then $(-1)^{|D|+1} = (-1)^{|D'|}$ and, by hypothesis, $\tau(G_D)(y) = \tau(G_{D'})(y)$. Hence

$$\Psi(y) = \sum_{D \in \mathcal{C}(G)_y} (-1)^{|D|} \tau(G_D)(y)$$

$$= \sum_{D \in \mathcal{B}_y} (-1)^{|D|} \tau(G_D)(y) + \sum_{D \in \mathcal{B}_y} (-1)^{|D'|} \tau(G_{D'})(y) = 0 \,,$$

as desired. $\qquad\qquad\square$

With the hypothesis of Theorem 9.19, notice if $\mathcal{S}$ is the set of subgroups of G, $A = \mathbb{Z}[\mathrm{Irr}(G)]$ and $f(H) = \tau(H)^G$, then we can apply Lemma 9.16 to conclude that Ψ can also be written as

$$\Psi = \sum_{C \in \mathcal{C}(G)} (-1)^{|C|} \frac{|G_C|}{|G|} \tau(G_C)^G = \sum_{C \in \mathcal{N}(G)} (-1)^{|C|} \frac{|G_C|}{|G|} \tau(G_C)^G \,,$$

for instance. Or

$$\Psi = \sum_{C \in \mathcal{C}(G)/G} (-1)^{|C|} \tau(G_C)^G = \sum_{C \in \mathcal{N}(G)/G} (-1)^{|C|} \tau(G_C)^G \,,$$

if we prefer to work with sets of representatives. Hence, the generalized characters that will appear until the end of this section can be defined over all the chains of p-subgroups, normal chains, or elementary chains.

Before coming back to AWC, let us prove a result of P. Webb.

Corollary 9.20 (Webb) *Let G be a finite group, and let $\mathcal{C}(G)$ be the set of chains of G. Then the generalized character*

$$\Psi = \sum_{C \in \mathcal{C}(G)} (-1)^{|C|} \frac{|G_C|}{|G|} (1_{G_C})^G$$

vanishes on the p-singular elements of G.

Proof It is enough to let $\tau(H)$ be the trivial character of G for every subgroup H of G, and apply Theorem 9.19. $\qquad\qquad\square$

For any group G, we define now the character

$$\gamma(G) = \sum_{\chi \in \mathrm{Irr}(G)} \chi \bar{\chi} \,.$$

If G is a group, recall that $l(G)$ is the number of conjugacy classes of G consisting of p-regular elements, $G_{p'}$ is the set of p-regular elements of G, and $k(G) = |\mathrm{Irr}(G)|$ is the number of conjugacy classes of G.

Lemma 9.21 *Let G be a finite group. Then the following hold.*

(a) If $g \in G$, then $\gamma(G)(g) = |\mathbf{C}_G(g)|$.
(b) $[\gamma(G), 1_G] = k(G)$.
(c)

$$\frac{1}{|G|} \sum_{x \in G_{p'}} \gamma(G)(x) = l(G).$$

Proof The first part is the second orthogonality relation. Also,

$$[\gamma(G), 1_G] = \sum_{\chi \in \mathrm{Irr}(G)} [\chi \bar{\chi}, 1_G] = \sum_{\chi \in \mathrm{Irr}(G)} [\chi, \chi] = |\mathrm{Irr}(G)| = k(G).$$

Finally, if we write $G_{p'} = K_1 \cup \cdots \cup K_l$, where the K_i's are the distinct conjugacy classes of G consisting of p-regular elements, then by (a) we have that

$$\frac{1}{|G|} \sum_{x \in G_{p'}} \gamma(G)(x) = \frac{1}{|G|} \sum_{x \in G_{p'}} |\mathbf{C}_G(x)| = \frac{1}{|G|} \sum_{j=1}^{l} \sum_{x \in K_j} |\mathbf{C}_G(x_j)| = l = l(G),$$

as desired.   □

Corollary 9.22 *Let G be a finite group, and let $\mathcal{C}(G)$ be the set of chains of G. Then the generalized character*

$$\Psi = \sum_{C \in \mathcal{C}(G)} (-1)^{|C|} \frac{|G_C|}{|G|} \gamma(G_C)^G$$

vanishes on the p-singular elements of G.

Proof We wish to apply Theorem 9.19. For every subgroup H of G, we let $\tau(H) = \gamma(H)$. If $g \in G$ and $h \in H$, notice that $\gamma(H^g)(h^g) = |\mathbf{C}_{H^g}(h^g)| = |\mathbf{C}_H(h)| = \gamma(H)(h)$. Also, if y is p-singular, C is a y-invariant chain and $C' = f_y(C)$ is as in Lemma 9.18, we have that $|\mathbf{C}_{G_C}(y)| = |\mathbf{C}_{G_{C'}}(y)|$ by Lemma 9.18(b). Hence $\gamma(G_C)(y) = \gamma(G_{C'})(y)$, and by Theorem 9.19, the proof of the corollary is complete.   □

Corollary 9.23 (Knörr–Robinson) *Let G be a finite group, and let $\mathcal{C}(G)$ be the set of chains of G. Then*

$$\sum_{C \in \mathcal{C}(G)} (-1)^{|C|} \frac{|G_C|}{|G|} k(G_C) = \sum_{C \in \mathcal{C}(G)} (-1)^{|C|} \frac{|G_C|}{|G|} l(G_C).$$

The same is true if C runs over $\mathcal{N}(G)$ or $\mathcal{E}(G)$.

Proof Let

$$\Psi = \sum_{C \in \mathcal{C}(G)} (-1)^{|C|} \frac{|G_C|}{|G|} \gamma(G_C)^G,$$

as in Corollary 9.22. Using Frobenius reciprocity, we have that

$$[\Psi, 1_G] = \sum_{C \in \mathcal{C}(G)} (-1)^{|C|} \frac{|G_C|}{|G|} [\gamma(G_C), 1_{G_C}] = \sum_{C \in \mathcal{C}(G)} (-1)^{|C|} \frac{|G_C|}{|G|} k(G_C),$$

using Lemma 9.21(b). Now, by Corollary 9.22, we have that Ψ vanishes on p-singular elements. Let ν be the class function of G that has the value 1 on p-regular elements and is zero on p-singular elements. Then

$$[\Psi, 1_G] = [\Psi, \nu] = \sum_{C \in \mathcal{C}(G)} (-1)^{|C|} \frac{|G_C|}{|G|} [\gamma(G_C)^G, \nu]$$

$$= \sum_{C \in \mathcal{C}(G)} (-1)^{|C|} \frac{|G_C|}{|G|} [\gamma(G_C), \nu_{G_C}] = \sum_{C \in \mathcal{C}(G)} (-1)^{|C|} \frac{|G_C|}{|G|} l(G_C),$$

by using Lemma 9.21(c).

As we have pointed out after Theorem 9.19, for the definition of Ψ it does not matter if we take all the chains, the elementary abelian chains, or the normal chains. The proof of the corollary is complete. ☐

Now, we are ready to prove the following crucial result.

Theorem 9.24 (Knörr–Robinson) *Let G be a finite group. Assume that AWC is true for all proper sections of G. Then AWC is true for G if and only if*

$$k_0(G) = \sum_{C \in \mathcal{N}(G)} (-1)^{|C|} \frac{|G_C|}{|G|} k(G_C).$$

Proof Apply Theorem 9.17 and Corollary 9.23. ☐

9.5 Dade's Ordinary Conjecture

Suppose that G is a finite group and let d be a nonnegative integer. We say that $\chi \in \mathrm{Irr}(G)$ has a **defect** d if

$$\left(\frac{|G|}{\chi(1)} \right)_p = p^d.$$

Also, we let $k_d(G)$ be the number of $\chi \in \mathrm{Irr}(G)$ with defect d. For instance, $k_0(G)$ is the number of defect zero characters of G, or $k_a(G) = |\mathrm{Irr}_{p'}(G)|$

if $|G|_p = p^a$. Of course, $k_d(G) = 0$ if $d > a$. As we have seen, the McKay conjecture counts $k_a(G)$ in terms of the normalizer of a Sylow p-subgroup, and the Alperin weight conjecture counts $k_0(G)$ as an alternating sum of quantities calculated in local subgroups.

Conjecture 9.25 (Dade's ordinary conjecture) *Let G be a finite group. Let $d > 0$. Then*

$$\sum_{C \in \mathcal{N}(G)} (-1)^{|C|} |G_C| k_d(G_C) = 0.$$

By Theorem 9.16 with $f(H) = k_d(H)$ (and dividing by $|G|$ in the expression above), notice that Dade's ordinary conjecture always holds if $\mathbf{O}_p(G) > 1$. Also, observe that Dade's ordinary conjecture is equivalent to

$$k_d(G) = \sum_{\substack{C \in \mathcal{N}(G) \\ |C| > 0}} (-1)^{|C|+1} \frac{|G_C|}{|G|} k_d(G_C)$$

or, if we prefer to use representatives,

$$k_d(G) = \sum_{\substack{C \in \mathcal{N}(G)/G \\ |C| > 0}} (-1)^{|C|+1} k_d(G_C).$$

If $\mathbf{O}_p(G) = 1$, then $G_C < G$ if C is nontrivial, and this gives a formula to count the number of irreducible characters of G of any defect $d > 0$ in terms of intersections of proper local subgroups of G. Notice too that by Theorem 9.16, we can use all the chains or the elementary abelian chains in Conjecture 9.25. Another observation is that Conjecture 9.25 is not true if $d = 0$, except if $k_0(G) = 0$.

Let us illustrate Dade's conjecture with an example. We let $G = \mathbf{S}_5$ and $p = 2$, and we calculate a set of representatives of the orbits of the elementary chains. To do that we fix a Sylow 2-subgroup P of G, say $P = \langle (1,2), (1,3,2,4) \rangle$. Now let $P_1 = \langle (1,2) \rangle$, $P_2 = \langle (1,3)(2,4) \rangle$, $K = \langle (1,2), (3,4) \rangle$ and finally let $L = \langle (1,3)(2,4), (1,4)(2,3) \rangle$. Then there are seven representatives of nontrivial elementary chains: $1 < P_1$ (with stabilizer D_{12}), $1 < P_2$ (with stabilizer D_8), $1 < K$ (with stabilizer D_8), $1 < L$ (with stabilizer S_4), $1 < P_1 < K$ (with stabilizer K), $1 < \langle (1,2)(3,4) \rangle < K$ (with stabilizer D_8), and $1 < P_2 < L$ (with stabilizer D_8). Since $k_2(D_{12}) = 4$, $k_2(D_8) = 1$, $k_2(S_4) = 1$ and $k_2(K) = 4$, we deduce that

$$k_2(S_5) = 4 + 1 + 1 + 1 - 4 - 1 - 1 = 1.$$

In fact, S_5 has a unique irreducible character with defect 2, which is the unique character of degree 6. The reader is invited to make the same calculation with characters of defect 1 and 3.

How does Dade's conjecture relate to AWC and to the McKay conjecture?

Theorem 9.26 *Assume that Conjecture 9.25 is true for every finite group G. Then AWC is true.*

Proof Let G be a finite group. We prove that AWC is true for G by induction on $|G|$. By induction, we know that AWC is true for every proper section of G. Notice that if $C \in \mathcal{N}(G)$ is not the trivial chain, then $\mathbf{O}_p(G_C) > 1$ (because all the members of the chain are normal in G_C). Hence, $k_0(G_C) = 0$ by Lemma 9.15(a). Since by hypothesis, Conjecture 9.25 holds for G, we have that

$$
\begin{aligned}
|G|k_0(G) &= \sum_{\substack{C \in \mathcal{N}(G) \\ |C|=0}} (-1)^{|C|}|G_C|k_0(G_C) + \sum_{\substack{C \in \mathcal{N}(G) \\ |C|>0}} (-1)^{|C|}|G_C|k_0(G_C) \\
&= \sum_{C \in \mathcal{N}(G)} (-1)^{|C|}|G_C|k_0(G_C) \\
&= \sum_{d \geq 0} \left(\sum_{C \in \mathcal{N}(G)} (-1)^{|C|}|G_C|k_d(G_C) \right) \\
&= \sum_{C \in \mathcal{N}(G)} (-1)^{|C|}|G_C| \left(\sum_{d \geq 0} k_d(G_C) \right) \\
&= \sum_{C \in \mathcal{N}(G)} (-1)^{|C|}|G_C|k(G_C) .
\end{aligned}
$$

By Theorem 9.24, we have that AWC is true for G. $\qquad\square$

Theorem 9.27 *Assume that Conjecture 9.25 is true for every finite group G. Then the McKay conjecture is true.*

Proof Let G be a finite group. We prove by induction on $|G|$ that

$$
|\mathrm{Irr}_{p'}(G)| = |\mathrm{Irr}_{p'}(\mathbf{N}_G(P))| ,
$$

where P is a Sylow p-subgroup of G. Write $|G|_p = p^a$. Of course, we may assume that $a > 0$.

Let $L = \mathbf{O}_p(G)$. First, we claim that we may assume that $L = 1$. Assume that $L > 1$, and let $\mu_1, \ldots, \mu_s$ be a complete set of representatives of the orbits of the action of $\mathbf{N}_G(P)$ on the linear P-invariant characters of L. Then, using Lemma 9.3, it is clear that

$$\mathrm{Irr}_{p'}(G) = \mathrm{Irr}_{p'}(G|\mu_1) \cup \cdots \cup \mathrm{Irr}_{p'}(G|\mu_s)$$

and

$$\mathrm{Irr}_{p'}(\mathbf{N}_G(P)) = \mathrm{Irr}_{p'}(\mathbf{N}_G(P)|\mu_1) \cup \cdots \cup \mathrm{Irr}_{p'}(\mathbf{N}_G(P)|\mu_s)$$

are disjoint unions. Let T_i be the stabilizer of μ_i in G. Now, notice that $\mathrm{Irr}_{p'}(T_i|\mu_i)$ is not empty if and only if μ_i extends to P. Indeed, if μ_i extends to some $\epsilon \in \mathrm{Irr}(P)$, then ϵ^{T_i} contains some p'-degree irreducible constituent, because $\epsilon^{T_i}(1) = |T_i : P|$ is not divisible by p. Conversely, if $\eta \in \mathrm{Irr}_{p'}(T_i|\mu_i)$, then η_P contains some linear constituent, which necessarily extends μ_i. Furthermore, notice that if μ_i extends to P, then by Lemma 5.11 we have that μ_i extends to T_i. Therefore $|\mathrm{Irr}_{p'}(T_i|\mu_i)| \neq 0$ if and only if $|\mathrm{Irr}_{p'}(\mathbf{N}_{T_i}(P)|\mu_i)| \neq 0$, and in this case, by using the Gallagher correspondence, we have that

$$|\mathrm{Irr}_{p'}(T_i|\mu_i)| = |\mathrm{Irr}_{p'}(T_i/L)| \ \text{ and } \ |\mathrm{Irr}_{p'}(\mathbf{N}_{T_i}(P)|\mu_i)| = |\mathrm{Irr}_{p'}(\mathbf{N}_{T_i}(P)/L)| \,.$$

Since we are assuming that $L > 1$, we have that

$$|\mathrm{Irr}_{p'}(T_i/L)| = |\mathrm{Irr}_{p'}(\mathbf{N}_{T_i}(P)/L)|$$

using induction and the fact that $\mathbf{N}_{T_i/L}(P/L) = \mathbf{N}_{T_i}(P)/L$. Thus $|\mathrm{Irr}_{p'}(T_i|\mu_i)| = |\mathrm{Irr}_{p'}(\mathbf{N}_{T_i}(P)|\mu_i)|$. Now, by the Clifford correspondence we have that

$$|\mathrm{Irr}_{p'}(G|\mu_i)| = |\mathrm{Irr}_{p'}(T_i|\mu_i)| \quad \text{and} \quad |\mathrm{Irr}_{p'}(\mathbf{N}_G(P)|\mu_i)| = |\mathrm{Irr}_{p'}(\mathbf{N}_{T_i}(P)|\mu_i)|.$$

We deduce that $|\mathrm{Irr}_{p'}(G|\mu_i)| = |\mathrm{Irr}_{p'}(\mathbf{N}_G(P)|\mu_i)|$, and therefore

$$|\mathrm{Irr}_{p'}(G)| = |\mathrm{Irr}_{p'}(\mathbf{N}_G(P))| \,.$$

This shows the claim. (Notice that this argument shows that if G is a minimal counterexample to the McKay conjecture, then $\mathbf{O}_p(G) = 1$.)

Let $C \in \mathcal{N}(G)$ be any normal chain. We claim that $k_a(G_C) \neq 0$ if and only if G_C contains a Sylow p-subgroup of G, and that in this case, $k_a(G_C) = |\mathrm{Irr}_{p'}(G_C)|$. Suppose that $\gamma \in \mathrm{Irr}(G_C)$ has defect a. This means that

$$|G_C|_p/\gamma(1)_p = p^a = |G|_p \,.$$

Since $|G_C|_p \leq p^a$, this forces $\gamma \in \mathrm{Irr}_{p'}(G_C)$, and $|G_C|_p = |G|_p$. By considering the trivial character, the converse is clear.

We divide the set of the nontrivial normal chains C with $|G_C|_p = |G|_p$ into three disjoint subsets: $\mathcal{B}$, which are the chains $1 < P$, where P is a Sylow p-subgroup of G; $\mathcal{D}$, which are the normal chains of length at least 2, whose final term is a Sylow p-subgroup of G; and $\mathcal{F}$, the rest of these normal chains. Notice that $\mathcal{F}$ is simply the set of nontrivial normal chains C whose final term is not a Sylow p-subgroup of G with $|G_C|_p = |G|_p$. We define a map f from $\mathcal{D}$ to $\mathcal{F}$ by removing the last term of the chain. Write $f(C) = \hat{C}$. Notice that this is a well-defined map. Indeed, if $1 = P_0 < P_1 < \cdots < P_n$ is a normal chain, $n \geq 2$, and $P_n \in \mathrm{Syl}_p(G)$, then $P_n \subseteq \mathbf{N}_G(P_1) \cap \cdots \cap \mathbf{N}_G(P_{n-1}) = G_{\hat{C}}$, and $\hat{C} \in \mathcal{F}$. Also, $P_n \in \mathrm{Syl}_p(G_{\hat{C}})$. Now we claim that f is surjective. Suppose that $F \in \mathcal{F}$ is a nontrivial normal chain $1 = P_0 < P_1 < \cdots < P_{n-1}$ with $|G_F|_p = |G|_p$, $|P_{n-1}| < |G|_p$, and $n \geq 2$. Since F is a normal chain, then $P_{n-1} \trianglelefteq G_F$, and therefore P_{n-1} is properly contained in every $P_n \in \mathrm{Syl}_p(G_F)$. Notice that every Sylow p-subgroup of G_F is necessarily a Sylow p-subgroup of G. Now the chain C defined by $1 = P_0 < P_1 < \cdots < P_{n-1} < P_n$ is a normal chain, with $n \geq 2$,

$$G_C = G_F \cap \mathbf{N}_G(P_n) = \mathbf{N}_{G_F}(P_n)$$

and $P_n \in \mathrm{Syl}_p(G_C)$. Thus $C \in \mathcal{D}$ and $f(C) = F$, as claimed.

Now, notice that $f(C_1) = f(C_2) = F$ if and only if C_1 and C_2 have the same length n, the same first $n - 1$ terms, and the last terms P_n of C_1 and Q_n of C_2 are G_F-conjugate. (This is because $P_n, Q_n \in \mathrm{Syl}_p(G_F)$.) There are exactly

$$|G_F : \mathbf{N}_{G_F}(P_n)| = |G_F : G_{C_1}|$$

such chains, and all of them are G_F-conjugate.

If $C \in \mathcal{D}$ and $f(C) = F$, notice now that $G_F < G$ because the first term P_1 of the nontrivial normal chain F is normal in G_F, and we are assuming that $\mathbf{O}_p(G) = 1$. Hence the McKay conjecture is true for G_F by induction. Thus, if P_n is the last term of C, then $P_n \in \mathrm{Syl}_p(G_F)$, $|G_F|_p = p^a$ and

$$k_a(G_F) = k_a(\mathbf{N}_{G_F}(P_n)) = k_a(G_C).$$

By Conjecture 9.25, we have that

$$|G|k_a(G) = \sum_{\substack{C \in \mathcal{N}(G), |C| > 0 \\ k_a(G_C) \neq 0}} (-1)^{|C|+1} |G_C| k_a(G_C) = |G|k_a(\mathbf{N}_G(P))$$

$$+ \sum_{C \in \mathcal{D}} (-1)^{|C|+1} |G_C| k_a(G_C) + \sum_{C \in \mathcal{F}} (-1)^{|C|+1} |G_C| k_a(G_C),$$

where the first summand is obtained by running over the chains in $\mathcal{B}$. Now,

$$\sum_{C \in \mathcal{D}} (-1)^{|C|+1} |G_C| k_a(G_C) = \sum_{F \in \mathcal{F}} \sum_{\substack{C \in \mathcal{D} \\ f(C) = F}} (-1)^{|C|+1} |G_C| k_a(G_C)$$

$$= \sum_{F \in \mathcal{F}} \left(\sum_{\substack{C \in \mathcal{D} \\ f(C)=F}} (-1)^{|C|+1} |G_C| \right) k_a(G_F)$$

$$= \sum_{F \in \mathcal{F}} |G_F| (-1)^{|F|} k_a(G_F),$$

by using that there are exactly $|G_F : G_C|$ chains $C \in \mathcal{D}$ with $f(C) = F$, all having the same length $|F| + 1$, all having the same $|G_C|$ (because they are conjugate), and all satisfying $k_a(G_F) = k_a(G_C)$. We deduce that $k_a(G) = k_a(\mathbf{N}_G(P))$. $\qquad\square$

As we have mentioned, Dade's conjecture predicts how to count $k_d(G)$ from the intersection of proper local subgroups of G, for any $d > 0$. This implies that the global invariant $k_d(G)$ is determined locally, a truly amazing fact. When $p^d = |G|_p$, then the McKay conjecture proposes a much simpler equation $k_d(G) = k_d(\mathbf{N}_G(P))$, if $|P| = p^d$. It seems, that in order to calculate $k_d(G)$ for other d, one needs to use chains of subgroups and alternating sums. Although, as we shall see, not for $d = 1$.

Lemma 9.28 *Let G be a finite group, and let $\chi \in \mathrm{Irr}(G)$ have defect 1. If $L = \mathbf{O}_p(G)$, then $|L| \leq p$.*

Proof Assume that $L > 1$. Let $\theta \in \mathrm{Irr}(L)$ be under χ. By Theorem 5.12, we have that $\chi(1)_p / \theta(1)$ divides $|G|_p / |L|$. By hypothesis, we have that $\chi(1)_p = |G|_p / p$, and therefore $|L| \leq p\theta(1)$. Since $L > 1$, then $\theta(1)^2 < |L|$ (for instance by the character degree formula for L), and we deduce that $|L| < p^2$, as desired. $\qquad\square$

Theorem 9.29 *Let G be a finite group. Assume that Conjecture 9.25 is true for G for $d = 1$. Then*

$$k_1(G) = \sum_Q \frac{|\mathbf{N}_G(Q)|}{|G|} k_1(\mathbf{N}_G(Q)),$$

where Q runs over the p-radical subgroups of order p of G.

Proof By Dade's conjecture, we have that

$$k_1(G) = \sum_{C \in \mathcal{N}(G), |C| > 0} (-1)^{|C|+1} \frac{|G_C|}{|G|} k_1(G_C).$$

Suppose that C is a normal chain $1 = P_0 < P_1 < \cdots < P_n$, with $n \geq 1$, such that $k_1(G_C) \neq 0$. Then $|\mathbf{O}_p(G_C)| \leq p$ by Lemma 9.28. Since C is a normal chain, we have that

$$P_n \subseteq \mathbf{N}_G(P_1) \cap \cdots \cap \mathbf{N}_G(P_n) = G_C \subseteq \mathbf{N}_G(P_n).$$

Thus $P_n \subseteq \mathbf{O}_p(G_C)$. Since $P_n > 1$, we deduce that $n = 1$, $|P_1| = p$ and $\mathbf{N}_G(P_1) = G_C$. Therefore

$$k_1(G) = \sum_{|Q|=p} \frac{|\mathbf{N}_G(Q)|}{|G|} k_1(\mathbf{N}_G(Q)).$$

Finally, if $k_1(\mathbf{N}_G(Q)) > 0$ and $|Q| = p$, then we have that $Q = \mathbf{O}_p(\mathbf{N}_G(Q))$ by Lemma 9.28, and the theorem follows. $\qquad\square$

As a matter of fact, Theorem 9.29 holds without any condition on G, and this is a reformulation of a deep theorem of Richard Brauer. This theorem was the origin of the so called *cyclic defect theory*, later completed by Dade, that describes the character theory of the blocks with a cyclic defect group.

What are these conjectures telling us, for instance, if $P \in \mathrm{Syl}_p(G)$ is cyclic?

Theorem 9.30 *Suppose that $P \in \mathrm{Syl}_p(G)$ is cyclic or generalized quaternion, and let Q be the unique subgroup of P of order p. Then Dade's conjecture holds for G if and only if $k_d(G) = k_d(\mathbf{N}_G(Q))$ for every $d > 0$. In this case, $k(G) - k_0(G) = k(\mathbf{N}_G(Q))$.*

Proof By hypothesis P has a unique subgroup of order p. By using the elementary abelian chains in the formulation of Conjecture 9.25, we only have to deal with the trivial chain and, up to G-conjugation, with the chain $1 < Q$. The last equation follows because $k_0(\mathbf{N}_G(Q)) = 0$ by Lemma 9.15(a). $\qquad\square$

In order to check Dade's ordinary conjecture for a particular finite group G, one soon realizes that $\mathcal{N}(G)$, or even $\mathcal{E}(G)$, contains too many chains (as we have seen in $\mathbf{S}_5$ with $p = 2$). If $\mathbf{O}_p(G) = 1$, which is the only relevant case, then a much smaller subset than $\mathcal{N}(G)$ only needs to be taken into account: the *radical chains*. A chain $1 = P_0 < P_1 < \cdots < P_n$ is **radical** if, for $i = 0, \ldots, n$, we have that $P_i = \mathbf{O}_p(G_{C_i})$, where C_i is the chain $1 = P_0 < P_1 < \cdots < P_i$. If $\mathcal{R}(G)$ is the set of radical chains of G, and $\mathbf{O}_p(G) = 1$, then it was remarked by Knörr and Robinson that

$$\sum_{C \in \mathcal{N}(G)} (-1)^{|C|} \frac{|G_C|}{|G|} f(G_C) = \sum_{C \in \mathcal{R}(G)} (-1)^{|C|} \frac{|G_C|}{|G|} f(G_C),$$

where f is any function as in Theorem 9.16.

9.6 Notes

The McKay conjecture was announced in 1971 in the Notices of the American Mathematical Society [McK71], for $p = 2$ and simple groups. Later this became a paper, in 1972 (see [McK72]). The first place where other primes and other types of groups are considered is in [Is73], where Isaacs proves the McKay conjecture if $|G : \mathbf{N}_G(P)|$ is odd and G is solvable. (In particular, for groups of odd order for any prime p, and for solvable groups and $p = 2$.) In 1975 J. L. Alperin announced the block version of the McKay conjecture in [Al75]. In 1976 J. B. Olsson proved the Alperin–McKay conjecture for symmetric groups (see [Ol76]) and for $\mathrm{GL}_n(q)$ (for p not dividing q). In 1978 T. R. Wolf proved the McKay conjecture for solvable groups [Wo78b], and in 1979 T. Okuyama and W. Wajima gave an argument that proved Alperin–McKay for p-solvable groups [OW79]. There are too many cases where the McKay and the Alperin–McKay conjectures have now been checked since their announcement to list them here, by the work of many mathematicians. (See the survey by G. Malle in [Ma17].) Specially relevant is, of course, that now we know that the McKay conjecture is true for $p = 2$ by work of G. Malle and B. Späth [MS16]. Also, we know that the McKay conjecture is true for finite simple groups. (The sporadic simple groups were checked in [Wils98].)

The final chapter of this book is dedicated to the reduction of the McKay conjecture to a question on simple groups, and we shall discuss this further later.

The Alperin weight conjecture was announced in [Al87], based on the idea of weights that is used in the modular representation theory of groups of Lie type in characteristic p. After the Green correspondence, T. Okuyama in [Ok81] was the first to realize the connection between Brauer characters of a finite p-solvable group G with vertex Q and those of $\mathbf{N}_G(Q)$, and this proved AWC for p-solvable groups. Later, J. L. Alperin and P. Fong proved AWC for symmetric groups and $\mathrm{GL}_n(q)$ for q not divisible by p in [AF90]. (J. L. Alperin told me once that a conjecture on finite groups is *safe* when it has been checked for solvable, symmetric, general linear groups and sporadic simple groups.)

The AWC was reduced to simple groups in [NT11], in work modeled upon the reduction in [IMN07]. Both AWC and Dade's conjectures have block versions, which we have chosen not to include in this book. (After all, our main subject here is ordinary character theory.) The reduction of these conjectures was achieved by B. Späth in [Sp13a] and [Sp17a].

At the time Knörr and Robinson gave their reformulation of AWC in terms of chains and alternating sums [KR89], simplicial complexes had already been studied by many, including S. Bouc, K. S. Brown, D. Quillen, J. Thévenaz and P. J. Webb. But the idea of relating this to blocks and AWC was, perhaps, totally unexpected. From this reformulation, a wealth of equivalent forms of AWC were proposed: by Thévenaz using functions defined on the posets of subgroups [Th92], or on equivariant K-theory [Th93]; by R. Boltje using contractible chain complexes [Bo03]; by M. Linckelmann using Bredon cohomology [Li05], etc. All this made it even more clear that AWC lies quite deep in the representation theory of finite groups and has unpredicted interconnections with other areas.

Dade's ordinary conjecture, announced in 1992 [Da92], was also inspired by the Knörr–Robinson reformulation of AWC. In order to get a conjecture whose proof could be reduced to simple groups, Dade started formulating a series of increasingly more general and complex conjectures – the *invariant*, the *extended*, the *projective* – and finally the *inductive conjecture*, about which he wrote: "With a great amount of work it can be shown to hold for all finite groups if it holds whenever G is a non-abelian finite simple group" (see [Da97]). Dade's reduction never appeared on print. Späth's reduction appeared in 2017.

The projective conjecture (see [Da94]) has received more attention than any other form of Dade's conjectures (probably because it was shown by Dade that it implied the Alperin–McKay conjecture). Dade's projective conjecture had a complicated statement in terms of twisted group algebras. This statement was greatly simplified by Robinson by only using characters lying over a fixed character of a central subgroup. Robinson also gave a proof of this conjecture for p-solvable groups in [Ro00]. (Greater detail on how a minimal counterexample would look was given later by C. Eaton and Robinson in [ER02].) Recently, A. Turull [Tu17] has given another proof of Dade's projective conjecture for p-solvable groups, which also takes into account many of the generalizations which we have previously mentioned in the chapter. It is impossible not to mention the work of J. An, J. B. Olsson, and K. Uno on checking Dade's conjectures and several strengthenings for several important families of groups.

Problems

(9.1) Let G be a finite group. Suppose that $L \leq \mathbf{Z}(G)$ has prime order. If the McKay conjecture is true for G, then show that the relative McKay conjecture is true for G with respect to L.
(*Hint:* Use Galois automorphisms.)

(9.2) If A is a group that acts by automorphisms on G, then write $\mathrm{Irr}_{p',A}(G)$ to denote the set of irreducible characters of G of degree not divisible by p which are fixed by A. If A is a p-group, then prove the following.
  (a) There is $P \in \mathrm{Syl}_p(G)$ which is fixed by A.
  (b) Let $K = \mathbf{O}^p(G)$ and let $K \subseteq M \trianglelefteq G$. Show that

$$|\mathrm{Irr}_{p',G}(M)| = \frac{t}{s}|\mathrm{Irr}_{p'}(G)|\,,$$

where $s = |\mathrm{Irr}(G/G'K)|$ and $t = |\mathrm{Irr}_G(M/M'K)|$.
  (c) Assume that the McKay conjecture is true. Then prove that

$$|\mathrm{Irr}_{p',A}(G)| = |\mathrm{Irr}_{p',A}(\mathbf{N}_G(P))|\,.$$

(9.3) Suppose that G is a finite group, and let $P \in \mathrm{Syl}_p(G)$. Suppose that $N \trianglelefteq G$. If the McKay conjecture is true, then prove that

$$|\mathrm{Irr}_{p',P}(N)| = |\mathrm{Irr}_{p',P}(\mathbf{N}_N(P))|\,.$$

(*Hint:* Use the previous problem and the relative Glauberman correspondence.)

(9.4) (*Isaacs*) Let G be a finite group, p a prime and $P \in \mathrm{Syl}_p(G)$. Suppose that $\chi \in \mathrm{Irr}_{p'}(G)$ is induced from some linear $\lambda \in \mathrm{Irr}(U)$, where $P \le U \le G$.
  (a) Show that

$$(\lambda_{\mathbf{N}_U(P)})^{\mathbf{N}_G(P)} \in \mathrm{Irr}(\mathbf{N}_G(P))\,.$$

  (b) If $\nu, \lambda \in \mathrm{Irr}(G)$ are linear, prove that $\nu_{\mathbf{N}_G(P)} = \lambda_{\mathbf{N}_G(P)}$ if and only if $\lambda = \nu$.
  (c) Assume that $\psi \in \mathrm{Irr}_{p'}(G)$ is induced from some linear $\nu \in \mathrm{Irr}(V)$, where $P \le V \le G$. If

$$(\lambda_{\mathbf{N}_U(P)})^{\mathbf{N}_G(P)} = (\nu_{\mathbf{N}_V(P)})^{\mathbf{N}_G(P)},$$

then show that $\chi = \psi$.
  (d) If all the characters in $\mathrm{Irr}_{p'}(G)$ are induced from linear characters and $|\mathrm{Irr}_{p'}(G)| = |\mathrm{Irr}_{p'}(\mathbf{N}_G(P))|$, show that the above procedure defines a canonical bijection $\mathrm{Irr}_{p'}(G) \to \mathrm{Irr}_{p'}(\mathbf{N}_G(P))$.
  (*Hint:* Use Problem 1.8.)

10

A Reduction Theorem for the McKay Conjecture

We formulate the **inductive McKay condition** for a non-abelian finite simple group. If all simple groups satisfy this condition, then we prove that the McKay conjecture is true.

10.1 Wreath Products

If H is a group and $m \geq 1$ is an integer, we usually denote by H^m the group $H \times \cdots \times H$ (m times). If $\mathcal{X}_i : H \to \mathrm{GL}_{d_i}(\mathbb{C})$ is a representation of H for $i = 1, \ldots, m$, then

$$(\mathcal{X}_1 \otimes \cdots \otimes \mathcal{X}_m)(x_1, \ldots, x_m) = \mathcal{X}_1(x_1) \otimes \cdots \otimes \mathcal{X}_m(x_m)$$

defines a representation of H^m with character $\psi_1 \times \cdots \times \psi_m$, where ψ_i is the character afforded by $\mathcal{X}_i$.

Now, any subgroup S of the symmetric group S_m naturally acts as automorphisms on H^m by setting

$$(x_1, \ldots, x_m)^\sigma = (x_{\sigma^{-1}(1)}, \ldots, x_{\sigma^{-1}(m)}),$$

for $x_i \in H$ and $\sigma \in S$. (Recall that in the symmetric group S_m, and in general when we compose maps, we use $\sigma\tau$ to denote the composition map $\tau \circ \sigma$. This is why we have to use σ^{-1} in the definition of the previous action.) In this way,

$$(x_1, \ldots, x_m)^{\sigma\tau} = ((x_1, \ldots, x_m)^\sigma)^\tau$$

for $x_i \in H$ and $\sigma, \tau \in S$. Notice too that if H is not the trivial group then the kernel of this action is trivial, so we may view S_m as a subgroup of $\mathrm{Aut}(H^m)$. As in Chapter 2, we may form the semidirect product $H^m \rtimes S$, which is called the **wreath product** of H with S and is denoted by $H \wr S$. Recall that in $H \wr S$, we multiply elements in the following way:

$$(x_1, \ldots, x_m)\sigma(y_1, \ldots, y_m)\tau = (x_1, \ldots, x_m)(y_1, \ldots, y_m)^{\sigma^{-1}}\sigma\tau$$
$$= (x_1 y_{\sigma(1)}, \ldots, x_m y_{\sigma(m)})\sigma\tau.$$

Also, if $\theta_i \in \mathrm{Irr}(H)$, $\theta = \theta_1 \times \cdots \times \theta_m \in \mathrm{Irr}(H^m)$, and $\sigma \in S$, then notice that

$$\theta^\sigma = \theta_{\sigma^{-1}(1)} \times \cdots \times \theta_{\sigma^{-1}(m)}.$$

Next, we recall some linear algebra. Let $V = \mathbb{C}^d$, where d is a positive integer, and let $W = V \otimes \cdots \otimes V$ (m times) be the tensor product complex vector space of dimension d^m. Suppose that $\mathcal{B}_d = \{e_1, \ldots, e_d\}$ is the canonical basis of $\mathbb{C}^d$. If $f_i \colon V \to V$ is a linear map with matrix A_i with respect to $\mathcal{B}_d$ for $i = 1, \ldots, m$, then the basis $\mathcal{B} = \{e_{j_1} \otimes \cdots \otimes e_{j_m} \mid 1 \le j_1 \le d, \ldots, 1 \le j_m \le d\}$ of W, with the appropriate order, is such that the matrix of $f_1 \otimes \cdots \otimes f_m$ with respect to $\mathcal{B}$ is $A_1 \otimes \cdots \otimes A_m$. Now, notice that if $\sigma \in S_m$, then σ uniquely defines a linear map $\tilde{\sigma} \colon W \to W$ satisfying

$$\tilde{\sigma}(v_1 \otimes \cdots \otimes v_m) = v_{\sigma^{-1}(1)} \otimes \cdots \otimes v_{\sigma^{-1}(m)},$$

for $v_i \in V$. The map $\sigma \mapsto \tilde{\sigma}$ defines a group homomorphism $S_m \to GL(W)$. If $\mathcal{X}_d(\sigma) \in GL_{d^m}(\mathbb{C})$ is the matrix of $\tilde{\sigma}$ with respect to $\mathcal{B}$, then we have constructed a representation

$$\mathcal{X}_d \colon S_m \to GL_{d^m}(\mathbb{C}).$$

Observe that $\tilde{\sigma}(f_1 \otimes \cdots \otimes f_m)\tilde{\sigma}^{-1} = f_{\sigma(1)} \otimes \cdots \otimes f_{\sigma(m)}$. Hence

$$\mathcal{X}_d(\sigma)(A_1 \otimes \cdots \otimes A_m)\mathcal{X}_d(\sigma)^{-1} = A_{\sigma(1)} \otimes \cdots \otimes A_{\sigma(m)}$$

for all matrices $A_i \in GL_d(\mathbb{C})$ and all $\sigma \in S_m$.

Lemma 10.1   *If $\mathcal{X}$ is a representation of H of degree d, then $\mathcal{Y} \colon H \wr S_m \to GL_{d^m}(\mathbb{C})$ given by*

$$\mathcal{Y}((x_1, \ldots, x_m)\sigma) = (\mathcal{X}(x_1) \otimes \cdots \otimes \mathcal{X}(x_m))\mathcal{X}_d(\sigma)$$

is a representation of $H \wr S_m$ that extends $\mathcal{X} \otimes \cdots \otimes \mathcal{X}$.

Proof   We have that

$$\mathcal{Y}((x_1, \ldots, x_m)\sigma)\mathcal{Y}((y_1, \ldots, y_m)\tau)$$
$$= (\mathcal{X}(x_1) \otimes \cdots \otimes \mathcal{X}(x_m))\mathcal{X}_d(\sigma)(\mathcal{X}(y_1) \otimes \cdots \otimes \mathcal{X}(y_m))\mathcal{X}_d(\tau)$$
$$= (\mathcal{X}(x_1) \otimes \cdots \otimes \mathcal{X}(x_m))\mathcal{X}_d(\sigma)(\mathcal{X}(y_1) \otimes \cdots \otimes \mathcal{X}(y_m))\mathcal{X}_d(\sigma^{-1})\mathcal{X}_d(\sigma\tau)$$
$$= (\mathcal{X}(x_1) \otimes \cdots \otimes \mathcal{X}(x_m))\big(\mathcal{X}(y_{\sigma(1)}) \otimes \cdots \otimes \mathcal{X}(y_{\sigma(m)})\big)\mathcal{X}_d(\sigma\tau)$$
$$= \big(\mathcal{X}(x_1)\mathcal{X}(y_{\sigma(1)}) \otimes \cdots \otimes \mathcal{X}(x_m)\mathcal{X}(y_{\sigma(m)})\big)\mathcal{X}_d(\sigma\tau)$$

for all $x_i, y_i \in H$ and $\sigma, \tau \in S_m$. $\qquad\square$

We denote by $\mathcal{X} \wr S_m$ the representation $\mathcal{Y}$ given by Lemma 10.1.

We use Lemma 10.1 to prove a well-known fact on extendability of characters in wreath products.

Corollary 10.2 *Let H be a group, $S \leq S_m$ and let $G = H \wr S$. If $\theta \in \operatorname{Irr}(H^m)$, then θ extends to the stabilizer G_θ of θ in G.*

Proof We may assume that $S = S_m$. Write $\theta = \theta_1 \times \cdots \times \theta_m$ and recall that

$$\theta^\sigma = \theta_{\sigma^{-1}(1)} \times \cdots \times \theta_{\sigma^{-1}(m)}.$$

Let $\Omega = \{1, \ldots, m\}$ and set $i \equiv j$ if $\theta_i = \theta_j$. This defines a partition of Ω. By replacing θ by some S_m-conjugate, we may assume that

$$\theta = (\eta_1 \times \cdots \times \eta_1) \times \cdots \times (\eta_k \times \cdots \times \eta_k),$$

where $\{\eta_1, \ldots, \eta_k\}$ are distinct, and η_i appears m_i times. Thus $m_1 + \cdots + m_k = m$. Notice that the stabilizer of θ in S_m is isomorphic to

$$S_{m_1} \times \cdots \times S_{m_k}.$$

Using that there is a natural isomorphism

$$H \wr (S_{m_1} \times \cdots \times S_{m_k}) \cong (H \wr S_{m_1}) \times \cdots \times (H \wr S_{m_k}),$$

we use Lemma 10.1 to easily conclude the proof. $\qquad\square$

10.2 Tensor Induction

Our next objective is to define tensor induction of characters. Although we shall not need it for the reduction of the McKay conjecture, tensor induction has turned out to be a useful tool for character theory, especially in the presence of non-abelian minimal normal subgroups of groups.

Let H be a subgroup of index m of a finite group G. If ψ is a character of H of degree d, we are going to define a new character $\psi^{\otimes G}$ of G of degree d^m. Let $\mathcal{T} = \{t_1, \ldots, t_m\}$ be a complete set of representatives of right cosets of H in G. We have that G acts on the set $\Omega = \{1, \ldots, m\}$ in the following way: if $x \in G$ and $i \in \Omega$, then there is a unique $i \cdot x \in \Omega$ such that

$$H t_i x = H t_{i \cdot x}.$$

This defines a group homomorphism

$$\rho: G \to S_m,$$

where $\rho(x)$ is the permutation $i \mapsto i \cdot x$. Therefore, given $x \in G$ and $i \in \Omega$, we have uniquely defined

$$h_{i,x} = (t_i x)(t_{i \cdot x})^{-1} \in H .$$

Lemma 10.3 *The map $f : G \to H \wr S_m$ given by*

$$f(x) = (h_{1,x}, \dots, h_{m,x}) \rho(x)$$

is a group homomorphism.

Proof It amounts to proving that

$$h_{i,x} h_{i \cdot x, y} = h_{i,xy}$$

for $x, y \in G$ and $i = 1, \dots, m$. But this easily follows by using that $t_i g = h_{i,g} t_{i \cdot g}$ for every $g \in G$, and multiplying the equation $t_i x = h_{i,x} t_{i \cdot x}$ by y. □

If $\mathcal{X}$ is a representation of H, then the **tensor-induced** representation $\mathcal{X}^{\otimes G}$ is the representation of G given by $(\mathcal{X} \wr S_m) \circ f$. It is straightforward to check that the similarity class of the tensor induced representation only depends on the class of $\mathcal{X}$, and is independent of the representatives for the right cosets chosen. If $\psi \in \mathrm{Char}(H)$ is the character afforded by $\mathcal{X}$, then we denote by $\psi^{\otimes G}$ the character of G afforded by $(\mathcal{X} \wr S_m) \circ f$. This is the **tensor-induced character** of ψ.

It is not a triviality to obtain the general formula for the values of the tensor induced character. However, there is an easy case which is enough for many applications.

Lemma 10.4 *Suppose that $H \leq G$ and that N is a normal subgroup of G contained in H. Let ψ be a character of H. For $n \in N$, we have that*

$$\psi^{\otimes G}(n) = \prod_{t \in \mathcal{T}} \psi(t n t^{-1}),$$

where $\mathcal{T}$ is any set of representatives of the right cosets of H in G.

Proof If $\mathcal{T} = \{t_1, \dots, t_m\}$ and $n \in N$, notice that $H t_i n = H t_i$ because $t_i n t_i^{-1} \in H$. Therefore $i \cdot n = i$ for $1 \leq i \leq m$. With the previous notation, we have that $h_{i,n} = t_i n t_i^{-1}$. If $\mathcal{X}$ is a representation that affords ψ, then

$$((\mathcal{X} \wr S_m) \circ f)(n) = \mathcal{X}(t_1 n t_1^{-1}) \otimes \cdots \otimes \mathcal{X}(t_m n t_m^{-1})$$

and this matrix has trace $\prod_{t \in \mathcal{T}} \psi(t n t^{-1})$. □

The following is a typical application of tensor induction, which is usually applied to non-abelian minimal normal subgroups of finite groups.

Corollary 10.5 *Suppose that $N \trianglelefteq G$, and that $N = S_1 \times \cdots \times S_t$ is the direct product of the set of subgroups $\Omega = \{S_1, \ldots, S_t\}$ of N which are transitively permuted by G by conjugation. Write $S = S_1$ and view $S/\mathbf{Z}(S) \trianglelefteq A = \mathrm{Aut}(S)$. Let $\theta = \theta_1 \times \cdots \times \theta_t \in \mathrm{Irr}(N)$ be G-invariant, where $\theta_i \in \mathrm{Irr}(S_i)$. Assume that $\mathbf{Z}(S) \subseteq \ker(\theta_1)$. If θ_1 extends to its stabilizer A_{θ_1} in A, then θ extends to G.*

Proof Let $H = \mathbf{N}_G(S)$, and notice that $N \subseteq H$. Since G acts transitively on Ω, for every j there is $x_j \in G$ such that $S_j = S^{x_j}$. Notice that $G = \bigcup_{j=1}^{t} H x_j$ is a disjoint union.

We claim that $\theta_i = \theta_1^{x_i}$ and that θ_1 is H-invariant. If $s \in S$ and $h \in H$, then $s^{h x_i} \in S_i$ and thus $\theta(s^{h x_i}) = \theta_i(s^{h x_i}) \prod_{j \neq i} \theta_j(1)$. Using that θ is G-invariant, we have that $\theta(s^{h x_i}) = \theta(s) = \theta_1(s) \prod_{j>1} \theta_j(1)$. It follows that

$$\theta_i^{x_i^{-1} h^{-1}}$$

is a rational multiple of θ_1. Since irreducible characters are linearly independent, these two characters coincide for every $h \in H$. The claim follows.

Let $C = \mathbf{C}_G(S)$. Now, $S/\mathbf{Z}(S)$ is naturally isomorphic to $SC/C \leq H/C \leq A$ and we can view $\theta_1 \in \mathrm{Irr}(SC/C)$. Since θ_1 extends to A_{θ_1}, we have that θ_1 extends to H. Let $\psi \in \mathrm{Irr}(H/C)$ be an extension of θ_1. Now, let

$$\rho = \psi^{\otimes G}$$

be the tensor induced character. By Lemma 10.4, we have that

$$\rho(n) = \prod_{j=1}^{t} \psi(x_j n x_j^{-1})$$

for $n \in N$. Write $T = \prod_{j=2}^{t} S^{x_j}$, so that $N = S \times T$. Notice that $T \leq C \leq \ker(\psi)$. Thus $\psi_N = \theta_1 \times 1_T$. If we write $n = s_1 \cdots s_t$ with $s_i \in S_i$, then we have that

$$\rho(n) = \prod_{j=1}^{t} \theta_1(s_j^{x_j^{-1}}) = \prod_{j=1}^{t} (\theta_1^{x_j})(s_j) = \theta(n).$$

It follows that ρ extends θ. $\qquad\square$

10.3 Central Products

In this short section, we discuss characters of central products.

Definition 10.6 *Suppose that G is a finite group, and let $U_1, \ldots, U_k$ be subgroups of G such that $[U_i, U_j] = 1$ for all $i \neq j$, where $k \geq 2$. Write $Z = \bigcap_{i=1}^{k} U_i$ and assume that $G = U_1 \cdots U_k$ and that $(\prod_{j \neq i} U_j) \cap U_i = Z$ for all i. Then we say that G is the* **central product** *of $U_1, \ldots, U_k$.*

Notice in this case that $Z \subseteq \mathbf{Z}(G)$, and that $G/Z = U_1/Z \times \cdots \times U_k/Z$ is a direct product. (We mention that there are different definitions of central products in the literature.)

Theorem 10.7 *Suppose that G is the central product of $U_1, \ldots, U_k$, and let $Z = \bigcap_{i=1}^{k} U_i$. Let $\lambda \in \mathrm{Irr}(Z)$. Given $\theta_i \in \mathrm{Irr}(U_i|\lambda)$ for $i = 1, \ldots, k$, then there exists a unique character $\chi \in \mathrm{Irr}(G)$, which we denote by $\theta_1 \cdot \ldots \cdot \theta_k$, that lies over every θ_i. In fact, the map*

$$\mathrm{Irr}(U_1|\lambda) \times \ldots \times \mathrm{Irr}(U_k|\lambda) \to \mathrm{Irr}(G|\lambda)$$

given by

$$(\theta_1, \ldots, \theta_k) \mapsto \theta_1 \cdot \ldots \cdot \theta_k$$

is a bijection. Furthermore $(\theta_1 \cdot \ldots \cdot \theta_k)(u_1 \cdots u_k) = \theta_1(u_1) \cdots \theta_k(u_k)$ for $u_i \in U_i$.

Proof Let $G^* = U_1 \times \cdots \times U_k$ be the direct product of the groups $U_1, \ldots, U_k$, and define $f \colon G^* \to G$ by $f(u_1, \ldots, u_k) = u_1 \cdots u_k$. Using that $[U_i, U_j] = 1$ for $i \neq j$, we easily check that f is a group homomorphism. Also, f is onto because $G = U_1 \cdots U_k$. Using that $G/Z = U_1/Z \times \cdots \times U_k/Z$, we see that

$$K = \ker(f) = \{(z_1, \ldots, z_k) \in Z^k \mid z_1 \cdots z_k = 1\}.$$

Now, if a is the isomorphism $G^*/K \to G$ associated with f, by Theorem 2.1 we have that a induces a bijection between the irreducible characters of G^* that have K in their kernel with the irreducible characters of G. Now, suppose that $\theta_i \in \mathrm{Irr}(U_i)$, and write $(\theta_i)_Z = e_i \lambda_i$, where $e_i = \theta_i(1)$ and $\lambda_i \in \mathrm{Irr}(Z)$ is linear. We easily check, by using elements of the form $(z, 1, \ldots, 1, z^{-1}, 1, \ldots, 1)$, that $\eta = \theta_1 \times \cdots \times \theta_k$ contains K in its kernel if and only if $\lambda_1 = \cdots = \lambda_k$. If $\theta_i \in \mathrm{Irr}(U_i|\lambda)$ for all i, for a given $\lambda \in \mathrm{Irr}(Z)$, then the image under a of η is the unique character $\chi \in \mathrm{Irr}(G)$ such that $\chi(u_1 \cdots u_k) = \theta_1(u_1) \cdots \theta_k(u_k)$ for all $u_i \in U_i$. Notice in this case that

$$\chi(u_i) = (\prod_{j \neq i} \theta_j(1))\theta_i(u_i)$$

for $u_i \in U_i$, and therefore χ lies over every θ_i, and, in particular, over λ. Let us write $\chi = \theta_1 \cdot \ldots \cdot \theta_k$.

Finally, if $\psi \in \mathrm{Irr}(G|\theta_i)$ for all i, then ψ lies over λ, and we may write $\psi = \mu_1 \cdot \ldots \cdot \mu_k$ for some $\mu_i \in \mathrm{Irr}(U_i|\lambda)$, by the first part of the proof. Then ψ_{U_i} is a multiple of μ_i, and therefore $\mu_i = \theta_i$ for all i and $\psi = \theta_1 \cdot \ldots \cdot \theta_k = \chi$. $\square$

Although Theorem 10.7 is useful in many instances, for the purposes of this chapter we shall only need a trivial case.

Corollary 10.8 *Suppose that $G = ZK$, where $Z \subseteq \mathbf{Z}(G)$. Let $\nu \in \mathrm{Irr}(Z)$ and let $\lambda = \nu_{K \cap Z}$. Then restriction defines a bijection $\mathrm{Irr}(G|\nu) \rightarrow \mathrm{Irr}(K|\lambda)$ whose inverse map is $\theta \mapsto \nu \cdot \theta$.*

Proof The fact that restriction is a bijection follows, for instance, from Lemma 6.8(d). Now, G is the central product of $Z = U_1$ and $K = U_2$. By Theorem 10.7, the map $\mathrm{Irr}(Z|\lambda) \times \mathrm{Irr}(K|\lambda) \mapsto \mathrm{Irr}(G|\lambda)$ given by $(\epsilon, \theta) \mapsto \epsilon \cdot \theta$ is a bijection, where $\epsilon \cdot \theta$ is the unique irreducible character χ of G that lies over ϵ and θ. Hence the map $\theta \mapsto \nu \cdot \theta$ is a bijection $\mathrm{Irr}(K|\lambda) \mapsto \mathrm{Irr}(G|\nu)$. Since $(\nu \cdot \theta)(k) = \nu(1)\theta(k) = \theta(k)$ for $k \in K$, the proof is finished. $\square$

10.4 Back to Projective Representations

As defined in Section 2 of Chapter 5, a projective representation of G is a map

$$\mathcal{P} \colon G \to \mathrm{GL}_n(\mathbb{C})$$

such that for every $x, y \in G$ there is $\alpha(x, y) \in \mathbb{C}^\times$ satisfying

$$\mathcal{P}(x)\mathcal{P}(y) = \alpha(x, y)\mathcal{P}(xy).$$

The function $\alpha \colon G \times G \to \mathbb{C}^\times$ is the factor set of $\mathcal{P}$. Recall that we write $\alpha \in \mathbf{Z}^2(G, \mathbb{C}^\times)$. In this chapter, we let $\mathrm{Proj}(G|\alpha)$ be the set of projective representations of G with factor set α. If $\alpha(x, y) = 1$ for all $x, y \in G$, then $\mathrm{Proj}(G|\alpha)$ is the set of complex representations of G.

There are two easy ways of constructing new projective representations from old ones. For instance, if $\mathcal{P}$ is a projective representation and $\mu \colon G \to \mathbb{C}^\times$ is any function, then $\mu\mathcal{P}$ defined by

$$(\mu \mathcal{P})(x) = \mu(x)\mathcal{P}(x)$$

for $x \in G$, is another projective representation of G. This representation has factor set $\mu \alpha$, where

$$(\mu \alpha)(x, y) = \mu(x)\mu(y)\mu(xy)^{-1}\alpha(x, y).$$

Also, if $M \in \mathrm{GL}_n(\mathbb{C})$ is any invertible matrix, then the map $M^{-1}\mathcal{P}M: G \to \mathrm{GL}_n(\mathbb{C})$ defined by

$$(M^{-1}\mathcal{P}M)(x) = M^{-1}\mathcal{P}(x)M$$

is a projective representation of G with factor set α. We say in this case that $\mathcal{P}$ and $M^{-1}\mathcal{P}M$ are **similar**. Furthermore, we say that $\mathcal{P}$ is **irreducible** if $\mathcal{P}$ is not similar to a projective representation of the form

$$\begin{pmatrix} * & 0 \\ 0 & * \end{pmatrix}.$$

If $\alpha \in \mathbf{Z}^2(G, \mathbb{C}^\times)$, then it is entirely possible to build a theory of projective representations of G with factor set α, as is done for ordinary representations. The role of the group algebra $\mathbb{C}G$ is then played by the so called **twisted group algebra** $\mathbb{C}^\alpha G$ (which is the complex vector space with basis $\bar{G} = \{\bar{g} \mid g \in G\}$ and with multiplication given by $\bar{g}\bar{h} = \alpha(g, h)\overline{gh}$ for $g, h \in G$).

The following is a trivial but useful fact whose proof is left to the reader.

Theorem 10.9 *Suppose that $a: G \to H$ is an isomorphism of groups. Let $\alpha \in \mathbf{Z}^2(G, \mathbb{C}^\times)$ and let $\alpha^a \in \mathbf{Z}^2(H, \mathbb{C}^\times)$ defined by $\alpha^a(h, k) = \alpha(h^{a^{-1}}, k^{a^{-1}})$ for $h, k \in H$. If $\mathcal{P} \in \mathrm{Proj}(G|\alpha)$ and $\mathcal{P}^a$ is defined by $\mathcal{P}^a(h) = \mathcal{P}(h^{a^{-1}})$, then the map $\mathcal{P} \mapsto \mathcal{P}^a$ is a bijection $\mathrm{Proj}(G|\alpha) \to \mathrm{Proj}(H|\alpha^a)$ that respects similarity and irreducibility.*

Recall from Chapter 5 that projective representations naturally appear in the presence of character triples. If (G, N, θ) is a character triple, we proved in Theorem 5.5 that there always exists a projective representation $\mathcal{P}$ such that $\mathcal{P}_N$ is an ordinary representation affording θ and such that $\mathcal{P}(ng) = \mathcal{P}(n)\mathcal{P}(g)$ and $\mathcal{P}(gn) = \mathcal{P}(g)\mathcal{P}(n)$ for $n \in N$ and $g \in G$. (The projective representations of G satisfying these conditions are called the projective representations associated with θ by Definition 5.2.)

We shall need a few more properties of this type of projective representation.

Lemma 10.10 *Suppose that (G, N, θ) is a character triple, and let $\mathcal{P}$ be a projective representation of G associated with θ with factor set α. Then*

(a) $\mathcal{P}(g)\mathcal{P}(n)\mathcal{P}(g)^{-1} = \mathcal{P}(gng^{-1})$ *for* $g \in G$, $n \in N$.

(b) *Let* $\mathcal{P}' \colon G \to \mathrm{GL}_{\theta(1)}(\mathbb{C})$ *be any function. Then* $\mathcal{P}'$ *is a projective representation associated with* θ *if and only if there is a matrix* $M \in \mathrm{GL}_{\theta(1)}(\mathbb{C})$, *and a function* $\mu \colon G \to \mathbb{C}^{\times}$ *with* $\mu(1) = 1$ *constant on cosets of* N *such that* $\mathcal{P}'(g) = \mu(g)M^{-1}\mathcal{P}(g)M$ *for all* $g \in G$.

(c) *Suppose that* $Z \trianglelefteq G$ *is contained in* $\ker(\theta)$. *Then the function* $\bar{\mathcal{P}} \colon G/Z \to \mathrm{GL}_{\theta(1)}(\mathbb{C})$ *defined by* $\bar{\mathcal{P}}(Zg) = \mathcal{P}(g)$ *is a well-defined projective representation associated with* $(G/Z, N/Z, \bar{\theta})$, *where* $\bar{\theta}(Zn) = \theta(n)$ *for* $n \in N$.

Proof If $g \in G$ and $n \in N$, then $\mathcal{P}(g)\mathcal{P}(n) = \mathcal{P}(gn) = \mathcal{P}(gng^{-1}g) = \mathcal{P}(gng^{-1})\mathcal{P}(g)$, and (a) follows.

Next, we prove (b). It is straightforward to check that any projective representation of the form $\mu M^{-1}\mathcal{P}M$ with $\mu(1) = 1$ and $\mu(gn) = \mu(g)$ for $g \in G$ and $n \in N$ is associated with θ. Suppose conversely that $\mathcal{P}'$ is a projective representation associated with θ. Since $(\mathcal{P}')_N$ affords θ, then there is a matrix $M_1 \in \mathrm{GL}_{\theta(1)}(\mathbb{C})$ such that

$$M_1^{-1}\mathcal{P}'(n)M_1 = \mathcal{P}(n)$$

for all $n \in N$. (This is because two ordinary representations affording the same character are similar.) So, replacing $\mathcal{P}'$ by $M_1^{-1}\mathcal{P}'M_1$, we may assume that $\mathcal{P}(n) = \mathcal{P}'(n)$ for all $n \in N$. Now, using (a), we have that

$$\mathcal{P}(g)\mathcal{P}(n)\mathcal{P}(g)^{-1} = \mathcal{P}'(g)\mathcal{P}(n)\mathcal{P}'(g)^{-1},$$

and we deduce that $\mathcal{P}(g)^{-1}\mathcal{P}'(g)$ is a scalar matrix by Schur's lemma (Theorem 1.7). Thus there is a function $\mu \colon G \to \mathbb{C}^{\times}$ such that $\mathcal{P}'(g) = \mu(g)\mathcal{P}(g)$ for all $g \in G$. Notice then that $\mu(n) = 1$ for $n \in N$ and that $\mu(gn) = \mu(g)$ for $g \in G$ and $n \in N$.

Finally, in (c), notice that $\mathcal{P}(z) = I_{\theta(1)}$, since $\mathcal{P}_N$ affords θ and $Z \subseteq \ker(\theta)$. For $z \in Z$, we have that $\mathcal{P}(gz) = \mathcal{P}(g)\mathcal{P}(z) = \mathcal{P}(g)$. The rest of (c) is straightforward. □

We end this section with a key result of A. H. Clifford relating ordinary representations and projective representations associated with a character triple. Although its proof is standard and not difficult (it only uses linear algebra and Schur's lemma), it is not a short proof either. Therefore, we only provide the reader with a reference for a proof.

We need to explain precisely its statement. Using that the Kronecker product of matrices defines a group homomorphism $\mathrm{GL}_n(\mathbb{C}) \times \mathrm{GL}_m(\mathbb{C}) \to \mathrm{GL}_{nm}(\mathbb{C})$, we check that if $\mathcal{Q} \in \mathrm{Proj}(G|\alpha)$ and $\mathcal{P} \in \mathrm{Proj}(G|\beta)$, then $\mathcal{Q} \otimes \mathcal{P}$ defined as

$$(\mathcal{Q} \otimes \mathcal{P})(g) = \mathcal{Q}(g) \otimes \mathcal{P}(g)$$

for $g \in G$, is a projective representation of G with factor set $\alpha\beta$, where

$$(\alpha\beta)(x, y) = \alpha(x, y)\beta(x, y)$$

for $x, y \in G$. In particular, if $\mathcal{Q} \in \mathrm{Proj}(G|\alpha^{-1})$, where α^{-1} is the factor set defined by

$$\alpha^{-1}(x, y) = \alpha(x, y)^{-1},$$

and $\mathcal{P} \in \mathrm{Proj}(G|\alpha)$, then we have that $\mathcal{Q} \otimes \mathcal{P}$ has trivial factor set, so is an ordinary representation of G.

In what follows, we shall view the projective representations of G/N as projective representations $\mathcal{Q}$ of G satisfying $\mathcal{Q}(gn) = \mathcal{Q}(g)$ for all $g \in G$ and $n \in N$. If $\beta \in \mathbf{Z}^2(G, \mathbb{C}^\times)$ is the factor set of such a representation, notice that $\beta(gn, hm) = \beta(g, h)$ for $g, h \in G$ and $n, m \in N$. If (G, N, θ) is a character triple, this is exactly what the factor set α of a projective representation associated with θ satisfies by Lemma 5.3(c). Hence, it makes perfect sense to write $\mathrm{Proj}(G/N|\alpha^{-1})$, as the set of projective representations $\mathcal{Q}$ of G such that $\mathcal{Q}(gn) = \mathcal{Q}(g)$ for $g \in G$ and $n \in N$, having factor set α^{-1}. If $\alpha(1, 1) = 1$ (as happens with factor sets coming from character triples by Lemma 5.3(a)) and $\mathcal{Q} \in \mathrm{Proj}(G/N|\alpha^{-1})$, then notice that $\mathcal{Q}(n)$ is the identity matrix for all $n \in N$.

We need one final bit of notation: if (G, N, θ) is a character triple, then let $\mathrm{Rep}(G|\theta)$ be the set of all the representations of G that afford characters in $\mathrm{Char}(G|\theta)$.

Theorem 10.11 (Clifford) *Suppose that (G, N, θ) is a character triple. Let $\mathcal{P}$ be a projective representation of G associated with θ with factor set α. Then the following hold.*

(a) *The map $\mathrm{Proj}(G/N|\alpha^{-1}) \rightarrow \mathrm{Rep}(G|\theta)$ defined by $\mathcal{Q} \mapsto \mathcal{Q} \otimes \mathcal{P}$ is injective.*

(b) *If $\chi \in \mathrm{Char}(G|\theta)$, then there is $\mathcal{Q} \in \mathrm{Proj}(G/N|\alpha^{-1})$ such that $\mathcal{Q} \otimes \mathcal{P}$ affords χ.*

(c) *If $\mathcal{Q} \in \mathrm{Proj}(G/N|\alpha^{-1})$, we have that $\mathcal{Q} \otimes \mathcal{P}$ is irreducible if and only if $\mathcal{Q}$ is irreducible.*

(d) *If $\mathcal{Q}, \mathcal{Q}' \in \mathrm{Proj}(G/N|\alpha^{-1})$, we have that $\mathcal{Q} \otimes \mathcal{P}$ and $\mathcal{Q}' \otimes \mathcal{P}$ are similar if and only if $\mathcal{Q}$ and $\mathcal{Q}'$ are similar.*

Proof From the previous discussion, we have that the map is well-defined. (Notice that $(\mathcal{Q} \otimes \mathcal{P})_N$ affords a multiple of θ, and therefore $\mathcal{Q} \otimes \mathcal{P} \in$

Rep($G|\theta$).) By using the definition of the Kronecker product of matrices, it follows that the map is one to one. Now, we use Theorems 8.16 and 8.18 of [Na98], replacing the field F by $\mathbb{C}$. Part (b) corresponds to Theorem 8.16. Parts (c) and (d) correspond to Theorem 8.18. □

10.5 Ordering Character Triples

Let us first motivate the reader on what we are pursuing now. In several problems in character theory we wish to compare certain characters of a finite group G with certain characters of a subgroup of G. In the McKay conjecture, for instance, we want to compare the characters of G of degree not divisible by p with the characters of $\mathbf{N}_G(P)$ of degree not divisible by p, where $P \in \mathrm{Syl}_p(G)$. Suppose that N is a normal subgroup of G and let $Q = P \cap N \in \mathrm{Syl}_p(N)$. By the Frattini argument, notice that $G = NH$, where $H = \mathbf{N}_G(Q)$. Suppose, say by induction or by any method, that given $\theta \in \mathrm{Irr}_{p'}(N)$ we can construct some $\theta' \in \mathrm{Irr}_{p'}(H \cap N)$ in a *nice* way. (In the Okuyama–Wajima hypotheses, we have a similar situation: N is a p'-group, P is a p-subgroup of G, $NP \trianglelefteq G$, $H = \mathbf{N}_G(P)$, θ is P-invariant, and $\theta' \in \mathrm{Irr}(N \cap H)$ is the P-Glauberman correspondent of θ.) Among these *nice* properties, we would like that $x \in H$ stabilizes θ if and only if x stabilizes θ' (that is, $H_\theta = H_{\theta'}$), and also that the triples (G_θ, N, θ) and $(H_{\theta'}, H \cap N, \theta')$ are isomorphic. If this is the case, then we will have a bijection $\mathrm{Irr}_{p'}(G|\theta) \to \mathrm{Irr}_{p'}(H|\theta')$, by using the Clifford correspondence and character triple isomorphisms. Since $\mathbf{N}_G(P) \subseteq H$, somehow induction will make the rest.

We start by defining an order relation between character triples. Suppose that (G, N, θ) and (H, M, φ) are character triples, where $H \le G$. We write

$$(G, N, \theta) \ge (H, M, \varphi)$$

if the following conditions are satisfied:

(a) $G = NH$.
(b) $N \cap H = M$.
(c) There exist projective representations $\mathcal{P}$ of G and $\mathcal{P}'$ of H associated with θ and φ respectively, with factor sets α and α', such that

$$\alpha'(h_1, h_2) = \alpha(h_1, h_2)$$

for all $h_1, h_2 \in H$.

In this case, we say that $(\mathcal{P}, \mathcal{P}')$ is **associated** with $(G, N, \theta) \ge (H, M, \varphi)$.

The following is a typical easy example of this kind of relation between character triples.

Theorem 10.12 *Suppose that (G, N, θ) and (H, M, φ) are character triples. Assume that $G = NH$ and $M = N \cap H$. If $\theta_M = \varphi$, then $(G, N, \theta) \geq (H, M, \varphi)$.*

Proof By Theorem 5.5, let $\mathcal{P}$ be any projective representation of G associated with θ. Then it is clear that $\mathcal{P}_H$ is a projective representation of H associated with φ, and of course the factor sets coincide on $H \times H$. $\square$

If $(G, N, \theta) \geq (H, M, \varphi)$, then we are going to show next that these two character triples are isomorphic.

Theorem 10.13 *Suppose that $(\mathcal{P}, \mathcal{P}')$ is associated with $(G, N, \theta) \geq (H, M, \varphi)$, and let α be the factor set of $\mathcal{P}$. Given $N \leq J \leq G$ and $\psi \in \mathrm{Irr}(J|\theta)$, we know by Theorem 10.11(b) that there exists $\mathcal{Q} \in \mathrm{Proj}(J/N|\alpha^{-1}{}_{J \times J})$ such that $\mathcal{Q} \otimes \mathcal{P}_J$ affords ψ. If we define*

$$\psi'(x) = \mathrm{trace}(\mathcal{Q}(x) \otimes \mathcal{P}'(x))$$

for $x \in J \cap H$, then the map $': \mathrm{Irr}(J|\theta) \rightarrow \mathrm{Irr}(J \cap H|\varphi)$ given by $\psi \mapsto \psi'$ defines an isomorphism between the character triples (G, N, θ) and (H, M, φ).

Proof Let $\psi \in \mathrm{Irr}(J|\theta)$. If $\mathcal{Q}, \mathcal{Q}' \in \mathrm{Proj}(J/N|\alpha^{-1}{}_{J \times J})$ are such that $\mathcal{Q} \otimes \mathcal{P}_J$ and $\mathcal{Q}' \otimes \mathcal{P}_J$ afford ψ, we know by Theorem 10.11(d), that $\mathcal{Q}$ and $\mathcal{Q}'$ are similar. Then so are $\mathcal{Q}_{J \cap H} \otimes \mathcal{P}'_{J \cap H}$ and $\mathcal{Q}'_{J \cap H} \otimes \mathcal{P}'_{J \cap H}$, and we conclude that ψ' is well defined. Since $\mathcal{Q}$ is irreducible (because ψ is irreducible), we have that $\mathcal{Q}_{J \cap H}$ is irreducible, by Theorem 10.9 applied to the natural isomorphism $J/N \rightarrow (J \cap H)/M$. Thus $\mathcal{Q}_{J \cap H} \otimes \mathcal{P}'_{J \cap H}$ is an irreducible representation (using Theorem 10.11(c)). Hence $\psi' \in \mathrm{Irr}(H \cap J|\varphi)$. A similar argument proves that $\psi \mapsto \psi'$ is a bijection. Suppose now that $\mu \in \mathrm{Irr}(J/N)$, and let $\mathcal{X}$ be a representation of J with $N \subseteq \ker(\mathcal{X})$ affording μ. Then $\mathcal{X} \otimes \mathcal{Q} \in \mathrm{Proj}(J/N|\alpha^{-1}{}_{J \times J})$ and $\mathcal{X} \otimes \mathcal{Q} \otimes \mathcal{P}_J$ affords $\mu\psi$. Then $(\mu\psi)'$ is the trace of $\mathcal{X} \otimes \mathcal{Q}_{J \cap H} \otimes \mathcal{P}'_{J \cap H}$ which is $\mu_{J \cap H}\psi'$. We easily check that our construction is compatible with restriction of characters. $\square$

We say that the character triple isomorphism from (G, N, θ) to (H, M, φ) obtained in Theorem 10.13 is **induced** by $(\mathcal{P}, \mathcal{P}')$.

10.6 Ordering Character Triples Centrally

There is one further property which is key in the applications that we have in mind and that gives the name to this section.

Let (G, N, θ) be a character triple and suppose that $\mathcal{P}$ is a projective representation associated with θ. If $c \in \mathbf{C}_G(N)$ and $n \in N$, then we have that

$$\mathcal{P}(n)\mathcal{P}(c) = \mathcal{P}(nc) = \mathcal{P}(cn) = \mathcal{P}(c)\mathcal{P}(n).$$

Since $\mathcal{P}_N$ is an ordinary irreducible representation, we deduce that $\mathcal{P}(c)$ is a scalar matrix, by Schur's lemma (Theorem 1.7). Therefore a projective representation $\mathcal{P}$ of G associated with θ uniquely defines a function $\mu \colon \mathbf{C}_G(N) \to \mathbb{C}^\times$. Furthermore, notice that $\mu_{\mathbf{Z}(N)} = \lambda$ is a linear ordinary character and that μ is a projective representation of $\mathbf{C}_G(N)$ associated with λ with factor set $\alpha_{\mathbf{C}_G(N) \times \mathbf{C}_G(N)}$.

Definition 10.14 *Let G be a finite group and suppose that $(G, N, \theta) \geq (H, M, \varphi)$. We shall write $(G, N, \theta) \geq_c (H, M, \varphi)$ if the following conditions hold.*

(a) $\mathbf{C}_G(N) \subseteq H$.
(b) There is $(\mathcal{P}, \mathcal{P}')$ associated with $(G, N, \theta) \geq (H, M, \varphi)$ and $\mu \colon \mathbf{C}_G(N) \to \mathbb{C}^\times$ such that $\mathcal{P}(x) = \mu(x)I_{\theta(1)}$ and $\mathcal{P}'(x) = \mu(x)I_{\varphi(1)}$ for all $x \in \mathbf{C}_G(N)$.

In this case, we will say that the character triple isomorphism induced by Theorem 10.13 is **central**, and that $(\mathcal{P}, \mathcal{P}')$ is **associated** with $(G, N, \theta) \geq_c (H, M, \varphi)$.

Notice that part (b) of Definition 10.14 makes perfect sense. Since $\mathbf{C}_G(N) \subseteq H$ by part (a), then we have that $\mathbf{C}_G(N) \subseteq \mathbf{C}_H(M)$, and therefore $\mathcal{P}'(x)$ is scalar for $x \in \mathbf{C}_G(N)$, by Schur's lemma. Notice too that $\mathbf{Z}(N) \subseteq M$ and that θ and φ lie over the same linear character of $\mathbf{Z}(N)$ if $(G, N, \theta) \geq_c (H, M, \varphi)$.

In summary, if we have two triples, (G, N, θ) and (H, M, φ), and we wish to prove that $(G, N, \theta) \geq_c (H, M, \varphi)$, then we need to check five conditions. Three of them are group theoretical: $G = NH$, $N \cap H = M$ and $\mathbf{C}_G(N) \subseteq H$. Then we have to find a projective representation $\mathcal{P}$ of G associated with θ and a projective representation $\mathcal{P}'$ associated with φ such that the factors sets coincide on $H \times H$ (thus, $(\mathcal{P}, \mathcal{P}')$ is associated with $(G, N, \theta) \geq (H, M, \varphi)$) and such that the scalar matrices $\mathcal{P}(x)$ and $\mathcal{P}'(x)$ have the same scalar for $x \in \mathbf{C}_G(N)$.

The following is an easy example of this situation.

Theorem 10.15 *Suppose that $N \trianglelefteq G$ and $N\mathbf{C}_G(N) = G$. Let $Z = \mathbf{Z}(N)$. Let $\theta \in \mathrm{Irr}(N)$ and write $\theta_Z = \theta(1)\lambda$, for some linear $\lambda \in \mathrm{Irr}(Z)$. Then $(G, N, \theta) \geq_c (\mathbf{C}_G(N), Z, \lambda)$.*

Proof The group theoretical conditions are obvious in this case. By Theorem 5.5, let $\mathcal{P}$ be a projective representation of G associated with θ. Write $H = \mathbf{C}_G(N)$. Now, let $h \in H$. Then

$$\mathcal{P}(h)\mathcal{P}(n) = \mathcal{P}(hn) = \mathcal{P}(nh) = \mathcal{P}(n)\mathcal{P}(h)$$

for all $n \in N$. By Schur's lemma, there is $\mu \colon H \to \mathbb{C}^\times$ such that $\mathcal{P}(h) = \mu(h)I_{\theta(1)}$ for $h \in H$. Notice that $\mu_Z = \lambda$ by taking traces in the previous equation for $h \in Z$. Now it is clear that $(\mathcal{P}, \mu)$ is associated with $(G, N, \theta) \geq_c (\mathbf{C}_G(N), Z, \lambda)$. □

Suppose that $(\mathcal{P}, \mathcal{P}')$ is associated with $(G, N, \theta) \geq (H, M, \varphi)$, and assume that $\mathbf{C}_G(N) \subseteq H$. There is a convenient way to check that the scalar in the matrices $\mathcal{P}(x)$ and $\mathcal{P}'(x)$ is the same, and therefore that $(\mathcal{P}, \mathcal{P}')$ is associated with $(G, N, \theta) \geq_c (H, M, \varphi)$.

Lemma 10.16 *Suppose that $(\mathcal{P}, \mathcal{P}')$ is associated with $(G, N, \theta) \geq (H, M, \varphi)$. Assume that $\mathbf{C}_G(N) \subseteq H$. For $N \leq J \leq G$ and $\psi \in \mathrm{Irr}(J|\theta)$, we let $\psi' \in \mathrm{Irr}(J \cap H|\varphi)$ be the image of ψ under the isomorphism induced by $(\mathcal{P}, \mathcal{P}')$. Then $(\mathcal{P}, \mathcal{P}')$ is associated with $(G, N, \theta) \geq_c (H, M, \varphi)$ if and only if $\mathrm{Irr}(\psi_{\mathbf{C}_J(N)}) = \mathrm{Irr}(\psi'_{\mathbf{C}_J(N)})$ for every $N \subseteq J \leq G$ and $\psi \in \mathrm{Irr}(J|\theta)$.*

Proof Suppose that $(\mathcal{P}, \mathcal{P}')$ is associated with $(G, N, \theta) \geq_c (H, M, \varphi)$ and let $\mu \colon \mathbf{C}_G(N) \to \mathbb{C}^\times$ be the function that satisfies $\mathcal{P}(c) = \mu(c)I_{\theta(1)}$ and $\mathcal{P}'(c) = \mu(c)I_{\varphi(1)}$ for $c \in \mathbf{C}_G(N)$. Let α be the factor set of $\mathcal{P}$. Let $\psi \in \mathrm{Irr}(J|\theta)$. Then there is $\mathcal{Q} \in \mathrm{Proj}(J/N|\alpha^{-1}{}_{J \times J})$ such that $\mathcal{Q} \otimes \mathcal{P}_J$ affords ψ and $\mathcal{Q}_{H \cap J} \otimes \mathcal{P}'_{H \cap J}$ affords ψ'. Let $c \in \mathbf{C}_J(N) \subseteq H$. Then $\mathcal{P}(c) = \mu(c)I_{\theta(1)}$ and

$$\psi(c) = \mathrm{trace}(\mathcal{Q}(c) \otimes \mathcal{P}(c)) = \theta(1)\mu(c)\mathrm{trace}(\mathcal{Q}(c)).$$

In the same way,

$$\psi'(c) = \mathrm{trace}(\mathcal{Q}(c) \otimes \mathcal{P}'(c)) = \varphi(1)\mu(c)\mathrm{trace}(\mathcal{Q}(c)),$$

and we deduce that $\varphi(1)\psi_{\mathbf{C}_J(N)} = \theta(1)\psi'_{\mathbf{C}_J(N)}$. Therefore, $\mathrm{Irr}(\psi_{\mathbf{C}_J(N)}) = \mathrm{Irr}(\psi'_{\mathbf{C}_J(N)})$.

We prove the converse. Let $c \in \mathbf{C}_G(N)$. Write $\mathcal{P}(c) = aI_{\theta(1)}$ and $\mathcal{P}'(c) = a'I_{\varphi(1)}$ for some scalars a and a'. We wish to prove that $a = a'$. Let $J = N\langle c \rangle$. Then $c \in \mathbf{Z}(J)$ and therefore $\mathbf{C}_J(N) = \mathbf{Z}(J)$. Let $\psi \in \mathrm{Irr}(J|\theta)$. We know that $\psi = \mathrm{trace}(\mathcal{Q} \otimes \mathcal{P}_J)$ and $\psi' = \mathrm{trace}(\mathcal{Q}_{J \cap H} \otimes \mathcal{P}'_{J \cap H})$ for some projective representation $\mathcal{Q}$ of J/N with factor set $\alpha^{-1}_{J \times J}$. In particular, $\psi(c) = a\theta(1)\mathrm{trace}(\mathcal{Q}(c))$ and $\psi'(c) = a'\varphi(1)\mathrm{trace}(\mathcal{Q}(c))$. By hypothesis, we can write $\psi_{\mathbf{C}_J(N)} = \theta(1)\nu$ and $\psi'_{\mathbf{C}_J(N)} = \varphi(1)\nu$, where $\nu \in \mathrm{Irr}(\mathbf{C}_J(N))$ is linear. In particular, $\psi(c)/\theta(1) = \nu(c) \neq 0$, and therefore $\mathrm{trace}(\mathcal{Q}(c)) \neq 0$. Since $\psi'(c)/\varphi(1) = \nu(c)$, we have that

$$a = \frac{\psi(c)}{\theta(1)\mathrm{trace}(\mathcal{Q}(c))} = \frac{\psi'(c)}{\varphi(1)\mathrm{trace}(\mathcal{Q}(c))} = a',$$

as desired. $\qquad\square$

It is now convenient to write the exact result that we are going to use in the final theorem of this chapter.

Theorem 10.17 *Let G be a finite group, and suppose that $(G, K, \theta) \geq_c$ (H, M, φ). Suppose that $Z \leq \mathbf{Z}(G)$, and let $\lambda \in \mathrm{Irr}(Z)$. Let p be a prime. Assume that θ and φ have degree not divisible by p and lie over $\lambda_{M \cap Z}$. Then*

$$|\mathrm{Irr}_{p'}(G|\theta \cdot \lambda)| = |\mathrm{Irr}_{p'}(H|\varphi \cdot \lambda)|.$$

Proof Let $(\mathcal{P}, \mathcal{P}')$ be associated with $(G, K, \theta) \geq_c (H, M, \varphi)$. Let

$$': \mathrm{Irr}(G|\theta) \to \mathrm{Irr}(H|\varphi)$$

be the bijection associated with the character triple isomorphism induced by $(\mathcal{P}, \mathcal{P}')$. By hypothesis, we have that $\mathbf{C}_G(K) \subseteq H$. In particular, $Z \cap K = Z \cap M$. By Lemma 10.16, we have that $\mathrm{Irr}(\psi_{\mathbf{C}_G(K)}) = \mathrm{Irr}(\psi'_{\mathbf{C}_G(K)})$, for $\psi \in \mathrm{Irr}(G|\theta)$. Since $Z \subseteq \mathbf{C}_G(K)$, then $\mathrm{Irr}(\psi_Z) = \mathrm{Irr}(\psi'_Z)$, and therefore ψ lies over λ if and only if ψ' lies over λ. Since it is clear that ψ lies over θ and λ if and only if ψ lies over $\theta \cdot \lambda$, and that ψ' lies over φ and λ if and only if ψ' lies over $\varphi \cdot \lambda$ (by Corollary 10.8, if we wish), then we have that $'$ sends $\mathrm{Irr}(G|\theta \cdot \lambda)$ onto $\mathrm{Irr}(H|\varphi \cdot \lambda)$. Now, by the remark after Definition 5.7 we have that

$$\psi(1)/\theta(1) = \psi'(1)/\varphi(1),$$

for $\psi \in \mathrm{Irr}(G|\theta)$. In particular, ψ has degree not divisible by p if and only if ψ' has degree not divisible by p. $\qquad\square$

The following is the main result concerning central isomorphisms of triples.

Theorem 10.18 (Späth) *Suppose that $(G, N, \theta) \geq_c (H, M, \varphi)$. Let $\epsilon: G \to \mathrm{Aut}(N)$ be the conjugation homomorphism $\epsilon(g) = \alpha_g$, where $\alpha_g(n) = n^g$. Suppose that $N \trianglelefteq \widehat{G}$, where $\widehat{G}$ is some finite group. Let $\hat{\epsilon}: \widehat{G} \to \mathrm{Aut}(N)$ be the corresponding conjugation homomorphism and assume that $\hat{\epsilon}(\widehat{G}) = \epsilon(G)$. Let*

$$\widehat{H} = \hat{\epsilon}^{-1}(\epsilon(H)).$$

Then

$$(\widehat{G}, N, \theta) \geq_c (\widehat{H}, M, \varphi).$$

Proof By definition, we have that

$$\widehat{H} = \{x \in \widehat{G} \mid \text{there exists } h \in H \text{ such that } n^x = n^h \text{ for every } n \in N\}.$$

Hence $\mathbf{C}_{\widehat{G}}(N) \subseteq \widehat{H}$ (by taking $h = 1$ for every $x \in \mathbf{C}_{\widehat{G}}(N)$).

Let us start by proving that $\widehat{G} = N\widehat{H}$. First of all notice that $\hat{\epsilon}(n) = \epsilon(n)$ for all $n \in N$. Let $x \in \widehat{G}$. Then $\hat{\epsilon}(x) = \epsilon(g)$ for some $g \in G$, by hypothesis. Since $G = HN$, write $g = hn$ for some $h \in H$ and $n \in N$. Then $\hat{\epsilon}(x) = \epsilon(hn) = \epsilon(h)\hat{\epsilon}(n)$. Thus $\hat{\epsilon}(xn^{-1}) = \epsilon(h)$ and $xn^{-1} \in \widehat{H}$. Therefore $\widehat{G} = N\widehat{H}$.

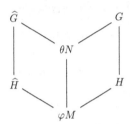

Next we prove that $N \cap \widehat{H} = M$. Let $n \in N \cap \widehat{H}$. Hence $\hat{\epsilon}(n) = \epsilon(h)$ for some $h \in H$. Thus $\epsilon(nh^{-1}) = 1$. Hence $nh^{-1} \in \mathbf{C}_G(N) \subseteq H$. Thus $n \in H$ and $n \in N \cap H = M$. Therefore $N \cap \widehat{H} \subseteq M$. If $n \in M = N \cap H$, then $\hat{\epsilon}(n) = \epsilon(h)$ for $h = n$ and thus $n \in \widehat{H}$. Hence $N \cap \widehat{H} = N \cap H = M$.

We claim now that θ is $\widehat{G}$-invariant and φ is $\widehat{H}$-invariant. This is because for $x \in \widehat{G}$ there exists $g \in G$ such that $n^x = n^g$ for every $n \in N$. Hence $\theta(n^x) = \theta(n^g) = \theta(n)$. The same argument proves that φ is $\widehat{H}$-invariant.

Suppose that $(\mathcal{P}, \mathcal{P}')$ is associated with $(G, N, \theta) \geq_c (H, M, \varphi)$. Hence $\mathcal{P}$ is a projective representation of G associated with θ and $\mathcal{P}'$ is a projective representation of H associated with φ, having factor sets α and α' such that

$$\alpha'(h_1, h_2) = \alpha(h_1, h_2)$$

for $h_i \in H$. Also, there is a map $\mu : \mathbf{C}_G(N) \to \mathbb{C}^\times$ such that $\mathcal{P}(c) = \mu(c)I_{\theta(1)}$ and $\mathcal{P}'(c) = \mu(c)I_{\varphi(1)}$ for $c \in \mathbf{C}_G(N)$.

We define a map $\bar{\epsilon} : G/\mathbf{C}_G(N) \to \widehat{G}/\mathbf{C}_{\widehat{G}}(N)$ given by

$$\bar{\epsilon}(x\mathbf{C}_G(N)) = y\mathbf{C}_{\widehat{G}}(N) \text{ if and only if } \epsilon(x) = \hat{\epsilon}(y).$$

In other words, $\bar{\epsilon}(x\mathbf{C}_G(N)) = y\mathbf{C}_{\widehat{G}}(N)$ if and only if x and y induce the same conjugation automorphism of N. We notice that $\bar{\epsilon}$ is a well-defined group isomorphism sending $N\mathbf{C}_G(N)/\mathbf{C}_G(N)$ onto $N\mathbf{C}_{\widehat{G}}(N)/\mathbf{C}_{\widehat{G}}(N)$. In fact,

$\bar{\epsilon}(n\mathbf{C}_G(N)) = n\mathbf{C}_{\widehat{G}}(N)$. Furthermore, the isomorphism $\bar{\epsilon}$ sends $H/\mathbf{C}_G(N)$ onto $\widehat{H}/\mathbf{C}_{\widehat{G}}(N)$.

Now, let $\mathcal{T} \subseteq H$ be a complete set of representatives of cosets of $M\mathbf{C}_G(N)$ in H with $1 \in \mathcal{T}$. Using that $G = NH$, by elementary group theory we have that $\mathcal{T}$ is a complete set of representatives of cosets of $N\mathbf{C}_G(N)$ in G. Hence, every $g \in G$ can be written as tcn for some $t \in \mathcal{T}$, $c \in \mathbf{C}_G(N)$ and $n \in N$. Notice that $t_1c_1n_1 = t_2c_2n_2$ if and only if there is some $z \in \mathbf{Z}(N)$ such that $t_1 = t_2$, $c_1 = c_2z$, and $n_1 = z^{-1}n_2$.

Now, since $\epsilon(G) = \hat{\epsilon}(\widehat{G})$, for every $t \in \mathcal{T}$ we can choose $\hat{t} \in \widehat{G}$ such that $\epsilon(t) = \hat{\epsilon}(\hat{t})$, with $\hat{1} = 1$. By definition, we have that

$$n^t = n^{\hat{t}}$$

for every $n \in N$ and $t \in \mathcal{T}$. In other words, $\bar{\epsilon}(t\mathbf{C}_G(N)) = \hat{t}\mathbf{C}_{\widehat{G}}(N)$.

Using that $\bar{\epsilon}$ is a group isomorphism, we have that $\widehat{\mathcal{T}} = \{\hat{t} \mid t \in \mathcal{T}\}$ is a complete set of representatives of right cosets of $N\mathbf{C}_{\widehat{G}}(N)$ in $\widehat{G}$. By the same reason, since $\mathcal{T}$ is a complete set of representatives of cosets of $M\mathbf{C}_G(N)$ in H, we have that $\widehat{\mathcal{T}}$ is a complete set of representatives of cosets of $M\mathbf{C}_{\widehat{G}}(N)$ in $\widehat{H}$.

Now, write $\theta_{\mathbf{Z}(N)} = \theta(1)\mu_0$, where $\mu_0 \in \mathrm{Irr}(\mathbf{Z}(N))$. Notice that $\mu_{\mathbf{Z}(N)} = \mu_0$. Let $\hat{\mu} \colon \mathbf{C}_{\widehat{G}}(N) \to \mathbb{C}^\times$ be any function such that $\hat{\mu}(cz) = \hat{\mu}(c)\mu_0(z)$ for $c \in \mathbf{C}_{\widehat{G}}(N)$ and $z \in \mathbf{Z}(N)$, and $\hat{\mu}(1) = 1$. (For instance, write $\mathbf{C}_{\widehat{G}}(N) = \bigcup_{j=1}^t x_j\mathbf{Z}(N)$, with $x_1 = 1$, and define $\hat{\mu}(x_jz) = \mu_0(z)$. See the paragraph before Lemma 5.4.)

Now, we define functions

$$\widehat{\mathcal{P}} \colon \widehat{G} \to \mathrm{GL}_{\theta(1)}(\mathbb{C}) \quad \text{and} \quad \widehat{\mathcal{P}}' \colon \widehat{H} \to \mathrm{GL}_{\varphi(1)}(\mathbb{C})$$

by

$$\widehat{\mathcal{P}}(\hat{t}nc) = \mathcal{P}(t)\mathcal{P}(n)\hat{\mu}(c) \quad \text{and} \quad \widehat{\mathcal{P}}'(\hat{t}mc) = \mathcal{P}'(t)\mathcal{P}'(m)\hat{\mu}(c),$$

where $t \in \mathcal{T}$, $n \in N$, $c \in \mathbf{C}_{\widehat{G}}(N)$ and $m \in M$.

Notice that if $\hat{t}nc = \hat{t}_1n_1c_1$, then $t = t_1$, $c = c_1z$ and $n = z^{-1}n_1$ for some $z \in \mathbf{Z}(N)$, by a previous argument. Then

$$\mathcal{P}(n_1)\hat{\mu}(c_1) = \mathcal{P}(n)\mu_0(z)\hat{\mu}(z^{-1}c) = \mathcal{P}(n)\hat{\mu}(c),$$

using our defining property for $\hat{\mu}$. So we see that $\widehat{\mathcal{P}}$ (and similarly $\widehat{\mathcal{P}}'$) are well-defined functions. Notice that $\widehat{\mathcal{P}}(n) = \mathcal{P}(n)$ and $\widehat{\mathcal{P}}'(m) = \mathcal{P}'(m)$ for $n \in N$ and $m \in M$.

We want to show that $\widehat{\mathcal{P}}$ and $\widehat{\mathcal{P}}'$ are projective representations of $\widehat{G}$ and $\widehat{H}$ associated with θ and φ, respectively, having factor sets $\hat{\alpha}$ and $\hat{\alpha}'$ that coincide on $\widehat{H} \times \widehat{H}$, and such that for $x \in \mathbf{C}_{\widehat{G}}(N)$ they are associated with the same scalar. The latter part is obvious by definition.

Let $x \in \widehat{G}$ and let $n \in N$. We prove that $\widehat{\mathcal{P}}(xn) = \widehat{\mathcal{P}}(x)\widehat{\mathcal{P}}(n)$. Write $x = \hat{t}mc$, where $t \in T$, $m \in N$ and $c \in \mathbf{C}_{\widehat{G}}(N)$. Then $xn = \hat{t}(mn)c$. Now

$$\widehat{\mathcal{P}}(xn) = \widehat{\mathcal{P}}(\hat{t}(mn)c) = \mathcal{P}(t)\mathcal{P}(mn)\hat{\mu}(c) = \mathcal{P}(t)\mathcal{P}(m)\hat{\mu}(c)\mathcal{P}(n) = \widehat{\mathcal{P}}(x)\widehat{\mathcal{P}}(n).$$

Next, we prove that $\widehat{\mathcal{P}}(nx) = \widehat{\mathcal{P}}(n)\widehat{\mathcal{P}}(x)$. Recall that by Lemma 10.10(a), we have that $\mathcal{P}(n)\mathcal{P}(t) = \mathcal{P}(t)\mathcal{P}(t^{-1}nt)$ for $t \in G$. Then

$$\begin{aligned}
\widehat{\mathcal{P}}(nx) = \widehat{\mathcal{P}}(n\hat{t}mc) &= \widehat{\mathcal{P}}(\hat{t}\hat{t}^{-1}n\hat{t}mc) \\
&= \mathcal{P}(t)\mathcal{P}(\hat{t}^{-1}n\hat{t})\mathcal{P}(m)\hat{\mu}(c) \\
&= \mathcal{P}(t)\mathcal{P}(t^{-1}nt)\mathcal{P}(m)\hat{\mu}(c) \\
&= \mathcal{P}(n)\mathcal{P}(t)\mathcal{P}(m)\hat{\mu}(c) \\
&= \widehat{\mathcal{P}}(n)\widehat{\mathcal{P}}(x).
\end{aligned}$$

Next we show that $\widehat{\mathcal{P}}$ is a projective representation of $\widehat{G}$. Suppose that $t_1, t_2, t_3 \in T$, $c_1, c_2, c_3 \in \mathbf{C}_{\widehat{G}}(N)$, and $n_1, n_2, n_3 \in N$ are such that

$$(\hat{t}_1n_1c_1)(\hat{t}_2n_2c_2) = \hat{t}_3n_3c_3.$$

Notice that

$$\bar{\epsilon}(t_1n_1t_2n_2\mathbf{C}_G(N)) = \hat{t}_1n_1\hat{t}_2n_2\mathbf{C}_{\widehat{G}}(N) = \hat{t}_3n_3\mathbf{C}_{\widehat{G}}(N) = \bar{\epsilon}(t_3n_3\mathbf{C}_G(N)).$$

Thus $t_1n_1t_2n_2 = t_3n_3c$, for some $c \in \mathbf{C}_G(N)$. Then

$$\begin{aligned}
\widehat{\mathcal{P}}(\hat{t}_1n_1c_1)\widehat{\mathcal{P}}(\hat{t}_2n_2c_2) &= \mathcal{P}(t_1n_1)\hat{\mu}(c_1)\mathcal{P}(t_2n_2)\hat{\mu}(c_2) \\
&= \mathcal{P}(t_1n_1t_2n_2)\alpha(t_1, t_2)\hat{\mu}(c_1)\hat{\mu}(c_2) \\
&= \mathcal{P}(t_3n_3c)\alpha(t_1, t_2)\hat{\mu}(c_1)\hat{\mu}(c_2) \\
&= \mathcal{P}(t_3n_3)\mathcal{P}(c)\alpha(t_3, c)^{-1}\alpha(t_1, t_2)\hat{\mu}(c_1)\hat{\mu}(c_2) \\
&= \mathcal{P}(t_3n_3)\hat{\mu}(c_3)\mu(c)\alpha(t_3, c)^{-1}\alpha(t_1, t_2)\hat{\mu}(c_1)\hat{\mu}(c_2)\hat{\mu}(c_3)^{-1} \\
&= \widehat{\mathcal{P}}(\hat{t}_3n_3c_3)\mu(c)\alpha(t_3, c)^{-1}\alpha(t_1, t_2)\hat{\mu}(c_1)\hat{\mu}(c_2)\hat{\mu}(c_3)^{-1}.
\end{aligned}$$

This implies that $\widehat{\mathcal{P}}$ is a projective representation of $\widehat{G}$ with factor set

$$\hat{\alpha}(\hat{t}_1n_1c_1, \hat{t}_2n_2c_2) = \mu(c)\alpha(t_3, c)^{-1}\alpha(t_1, t_2)\hat{\mu}(c_1)\hat{\mu}(c_2)\hat{\mu}(c_3)^{-1},$$

where c is any element of $\mathbf{C}_G(N)$ satisfying the equation $t_1n_1t_2n_2 = t_3n_3c$. The same argument, substituting elements in N by elements in M, shows that $\widehat{\mathcal{P}}'$ is a projective representation of $\widehat{H}$ with factor set

$$\hat{\alpha}'(\hat{t}_1m_1c_1, \hat{t}_2m_2c_2) = \mu(c)\alpha'(t_3, c)^{-1}\alpha'(t_1, t_2)\hat{\mu}(c_1)\hat{\mu}(c_2)\hat{\mu}(c_3)^{-1},$$

where $c_3 \in \mathbf{C}_G(N)$ satisfies

$$(\hat{t}_1 m_1 c_1)(\hat{t}_1 m_2 c_2) = \hat{t}_3 m_3 c_3$$

and c is any element of $\mathbf{C}_G(N)$ satisfying the equation $t_1 m_1 t_2 m_2 = t_3 m_3 c$. Since $\mathcal{T}$ and $\mathbf{C}_G(N)$ are contained in H, and α and α' coincide on $H \times H$, it is now clear that the factor sets $\hat{\alpha}$ and $\hat{\alpha}'$ coincide on $\widehat{H} \times \widehat{H}$. This finishes the proof. $\qquad \square$

10.7 Building Central Isomorphisms

The next results, of a technical nature, will be applied later.

Lemma 10.19 *Suppose that $(G, N, \theta) \geq_c (H, M, \varphi)$. Let $Z \subseteq \ker(\theta) \cap \ker(\varphi)$ be normal in G. Write $\bar{N} = N/Z$, $\bar{G} = G/Z$, $\bar{H} = H/Z$ and $\bar{M} = M/Z$. Also, write $\bar{\theta} \in \mathrm{Irr}(N/Z)$ and $\bar{\varphi} \in \mathrm{Irr}(H/Z)$ for the corresponding characters that satisfy $\bar{\theta}(nZ) = \theta(n)$ and $\bar{\varphi}(mZ) = \varphi(m)$. Assume that $\mathbf{C}_{\bar{G}}(\bar{N}) = \mathbf{C}_G(N)Z/Z$. Then $(\bar{G}, \bar{N}, \bar{\theta}) \geq_c (\bar{H}, \bar{M}, \bar{\varphi})$.*

Proof Clearly, we have that $\bar{G} = \bar{N}\bar{H}$, $\bar{M} = \bar{N} \cap \bar{H}$, and $\mathbf{C}_{\bar{G}}(\bar{N}) \subseteq \bar{H}$, by hypothesis. Suppose that $(\mathcal{P}, \mathcal{P}')$ is associated with $(G, N, \theta) \geq_c (H, M, \varphi)$ and let $\mu \colon \mathbf{C}_G(N) \to \mathbb{C}^\times$ be satisfying $\mathcal{P}(c) = \mu(c) I_{\theta(1)}$ and $\mathcal{P}'(c) = \mu(c) I_{\varphi(1)}$ for $c \in \mathbf{C}_G(N)$. Now, using Lemma 10.10(c), it is straightforward to check that $\overline{\mathcal{P}}(gZ) = \mathcal{P}(g)$ and $\overline{\mathcal{P}'}(hZ) = \mathcal{P}'(h)$ define projective representations of G/Z and H/Z such that $(\overline{\mathcal{P}}, \overline{\mathcal{P}'})$ is associated with $(\bar{G}, \bar{N}, \bar{\theta}) \geq_c (\bar{H}, \bar{M}, \bar{\varphi})$. $\qquad \square$

Lemma 10.20 *Suppose that $(G_i, N_i, \theta_i) \geq_c (H_i, M_i, \varphi_i)$ for $i = 1, 2$. Then $(G_1 \times G_2, N_1 \times N_2, \theta_1 \times \theta_2) \geq_c (H_1 \times H_2, M_1 \times M_2, \varphi_1 \times \varphi_2)$.*

Proof The group theoretical conditions are trivial to check. Suppose that $(\mathcal{P}_i, \mathcal{P}'_i)$ is associated with $(G_i, N_i, \theta_i) \geq_c (H_i, M_i, \varphi_i)$ for $i = 1, 2$. Define

$$(\mathcal{P}_1 \otimes \mathcal{P}_2)(g_1, g_2) = \mathcal{P}_1(g_1) \otimes \mathcal{P}(g_2)$$

for $g_1 \in G_1, g_2 \in G_2$, and check that this is a projective representation with factor set

$$(\alpha_1 \times \alpha_2)((x_1, y_1), (x_2, y_2)) = \alpha_1(x_1, y_1)\alpha_2(x_2, y_2),$$

where α_i is factor set of $\mathcal{P}_i$. Then it is routine to verify that $(\mathcal{P}_1 \otimes \mathcal{P}_2, \mathcal{P}'_1 \otimes \mathcal{P}'_2)$ is associated with

$$(G_1 \times G_2, N_1 \times N_2, \theta_1 \times \theta_2) \geq_c (H_1 \times H_2, M_1 \times M_2, \varphi_1 \times \varphi_2),$$

as desired.   □

If G is a group and $m \geq 1$ is an integer, recall from the first section in this chapter that the symmetric group S_m naturally acts as automorphisms on $G^m = G \times \cdots \times G$ by letting

$$(g_1, \ldots, g_m)^\sigma = (g_{\sigma^{-1}(1)}, \ldots, g_{\sigma^{-1}(m)}),$$

for $g_i \in G$ and $\sigma \in \mathsf{S}_m$. If $\theta_i \in \mathrm{Irr}(G)$, we have that

$$(\theta_1 \times \cdots \times \theta_m)^\sigma = \theta_{\sigma^{-1}(1)} \times \cdots \times \theta_{\sigma^{-1}(m)}.$$

Also, for every $d \geq 1$ we have constructed a representation

$$\mathcal{X}_d : \mathsf{S}_m \to \mathrm{GL}_{d^m}(\mathbb{C})$$

such that

$$\mathcal{X}_d(\sigma)(A_1 \otimes \cdots \otimes A_m)\mathcal{X}_d(\sigma)^{-1} = A_{\sigma(1)} \otimes \cdots \otimes A_{\sigma(m)}$$

for all matrices $A_i \in \mathrm{GL}_d(\mathbb{C})$.

Theorem 10.21 *Suppose that* $(G, N, \theta) \geq_c (H, M, \varphi)$. *Let* $m \geq 1$ *be an integer, and let* $\eta = \theta \times \cdots \times \theta \in \mathrm{Irr}(N^m)$ *and* $\gamma = \varphi \times \cdots \times \varphi \in \mathrm{Irr}(M^m)$. *Then*

$$(G \wr \mathsf{S}_m, N^m, \eta) \geq_c (H \wr \mathsf{S}_m, M^m, \gamma).$$

Proof Write $G \wr \mathsf{S}_m = G^m \rtimes \mathsf{S}_m = G^m \mathsf{S}_m$. It is clear that we may assume that $N > 1$. If $1 \neq n \in N$, notice that $(n, 1, \ldots, 1)^{(g_1, \ldots, g_m)\sigma} = (n, 1, \ldots, 1)$ if and only if $\sigma(1) = 1$ and $n^{g_1} = n$. Using this argument with n in the ith position, we check that

$$\mathbf{C}_{G^m \mathsf{S}_m}(N^m) = \mathbf{C}_G(N)^m \subseteq H^m \subseteq H^m \mathsf{S}_m.$$

Of course, $G^m \mathsf{S}_m = (N^m)(H^m \mathsf{S}_m)$ and $M^m = N^m \cap (H^m \mathsf{S}_m)$.

Let $u = \theta(1)$ and $v = \varphi(1)$. Suppose now that $(\mathcal{P}, \mathcal{P}')$ is associated with $(G, N, \theta) \geq_c (H, M, \varphi)$ and let $\mu \colon \mathbf{C}_G(N) \to \mathbb{C}^\times$ be the function satisfying $\mathcal{P}(c) = \mu(c)I_u$ and $\mathcal{P}'(c) = \mu(c)I_v$ for $c \in \mathbf{C}_G(N) \subseteq H$. If $\sigma \in \mathsf{S}_m$, we see that $\eta^\sigma = \eta$ and $\gamma^\sigma = \gamma$.

Next, we define a projective representation $\tilde{\mathcal{P}}$ of $G^m \mathsf{S}_m$ associated with η by setting

$$\tilde{\mathcal{P}}((x_1, \ldots, x_m)\sigma) = (\mathcal{P}(x_1) \otimes \cdots \otimes \mathcal{P}(x_m)) \, \mathcal{X}_u(\sigma).$$

We have that

$$\tilde{\mathcal{P}}((x_1, \ldots, x_m)\sigma)\tilde{\mathcal{P}}((y_1, \ldots, y_m)\tau)$$
$$= (\mathcal{P}(x_1) \otimes \cdots \otimes \mathcal{P}(x_m)) \, \mathcal{X}_u(\sigma) \, (\mathcal{P}(y_1) \otimes \cdots \otimes \mathcal{P}(y_m)) \, \mathcal{X}_u(\tau)$$
$$= (\mathcal{P}(x_1) \otimes \cdots \otimes \mathcal{P}(x_m)) \, \mathcal{X}_u(\sigma) \, (\mathcal{P}(y_1) \otimes \cdots \otimes \mathcal{P}(y_m)) \, \mathcal{X}_u(\sigma^{-1}) \mathcal{X}_u(\sigma\tau)$$
$$= (\mathcal{P}(x_1) \otimes \cdots \otimes \mathcal{P}(x_m)) \, \big(\mathcal{P}(y_{\sigma(1)}) \otimes \cdots \otimes \mathcal{P}(y_{\sigma(m)})\big) \, \mathcal{X}_u(\sigma\tau)$$
$$= \big(\mathcal{P}(x_1)\mathcal{P}(y_{\sigma(1)}) \otimes \cdots \otimes \mathcal{P}(x_m)\mathcal{P}(y_{\sigma(m)})\big) \, \mathcal{X}_u(\sigma\tau).$$

From this equation we easily check that $\tilde{\mathcal{P}}$ is a projective representation associated with $(G^m \mathsf{S}_m, N^m, \eta)$ with factor set

$$\tilde{\alpha}((x_1, \ldots, x_m)\sigma, (y_1, \ldots, y_m)\tau) = \prod_i \alpha(x_i, y_{\sigma(i)}),$$

where α is the factor set of $\mathcal{P}$. By doing the same construction $\tilde{\mathcal{P}}'$ for $\mathcal{P}'$, we check that $(\tilde{\mathcal{P}}, \tilde{\mathcal{P}}')$ is associated with

$$(G^m \mathsf{S}_m, N^m, \eta) \geq (H^m \mathsf{S}_m, M^m, \gamma).$$

Finally, suppose that $(c_1, \ldots, c_m) \in \mathbf{C}_{G^m \mathsf{S}_m}(N^m) = \mathbf{C}_G(N)^m$. Then

$$\tilde{\mathcal{P}}(c_1, \ldots, c_m) = \prod_i \mu(c_i) I_{u^m},$$

and

$$\tilde{\mathcal{P}}'(c_1, \ldots, c_m) = \prod_i \mu(c_i) I_{v^m},$$

and this finishes the proof of the theorem. □

If G is a finite group and $m \geq 1$, the group $\mathrm{Aut}(G^m)$ will be relevant in some of what we are going to do next. We already know that if G is nontrivial, then we can view S_m as a subgroup of $\mathrm{Aut}(G^m)$. In fact, as the reader can check, we can view $\mathrm{Aut}(G) \wr \mathsf{S}_m$ as a subgroup of $\mathrm{Aut}(G^m)$: if $(\sigma_1, \ldots, \sigma_m) \in \mathrm{Aut}(G)^m$ and $\sigma \in \mathsf{S}_m$, then we shall denote by $(\sigma_1, \ldots, \sigma_m)\sigma$ the automorphism of G^m given by

$$(x_1, \ldots, x_m)^{(\sigma_1, \ldots, \sigma_m)\sigma} = (x_1^{\sigma_1}, \ldots, x_m^{\sigma_m})^{\sigma}.$$

If $\theta_i \in \mathrm{Irr}(G)$, then we check that

$$(\theta_1 \times \cdots \times \theta_m)^{(\sigma_1, \ldots, \sigma_m)\sigma} = (\theta_{\sigma^{-1}(1)})^{\sigma_{\sigma^{-1}(1)}} \times \cdots \times (\theta_{\sigma^{-1}(m)})^{\sigma_{\sigma^{-1}(m)}}.$$

In some cases as in Lemma 10.24 below, we even have that the equality $\mathrm{Aut}(G^m) = \mathrm{Aut}(G) \wr S_m$ holds.

If $H \leq G$, we denote by $\mathrm{Aut}(G)_H$ the subgroup of automorphisms $\alpha \in \mathrm{Aut}(G)$ such that $\alpha(H) = H$. If A is any subgroup of $\mathrm{Aut}(G)_H$, we are going to work in the group $G \rtimes A$. Since H is A-invariant, notice that $H \rtimes A \leq G \rtimes A$. If $m \geq 1$, we also have that $\tilde{A} = A \wr S_m \leq \mathrm{Aut}(G^m)_{H^m}$, and again we can form the group $G^m \rtimes \tilde{A}$ and its subgroup $H^m \rtimes \tilde{A}$.

Corollary 10.22 *Suppose that H is a subgroup of a finite group G, and let $A \leq \mathrm{Aut}(G)_H$. Assume that $\Omega \colon \mathrm{Irr}_{p'}(G) \to \mathrm{Irr}_{p'}(H)$ is an A-equivariant bijection. Write $\eta' = \Omega(\eta)$ for $\eta \in \mathrm{Irr}_{p'}(G)$. Assume that*

$$(G \rtimes A_\eta, G, \eta) \geq_c (H \rtimes A_\eta, H, \eta')$$

for every $\eta \in \mathrm{Irr}_{p'}(G)$, where A_η is the stabilizer of η in A. Let $m \geq 1$ be an integer, and let $\tilde{A} = A \wr S_m \leq \mathrm{Aut}(G^m)_{H^m}$. Then $\Omega \colon \mathrm{Irr}_{p'}(G^m) \to \mathrm{Irr}_{p'}(H^m)$, defined by $\Omega(\theta) = \theta' = \theta_1' \times \cdots \times \theta_m' \in \mathrm{Irr}_{p'}(H^m)$ for $\theta = \theta_1 \times \cdots \times \theta_m \in \mathrm{Irr}_{p'}(G^m)$, is a $\tilde{A}$-equivariant bijection, and

$$(G^m \rtimes \tilde{A}_\theta, G^m, \theta) \geq_c (H^m \rtimes \tilde{A}_\theta, H^m, \theta')$$

for $\theta \in \mathrm{Irr}_{p'}(G^m)$.

Proof Observe that $A_\eta = A_{\eta'}$ for $\eta \in \mathrm{Irr}_{p'}(G)$ by hypothesis. It is straightforward to check that $\Omega \colon \mathrm{Irr}_{p'}(G^m) \to \mathrm{Irr}_{p'}(H^m)$ is a $\tilde{A}$-equivariant bijection, using that $\Omega \colon \mathrm{Irr}_{p'}(G) \to \mathrm{Irr}_{p'}(H)$ is A-equivariant. (We denote both bijections by the same letter because there is no risk of confusion.) Hence $\tilde{A}_\theta = \tilde{A}_{\theta'}$ for $\theta \in \mathrm{Irr}_{p'}(G^m)$.

Suppose now that $\theta = \theta_1 \times \cdots \times \theta_m \in \mathrm{Irr}_{p'}(G^m)$. Assume that $(\theta_i)^a = \theta_j$ for some i, j and $a \in A$. Then $(\theta_i')^a = \theta_j'$. Since $A^m \subseteq \tilde{A}$, by conjugating by an appropriate element of A^m and applying Problem 10.1, it is no loss to assume that θ_i and θ_j are actually equal. Thus we may assume that θ_i and θ_j are A-conjugate if and only if they are equal. We now use the same idea as in the proof of Corollary 10.2. In $\{1, \ldots, m\}$, we set $i \equiv j$ if $\theta_i = \theta_j$. This defines a partition of $\{1, \ldots, m\}$. Since $S_m \subseteq \tilde{A}$, again by Problem 10.1 it is no loss if we replace θ by some S_m-conjugate. Hence, we may assume that

$$\theta = (\eta_1 \times \cdots \times \eta_1) \times \cdots \times (\eta_k \times \cdots \times \eta_k),$$

where $\{\eta_1, \ldots, \eta_k\}$ are the distinct θ_is, η_i appears m_i times, and $m_1 + \cdots + m_k = m$. Now, using that distinct η_i's are not A-conjugate, we check that

$$\tilde{A}_\theta = (A_{\eta_1} \wr S_{m_1}) \times \cdots \times (A_{\eta_k} \wr S_{m_k}).$$

By hypothesis, $(G \rtimes A_{\eta_i}, G, \eta_i) \geq_c (H \rtimes A_{\eta_i}, H, \eta_i')$, and therefore

$$((G \rtimes A_{\eta_i}) \wr S_{m_i}, G^{m_i}, \eta_i \times \cdots \times \eta_i) \geq_c ((H \rtimes A_{\eta_i}) \wr S_{m_i}, H^{m_i}, \eta_i' \times \cdots \times \eta_i')$$

by Theorem 10.21. Now, notice that there is a natural isomorphism

$$f \colon G^m \rtimes \tilde{A}_\theta \to (G \rtimes A_{\eta_1}) \wr S_{m_1} \times \cdots \times (G \rtimes A_{\eta_k}) \wr S_{m_k},$$

that sends $H^m \rtimes \tilde{A}_\theta$ onto $(H \rtimes A_{\eta_1}) \wr S_{m_1} \times \cdots \times (H \rtimes A_{\eta_k}) \wr S_{m_k}$, G^m onto $G^{m_1} \times \cdots \times G^{m_k}$ and H^m onto $H^{m_1} \times \cdots \times H^{m_k}$. Also f maps the character θ to $(\eta_1 \times \cdots \times \eta_1) \times \cdots \times (\eta_k \times \cdots \times \eta_k)$, and θ' to $(\eta_1' \times \cdots \times \eta_1') \times \cdots \times (\eta_k' \times \cdots \times \eta_k')$. By using Lemma 10.20, we easily conclude the proof. □

10.8 The Inductive McKay Condition

We can now define when a non-abelian simple group S satisfies the *inductive McKay condition*. In order to do that, we shall use a universal covering group X of S, and $\mathrm{Aut}(X)$. (For the reader not used to universal covering groups, we have written in Appendix B the results that we are going to use, with some proofs.)

Definition 10.23 (Späth) *Let p be a prime. Let S be a non-abelian finite simple group, and let X be a universal covering group of S. Let $R \in \mathrm{Syl}_p(X)$. Let $A = \mathrm{Aut}(X)_R$ be the group of automorphisms of X that fix R setwise. We say that S satisfies the **inductive McKay condition for** p if there exists an A-equivariant bijection $\Omega \colon \mathrm{Irr}_{p'}(X) \to \mathrm{Irr}_{p'}(\mathbf{N}_X(R))$ such that for every $\theta \in \mathrm{Irr}_{p'}(X)$, we have that*

$$(X \rtimes A_\theta, X, \theta) \geq_c (\mathbf{N}_X(R) \rtimes A_\theta, \mathbf{N}_X(R), \Omega(\theta)).$$

(In truth, the inductive McKay condition only requires the existence of an A-invariant subgroup $\mathbf{N}_X(R) \subseteq M < X$ and an A-equivariant bijection $\mathrm{Irr}_{p'}(X) \to \mathrm{Irr}_{p'}(M)$, satisfying the corresponding central isomorphism of triples with $\mathbf{N}_X(R)$ replaced by M. The proof of the reduction theorem below is just the same, but we shall avoid this technicality in what follows. In simple groups, it is sometimes much more convenient to work with some appropriate *Malle intermediate subgroup M* rather than with $\mathbf{N}_X(R)$.)

Recall that a finite group G is **quasisimple** if G is perfect and $G/\mathbf{Z}(G)$ is simple. (By Theorem B.4, we know that X is quasisimple in Definition 10.23.)

Lemma 10.24 *Suppose that G is quasisimple, and let $n \geq 1$.*

(a) We have that $\mathrm{Aut}(G^n) = \mathrm{Aut}(G) \wr \mathsf{S}_n$.

(b) If $H \leq G$, then $\mathrm{Aut}(G^n)_{H^n} = \mathrm{Aut}(G)_H \wr \mathsf{S}_n$.

(c) If $G^n \subseteq W \leq G^n \rtimes \mathrm{Aut}(G^n)$ and $Z \subseteq \mathbf{Z}(G^n)$ is W-invariant, then $\mathbf{C}_{W/Z}(G^n/Z) = \mathbf{Z}(G^n)/Z$.

Proof If G is any finite group, and $n \geq 1$, then we know that $\mathrm{Aut}(G) \wr \mathsf{S}_n \leq \mathrm{Aut}(G^n)$. Now, let $G_i = 1 \times \cdots \times G \times \cdots \times 1$, where G appears in the ith position. Notice that G_i is a normal quasisimple subgroup of G^n. We claim that there are no other normal quasisimple subgroups of G^n. If $X \trianglelefteq \trianglelefteq G^n$ is quasisimple and $X \neq G_i$, then $[X, G_i] = 1$ (because distinct components commute, by Theorem 9.4 of [Is08]). If $X \neq G_i$ for all i, then $X \subseteq \mathbf{Z}(G^n)$, and this is a contradiction because X is perfect. This proves the claim. Now, if $\alpha \in \mathrm{Aut}(G^n)$, then it follows that α permutes the set $\{G_1, \ldots, G_n\}$. Hence, there is $\sigma \in \mathsf{S}_n$ such that $(G_i)^\alpha = G_{\sigma(i)} = G_i^\sigma$, where in the latter, we are viewing $\sigma \in \mathrm{Aut}(G^n)$ as the automorphism $(g_1, \ldots, g_n)^\sigma = (g_{\sigma^{-1}(1)}, \ldots, g_{\sigma^{-1}(n)})$. Let $\alpha_i = (\alpha\sigma^{-1})|_{G_i} \in \mathrm{Aut}(G_i) = \mathrm{Aut}(G)$. It is straightforward to check that $\alpha = (\alpha_1, \ldots, \alpha_n)\sigma$. Part (b) easily follows now.

For (c), notice that it suffices to show that if $G^n \subseteq W \leq G^n \rtimes \mathrm{Aut}(G^n)$, then

$$\mathbf{C}_{W/\mathbf{Z}(G)^n}(G^n/\mathbf{Z}(G)^n)$$

is trivial. We start by claiming that if $\alpha \in \mathrm{Aut}(G)$ is such that $g\mathbf{Z}(G) = g^\alpha\mathbf{Z}(G)$ for every $g \in G$, then α is the identity. Notice that for every $g \in G$ there is a unique $z_g \in \mathbf{Z}(G)$ such that $g^\alpha = gz_g$. Then the map $g \mapsto z_g$ is a group homomorphism. Since G is perfect, we deduce that $z_g = 1$ for all $g \in G$, and the claim is proven. Now, for the general case, if $(\alpha_1, \ldots, \alpha_n)\sigma$ fixes $(g_1, \ldots, g_n)\mathbf{Z}(G)^n$ for every $g_i \in G$, by choosing $x \in G - \mathbf{Z}(G)$ and considering the element $(1, \ldots, x, \ldots, 1)$ in the ith position, we see that σ is the identity. Now, each α_i is the identity by the previous claim. $\square$

Next is the crucial result. In its proof, we are going to use universal extensions of perfect groups. (Specifically, we shall use Corollaries B.8 and B.10 in Appendix B.)

Theorem 10.25 *Suppose that $K \trianglelefteq G$, where K is perfect and $K/\mathbf{Z}(K) \cong S^n$, where S is a non-abelian simple group. Let $Q \in \mathrm{Syl}_p(K)$. If S satisfies the inductive McKay condition, then there exists a $\mathbf{N}_G(Q)$-equivariant bijection*

$$\Omega \colon \mathrm{Irr}_{p'}(K) \to \mathrm{Irr}_{p'}(\mathbf{N}_K(Q))$$

such that for every $\theta \in \operatorname{Irr}_{p'}(K)$ *and* $\theta' = \Omega(\theta)$ *we have that*

$$(G_\theta, K, \theta) \geq_c (\mathbf{N}_G(Q)_{\theta'}, \mathbf{N}_K(Q), \theta').$$

Proof We have that $G = K\mathbf{N}_G(Q)$ by the Frattini argument. Notice that if the theorem is true for Q, then it is true for Q^k for every $k \in K$. (The map $\Omega_k(\theta) = \Omega(\theta)^k$ is $\mathbf{N}_G(Q)^k$-equivariant and we apply Problem 10.1 to get the conclusion.) Hence, in order to prove the theorem, we may choose any Sylow p-subgroup of the normal subgroup in question.

The idea of the proof is the following. First, we are going to prove the theorem for the group $\widehat{G} = K \rtimes \operatorname{Aut}(K)_Q$, which of course contains K as a normal subgroup and satisfies the hypotheses. Then we shall use Theorem 10.18 to relate $\widehat{G}$ with G via their conjugation homomorphisms into $\operatorname{Aut}(K)$.

By Corollary B.10, we know that if X is a universal covering group of S, then X^n is a universal covering of K. Hence, let (X^n, π) be a central universal extension of K. Let $Z = \ker(\pi) \subseteq \mathbf{Z}(X^n) = \mathbf{Z}(X)^n$. By definition, π is onto. Let $\bar\pi \colon X^n/Z \to K$ be the associated isomorphism. Now, let us fix $R \in \operatorname{Syl}_p(X)$. Then $R^n \in \operatorname{Syl}_p(X^n)$ and $R^nZ/Z \in \operatorname{Syl}_p(X^n/Z)$. Using that Z is central, we have that

$$\mathbf{N}_{X^n/Z}(R^nZ/Z) = \mathbf{N}_{X^n}(R^n)/Z = \mathbf{N}_X(R)^n/Z.$$

Also, we may assume that $\bar\pi(R^nZ/Z) = Q$ (by replacing Q if necessary).

We are going to prove the theorem for $(X^n/Z) \rtimes \operatorname{Aut}(X^n/Z)_{R^nZ/Z}$ (which is isomorphic to $K \rtimes \operatorname{Aut}(K)_Q$).

One more feature of universal central extensions that we need now is that the natural map $\operatorname{Aut}(X^n)_Z \to \operatorname{Aut}(X^n/Z)$ given by $\sigma \mapsto \tilde\sigma$, where $\tilde\sigma(yZ) = \sigma(y)Z$ and $y \in X^n$, is a group isomorphism. (This follows from Corollary B.8, using that X^n is a universal covering of X^n/Z. In general, the fact that this homomorphism is injective follows from the fact that X^n is perfect, using the argument in Lemma 10.24(c). To prove that it is surjective, one needs to use universal central extensions.) Notice that this isomorphism sends $\operatorname{Aut}(X^n)_{R^n,Z}$ onto $\operatorname{Aut}(X^n/Z)_{R^nZ/Z}$, where $\operatorname{Aut}(X^n)_{R^n,Z}$ are the elements of $\operatorname{Aut}(X^n)_Z$ that send R^n to R^n. Notice too that $\operatorname{Aut}(X^n)_{R^n,Z}$ acts on X^n/Z the same way as $\operatorname{Aut}(X^n/Z)_{R^nZ/Z}$ does. Therefore the groups $(X^n/Z) \rtimes \operatorname{Aut}(X^n/Z)_{R^nZ/Z}$ and $(X^n/Z) \rtimes \operatorname{Aut}(X^n)_{R^n,Z}$ are naturally isomorphic. Hence, we are going to prove now the theorem for the group

$$(X^n/Z) \rtimes \operatorname{Aut}(X^n)_{R^n,Z}.$$

If $B = \operatorname{Aut}(X^n)_{R^n,Z}$, then we wish to find a B-equivariant bijection

$$\Omega \colon \operatorname{Irr}_{p'}(X^n/Z) \to \operatorname{Irr}_{p'}(\mathbf{N}_X(R)^n/Z)$$

such that

$$((X^n/Z) \rtimes B_\theta, X^n/Z, \theta) \geq_c ((\mathbf{N}_X(R)^n/Z) \rtimes B_\theta, \mathbf{N}_X(R)^n/Z, \Omega(\theta))$$

for all $\theta \in \mathrm{Irr}_{p'}(X^n/Z)$.

Let $A = \mathrm{Aut}(X)_R$. Notice that $A = \mathrm{Aut}(X)_{\mathbf{N}_X(R)}$, because R and $\mathbf{N}_X(R)$ uniquely determine one another. Since S satisfies the inductive McKay condition by hypothesis, there is an A-equivariant bijection $\Omega \colon \mathrm{Irr}_{p'}(X) \to \mathrm{Irr}_{p'}(\mathbf{N}_X(R))$ such that

$$(X \rtimes A_\gamma, X, \gamma) \geq_c (\mathbf{N}_X(R) \rtimes A_\gamma, \mathbf{N}_X(R), \gamma')$$

for $\gamma \in \mathrm{Irr}_{p'}(X)$, and where $\Omega(\gamma) = \gamma'$. Let $\tilde{A} = A \wr S_n$, and notice that $\tilde{A} = \mathrm{Aut}(X^n)_{R^n}$ by Lemma 10.24(b). Next we apply Corollary 10.22. First, by Corollary 10.22, we have that the map $\Omega \colon \mathrm{Irr}_{p'}(X^n) \to \mathrm{Irr}_{p'}(\mathbf{N}_X(R)^n)$, defined by

$$\Omega(\gamma_1 \times \cdots \times \gamma_n) = \gamma_1' \times \cdots \times \gamma_n'$$

for $\gamma_i \in \mathrm{Irr}_{p'}(X)$, is a $\tilde{A}$-equivariant bijection. Furthermore

$$(X^n \rtimes \tilde{A}_\theta, X^n, \theta) \geq_c (\mathbf{N}_X(R)^n \rtimes \tilde{A}_\theta, \mathbf{N}_X(R)^n, \theta')$$

for $\theta = \gamma_1 \times \cdots \times \gamma_n \in \mathrm{Irr}_{p'}(X^n)$, and where $\theta' = \Omega(\theta)$. Since this is a central isomorphism and $Z \subseteq \mathbf{Z}(X^n)$, notice that $\mathrm{Irr}(\theta_Z) = \mathrm{Irr}(\theta_Z')$, by Lemma 10.16. In particular, $Z \subseteq \ker(\theta)$ if and only if $Z \subseteq \ker(\theta')$. Thus $\Omega \colon \mathrm{Irr}_{p'}(X^n/Z) \to \mathrm{Irr}_{p'}(\mathbf{N}_X(R)^n/Z)$ is an $\tilde{A}_Z$-equivariant bijection. Notice that $\tilde{A}_Z = B$. Finally, if $\theta \in \mathrm{Irr}_{p'}(X^n/Z)$, then $B_\theta \leq \tilde{A}_\theta$, and thus

$$(X^n \rtimes B_\theta, X^n, \theta) \geq_c (\mathbf{N}_X(R)^n \rtimes B_\theta, \mathbf{N}_X(R)^n, \theta').$$

Using Lemma 10.19 and Lemma 10.24(c), we conclude that

$$((X^n/Z) \rtimes B_\theta, X^n/Z, \theta) \geq_c ((\mathbf{N}_X(R)^n/Z) \rtimes B_\theta, \mathbf{N}_X(R)^n/Z, \theta'),$$

as desired. (Here, we are identifying $(X^n/Z) \rtimes B_\theta$ with $(X^n \rtimes B_\theta)/Z$, which we can certainly do.) This completes the proof that the theorem is true for $\hat{G} = K \rtimes \mathrm{Aut}(K)_Q$. In other words, we have shown that there is a bijection

$$\Omega \colon \mathrm{Irr}_{p'}(K) \to \mathrm{Irr}_{p'}(\mathbf{N}_K(Q)),$$

which is $\mathrm{Aut}(K)_Q$-equivariant, and that satisfies

$$(K \rtimes \mathrm{Aut}(K)_{Q,\theta}, K, \theta) \geq_c (\mathbf{N}_K(Q) \rtimes \mathrm{Aut}(K)_{Q,\theta}, \mathbf{N}_K(Q), \theta')$$

for $\theta \in \mathrm{Irr}_{p'}(K)$, where $\theta' = \Omega(\theta)$. (Here $\mathrm{Aut}(K)_{Q,\theta}$ is the stabilizer in $\mathrm{Aut}(K)_Q$ of θ.) That is

$$(\hat{G}_\theta, K, \theta) \geq_c (\mathbf{N}_{\hat{G}}(Q)_\theta, \mathbf{N}_K(Q), \theta').$$

Our final objective is to apply Theorem 10.18 to our situation. Let $\epsilon\colon G \to$ $\mathrm{Aut}(K)$ and $\hat{\epsilon}\colon \widehat{G} \to \mathrm{Aut}(K)$ be the corresponding conjugation homomorphisms. Notice that $\epsilon(\mathbf{N}_G(Q)) \subseteq \mathrm{Aut}(K)_Q \subseteq \hat{\epsilon}(\widehat{G})$. In particular, we have that Ω is $\mathbf{N}_G(Q)$-equivariant. Also, since $\epsilon(K) = \hat{\epsilon}(K)$, by the Frattini argument $\epsilon(G) = \epsilon(K\mathbf{N}_G(Q)) \subseteq \hat{\epsilon}(\widehat{G})$.

Finally, let $\theta \in \mathrm{Irr}_{p'}(K)$, let $V = \epsilon(G_\theta)$ and let

$$\widehat{V} = \hat{\epsilon}^{-1}(V) = \{x \in \widehat{G} \mid \hat{\epsilon}(x) = \epsilon(g) \text{ for some } g \in G_\theta\}.$$

We have that $\hat{\epsilon}(\widehat{V}) = V = \epsilon(G_\theta)$, using that $\epsilon(G_\theta) \subseteq \hat{\epsilon}(\widehat{G})$. If $x \in \widehat{V}$, then x acts on K as some $g \in G_\theta$ acts on K. Thus $(\widehat{V}, K, \theta)$ is a character triple. Since $\widehat{V} \subseteq \widehat{G}_\theta$, we have that

$$(\widehat{V}, K, \theta) \geq_c (\mathbf{N}_{\widehat{V}}(Q), \mathbf{N}_K(Q), \theta').$$

If we prove that

$$\mathbf{N}_{G_\theta}(Q) = \epsilon^{-1}(\hat{\epsilon}(\mathbf{N}_{\widehat{V}}(Q))$$

then by Corollary 10.18, we will conclude that

$$(G_\theta, K, \theta) \geq_c (\mathbf{N}_{G_\theta}(Q), \mathbf{N}_K(Q), \theta').$$

(Here the roles of $\hat{\epsilon}$ and ϵ are reversed with respect to Theorem 10.18.) However, $W = \epsilon^{-1}(\hat{\epsilon}(\mathbf{N}_{\widehat{V}}(Q))$ is exactly the set of $g \in G$ that act on K in the same way that some $x \in \mathbf{N}_{\widehat{V}}(Q)$ acts. Since x normalizes Q and fixes θ, we get that $g \in \mathbf{N}_{G_\theta}(Q)$. Thus $W \subseteq \mathbf{N}_{G_\theta}(Q)$. Conversely, if $g \in \mathbf{N}_{G_\theta}(Q)$, then $\epsilon(g) \in V$ by definition, and thus $\epsilon(g) = \hat{\epsilon}(x)$ for some $x \in \widehat{V}$. Now, g and x act on K in the same way. Therefore $x \in \mathbf{N}_{\widehat{V}}(Q)$ and $g \in W$. This finishes the proof of the theorem. $\qquad\square$

Recall that a non-abelian simple group S is **involved** in a group G if there exists a subgroup $K \leq G$ and $L \lhd K$ such that $K/L \cong S$. Notice that a group involved in a subgroup of G is involved in G, and that a group involved in a factor group of G is also involved in G.

The following constitutes the main result of this chapter. The final conclusion is obtained by setting $Z = 1$.

Theorem 10.26 (Isaacs–Malle–Navarro) *Let G be a finite group, let p be a prime, and let $P \in \mathrm{Syl}_p(G)$. Suppose that $Z \trianglelefteq G$ and let $\lambda \in \mathrm{Irr}(Z)$ be P-invariant of degree not divisible by p. Assume that all the non-abelian simple groups of order divisible by p involved in G/Z satisfy the inductive McKay condition (with respect to p). Then*

$$|\mathrm{Irr}_{p'}(G|\lambda)| = |\mathrm{Irr}_{p'}(\mathbf{N}_G(P)Z|\lambda)|.$$

Proof We argue by induction on $|G : Z|$. Let $N = \mathbf{N}_G(P)$. If G_λ is the stabilizer of λ in G, using that $P \subseteq G_\lambda$ and the Clifford correspondence, we may assume that λ is G-invariant. Now, we are going to use Corollary 5.9 and character triple isomorphisms. Suppose that (G^*, Z^*, λ^*) is a character triple isomorphic to (G, Z, λ), where $Z^* \subseteq \mathbf{Z}(G^*)$. We use our notation for character triple isomorphisms, namely, if $Z \le H \le G$ then H^*/Z^* is the group $(H/Z)^*$ under the isomorphism $^*: G/Z \to G^*/Z^*$. By elementary group theory, notice that PZ/Z is a Sylow p-subgroup of G/Z and that $\mathbf{N}_{G/Z}(PZ/Z) = NZ/Z$. Now, $(PZ)^*/Z^* = (PZ/Z)^*$ is a Sylow p-subgroup of G^*/Z^*. Since Z^* is central, then there is a unique Sylow p-subgroup P^* of G^* such that $(PZ)^* = P^*Z^*$. By uniqueness, $\mathbf{N}_{G^*}(P^*) = \mathbf{N}_{G^*}(P^*Z^*)$, and therefore

$$(NZ)^*/Z^* = \mathbf{N}_{G/Z}(PZ/Z)^* = \mathbf{N}_{G^*/Z^*}(P^*Z^*/Z^*) = \mathbf{N}_{G^*}(P^*)/Z^*.$$

By the definition of character triple isomorphisms, there are bijections

$$^*: \mathrm{Irr}(G|\lambda) \to \mathrm{Irr}(G^*|\lambda^*) \quad \text{and} \quad ^*: \mathrm{Irr}(NZ|\lambda) \to \mathrm{Irr}(\mathbf{N}_{G^*}(P^*)|\lambda^*)$$

that preserve ratios of degrees. Since $\lambda(1)$ is not divisible by p, we conclude that $|\mathrm{Irr}_{p'}(G|\lambda)| = |\mathrm{Irr}_{p'}(G^*|\lambda^*)|$ and $|\mathrm{Irr}_{p'}(NZ|\lambda)| = |\mathrm{Irr}_{p'}(\mathbf{N}_{G^*}(P^*)|\lambda^*)|$. Hence, by working in G^*, it is no loss to assume that $Z \subseteq \mathbf{Z}(G)$. In particular, $Z \subseteq N$.

Now, let L/Z be a chief factor of G. Let $\mathcal{A}$ be a complete set of representatives of the orbits of the action of N on the P-invariant characters in $\mathrm{Irr}_{p'}(L|\lambda)$. Then, using Lemma 9.3, we have that

$$\mathrm{Irr}_{p'}(G|\lambda) = \bigcup_{\theta \in \mathcal{A}} \mathrm{Irr}_{p'}(G|\theta) \text{ and } \mathrm{Irr}_{p'}(LN|\lambda) = \bigcup_{\theta \in \mathcal{A}} \mathrm{Irr}_{p'}(LN|\theta)$$

are disjoint unions. By induction, we have that

$$|\mathrm{Irr}_{p'}(G|\lambda)| = \sum_{\theta \in \mathcal{A}} |\mathrm{Irr}_{p'}(G|\theta)| = \sum_{\theta \in \mathcal{A}} |\mathrm{Irr}_{p'}(LN|\theta)| = |\mathrm{Irr}_{p'}(LN|\lambda)|.$$

If $LN < G$, then by induction $|\mathrm{Irr}_{p'}(LN|\lambda)| = |\mathrm{Irr}_{p'}(N|\lambda)|$, and we are done. Therefore we may assume that $LN = G$, or equivalently, that $LP \trianglelefteq G$.

Suppose first that L/Z is a p-group. Then LP/Z is a normal Sylow p-subgroup of G/Z. Since $Z \subseteq \mathbf{Z}(G)$ and $LP = ZP$, then $P \trianglelefteq G$, and the theorem is proven in this case.

Suppose now that L/Z is a p'-group. Then the Sylow p-subgroup Z_p of Z, is a central Sylow p-subgroup of L. By elementary group theory, we can write $L = K \times Z_p$, where K is a normal p-complement of L. Then $K \trianglelefteq G$. Observe that $KP = LP \trianglelefteq G$. Also, $Z = (Z \cap K) \times Z_p$. Write $\lambda = \lambda_{p'} \times \lambda_p$, where $\lambda_p \in \mathrm{Irr}(Z_p)$ and $\lambda_{p'} \in \mathrm{Irr}(Z \cap K)$. Let $\mathcal{B}$ be a complete set of representatives of the orbits of the action of N on the P-invariant characters in $\mathrm{Irr}(K|\lambda_{p'})$. Let $': \mathrm{Irr}_P(K) \to \mathrm{Irr}(C)$ be the P-Glauberman correspondence, where $C = \mathbf{C}_K(P)$. Since θ' is an irreducible constituent of θ_C for $\theta \in \mathrm{Irr}_P(K)$, notice that $': \mathrm{Irr}_P(K|\lambda_{p'}) \to \mathrm{Irr}(C|\lambda_{p'})$ is a bijection, where $\mathrm{Irr}_P(K|\lambda_{p'})$ is the set of P-invariant characters in $\mathrm{Irr}(K|\lambda_{p'})$. By using Lemma 2.10, we have that $\mathcal{B}' = \{\theta' \mid \theta \in \mathcal{B}\}$ is a complete set of representatives of the orbits of the action of N on $\mathrm{Irr}(C|\lambda_{p'})$. Now

$$\mathrm{Irr}_{p'}(G|\lambda) = \bigcup_{\theta \in \mathcal{B}} \mathrm{Irr}_{p'}(G|\theta \times \lambda_p)$$

and

$$\mathrm{Irr}_{p'}(N|\lambda) = \bigcup_{\theta \in \mathcal{B}} \mathrm{Irr}_{p'}(N|\theta' \times \lambda_p)$$

are disjoint unions, using Lemma 9.3. By Theorem 8.11, we know that

$$|\mathrm{Irr}_{p'}(G|\theta \times \lambda_p)| = |\mathrm{Irr}_{p'}(N|\theta' \times \lambda_p)|$$

for $\theta \in \mathcal{B}$, and this proves the theorem in this case.

Finally, we may assume that $L/Z \cong S^n$, where S is a non-abelian simple group of order divisible by p, which by hypothesis satisfies the inductive McKay condition. Let $K = L'$ and $Z_1 = K \cap Z = \mathbf{Z}(K)$, and notice that K is perfect and $K/\mathbf{Z}(K) \cong S^n$. In fact, $L = KZ$ is a central product. (In Theorem C.1, in Appendix C, there are a few lines explaining all these group theoretical facts.) Let $Q = K \cap P \in \mathrm{Syl}_p(K)$. We have that $K\mathbf{N}_G(Q) = G$ by the Frattini argument. Also, $N \subseteq \mathbf{N}_G(Q)$.

By Corollary 10.8 we can write $\mathrm{Irr}(L|\lambda) = \{\theta \cdot \lambda \mid \theta \in \mathrm{Irr}(K|\lambda_1)\}$, where $\lambda_1 = \lambda_{Z_1}$. Recall that $(\theta \cdot \lambda)(1) = \theta(1)$. Hence, can write the set $\mathcal{A}$ (of representatives of the orbits of the action of N on the P-invariant characters in $\mathrm{Irr}_{p'}(L|\lambda)$) as

$$\mathcal{A} = \{\theta \cdot \lambda \mid \theta \in \mathcal{C}\},$$

where $\mathcal{C}$ is a complete set of representatives of the orbits of the action of N on the P-invariant characters in $\mathrm{Irr}_{p'}(K|\lambda_1)$. Thus

$$\mathrm{Irr}_{p'}(G|\lambda) = \bigcup_{\theta \in \mathcal{C}} \mathrm{Irr}_{p'}(G|\theta \cdot \lambda)$$

is a disjoint union.

By Theorem 10.25, there exists an $\mathbf{N}_G(Q)$-equivariant bijection $\Omega: \mathrm{Irr}_{p'}(K) \to \mathrm{Irr}_{p'}(\mathbf{N}_K(Q))$ such that for every $\theta \in \mathrm{Irr}_{p'}(K)$ we have that

$$(G_\theta, K, \theta) \geq_c (\mathbf{N}_G(Q)_{\Omega(\theta)}, \mathbf{N}_K(Q), \Omega(\theta)).$$

Also, since Ω is $\mathbf{N}_G(Q)$-equivariant, we have that the stabilizers in $\mathbf{N}_G(Q)$ of θ and $\Omega(\theta)$ are the same. Therefore

$$G_\theta \cap \mathbf{N}_G(Q) = \mathbf{N}_G(Q)_{\Omega(\theta)}.$$

Also, by using that $N \subseteq \mathbf{N}_G(Q)$ and that the map Ω is $\mathbf{N}_G(Q)$-equivariant, we have that $\mathcal{A}' = \{\Omega(\theta) \cdot \lambda \mid \theta \in \mathcal{C}\}$ is a complete set of representatives of the orbits of the action of N on the P-invariant characters in $\mathrm{Irr}_{p'}(\mathbf{N}_L(Q)|\lambda)$. (We are using that $\Omega(\theta)$ lies over λ_1 if and only if θ lies over λ_1 and that $\mathbf{N}_L(Q)$ is the central product of $\mathbf{N}_K(Q)$ and Z.) Hence

$$\mathrm{Irr}_{p'}(\mathbf{N}_G(Q)|\lambda) = \bigcup_{\theta \in \mathcal{C}} \mathrm{Irr}_{p'}(\mathbf{N}_G(Q)|\Omega(\theta) \cdot \lambda)$$

is also a disjoint union.

By Theorem 10.17, we have that

$$|\mathrm{Irr}_{p'}(G_\theta|\theta \cdot \lambda)| = |\mathrm{Irr}_{p'}(\mathbf{N}_G(Q)_{\Omega(\theta)}|\Omega(\theta) \cdot \lambda)|.$$

Now, using that $G_{\theta \cdot \lambda} = G_\theta$, $\mathbf{N}_G(Q)_{\Omega(\theta) \cdot \lambda} = \mathbf{N}_G(Q)_{\Omega(\theta)}$ and the Clifford correspondence, we have that

$$
\begin{aligned}
|\mathrm{Irr}_{p'}(\mathbf{N}_G(Q)|\lambda)| &= \sum_{\theta \in \mathcal{C}} |\mathrm{Irr}_{p'}(\mathbf{N}_G(Q)|\Omega(\theta) \cdot \lambda)| \\
&= \sum_{\theta \in \mathcal{C}} |\mathrm{Irr}_{p'}(\mathbf{N}_G(Q)_{\Omega(\theta)}|\Omega(\theta) \cdot \lambda)| \\
&= \sum_{\theta \in \mathcal{C}} |\mathrm{Irr}_{p'}(G_\theta|\theta \cdot \lambda)| \\
&= \sum_{\theta \in \mathcal{C}} |\mathrm{Irr}_{p'}(G|\theta \cdot \lambda)| \\
&= |\mathrm{Irr}_{p'}(G|\lambda)|.
\end{aligned}
$$

Since L/Z does not have a normal Sylow p-subgroup, we have that $\mathbf{N}_G(Q) < G$. Now, $P \in \mathrm{Syl}_p(\mathbf{N}_G(Q))$, and by induction we have that

$$|\mathrm{Irr}_{p'}(\mathbf{N}_G(Q)|\lambda)| = |\mathrm{Irr}_{p'}(N|\lambda)|.$$

The proof of the theorem is finished. □

10.9 Notes

As we have mentioned before, E. C. Dade announced a reduction of his projective conjecture to a question on simple groups. In particular, this would provide a reduction of the McKay conjecture to a question on simple groups. In 2007, I. M. Isaacs, G. Malle and I gave a reduction of the McKay conjecture which was specifically tailored for the McKay problem (see [IMN07]). For instance, the introduction of the Malle intermediate subgroup is a trick that only a simple group specialist could have devised. Since then, many simple groups have been proved to satisfy the inductive McKay condition. The treatment of the alternating and the sporadic groups was completed in [Ma08], and groups of Lie type in defining characteristic in [Sp12]. Further groups of Lie type have been treated by M. Cabanes and B. Späth, [CS17a] and [CS17b], among other references. The absolute highlight of the subject is the proof that every simple group satisfies the inductive McKay condition for the prime $p = 2$ by G. Malle and B. Späth in [MS16], hence proving the McKay conjecture for this prime.

Problems

(10.1) Suppose that $N \trianglelefteq G$, $H \leq G$ and $G = NH$. Let $M = N \cap H$. Let $N \subseteq K \leq G$, and assume that $(K, N, \theta) \geq_c (K \cap H, M, \varphi)$. If $h \in H$, then prove that $(K^h, N, \theta^h) \geq_c (K^h \cap H, M, \varphi^h)$.

(10.2) Suppose that $(G, N, \theta) \geq (H, M, \varphi)$ and $(H, M, \varphi) \geq (K, L, \delta)$. Show that $(G, N, \theta) \geq (K, L, \delta)$. Prove the same for central isomorphisms.

(10.3) Suppose that $(\mathcal{P}, \mathcal{P}')$ is associated with $(G, N, \theta) \geq (H, M, \varphi)$. Then $(\mathcal{Q}, \mathcal{Q}')$ is associated with $(G, N, \theta) \geq (H, M, \varphi)$ if and only if there is a function $\mu : G \to N$ with $\mu(1) = 1$ and $\mu(gn) = \mu(g)$ for $g \in G$ and $n \in N$, a linear character $\lambda \in \mathrm{Irr}(H/M)$ and invertible matrices A and B such that $\mathcal{Q}(g) = \mu(g)A^{-1}\mathcal{P}(g)A$ and $\mathcal{Q}'(h) = \lambda(h)\mu(h)B^{-1}\mathcal{P}'(h)B$ for all $g \in G$ and $h \in H$.

(10.4) *(Späth)* Show that the character triple isomorphism obtained in Theorem 10.13 is strong.
    (Hint: Use Lemma 10.10(b).)

Appendix A

In several places in this book we have used the Classification of Finite Simple Groups (CFSG).

Theorem A.1 (CFSG)   *Let S be a non-abelian finite simple group. Then one of the following holds.*

(a) S is isomorphic to the alternating group A_n, for some $n \geq 5$.
(b) S is isomorphic to a simple group of Lie type.
(c) S is isomorphic to one of the 26 sporadic simple groups.

The representation theory of finite simple groups is a vast fast-growing field. A classic reference for symmetric and alternating groups is [JK81], and for simple groups of Lie type [Car89]. For the sporadic groups, it is common to use the ATLAS, the computer programs GAP and Magma, and their websites.

We have used the CFSG in Theorem 6.7, in Theorem 7.3, and in Theorem 8.13.

Appendix B

In the proof of the Howlett–Isaacs theorem (Theorem 8.13) we have specifically used Schur multipliers of simple groups, while in the reduction of the McKay conjecture we have used universal covering groups of simple groups. In character theory, Schur multipliers usually appear when we are dealing with *projective statements* over characters of normal subgroups. (Frequently, the central subgroups appear after we have used the theory of character triple isomorphisms.) As we shall see, each finite simple group S uniquely determines (up to isomorphism) a perfect finite group $\widehat{S}$ with $\widehat{S}/\mathbf{Z}(\widehat{S}) \cong S$, such that whenever G is perfect and $G/\mathbf{Z}(G) \cong S$, then $G \cong \widehat{S}/Z$ for some $Z \subseteq \mathbf{Z}(\widehat{S})$. The **Schur multiplier** of S is $\mathbf{M}(S) = \mathbf{Z}(\widehat{S})$. A character theoretical introduction to Schur multipliers can be found in Chapter 11 of [Is06].

In the reduction of the McKay conjecture, we have used another relationship between S and $\widehat{S}$, namely the connection between their automorphism groups. Although this is standard, let us write down next a few sketches of the relevant proofs, for the sake of completeness. We are following [As00].

Definition B.1 *Let G be a group. A* **central extension** *of G is a pair (H, π), where H is a group and $\pi : H \to G$ is an onto homomorphism such that $\ker(\pi) \subseteq \mathbf{Z}(H)$. If (G_1, π_1) and (G_2, π_2) are central extensions of G, we say that $f : (G_1, \pi_1) \to (G_2, \pi_2)$ is a* **morphism of central extensions** *if $f : G_1 \to G_2$ is a group homomorphism such that $\pi_1 = f\pi_2$.*

(Recall that in this book, maps are composed from the left.)

Definition B.2 *A central extension $(\widehat{G}, \pi)$ of G is* **universal** *if for each central extension (H, δ) of G there exists a unique morphism $f : (\widehat{G}, \pi) \to (H, \delta)$ of central extensions. In this case, we say that $\widehat{G}$ is a* **universal covering group** *of G and $\ker(\pi)$ a* **Schur multiplier** *of G.*

From the definition it is easy to check that, if they exist, universal covering groups and Schur multipliers are unique up to isomorphism.

Lemma B.3 *Suppose that* $(\widehat{G}_1, \pi_1)$ *and* $(\widehat{G}_2, \pi_2)$ *are universal central extensions of* G. *Then there exists a unique isomorphism* $\alpha : \widehat{G}_1 \to \widehat{G}_2$ *such that* $\alpha \pi_2 = \pi_1$. *Also* α *sends* $\ker(\pi_1)$ *isomorphically onto* $\ker(\pi_2)$.

Proof See 33.1 of [As00]. $\qquad\qquad\qquad\qquad\qquad\qquad\qquad\qquad\quad$ □

The main result is the following.

Theorem B.4 *If* G *is a finite perfect group, then* G *possesses a universal central extension* $(\widehat{G}, \pi)$. *Also,* $\widehat{G}$ *is perfect. In particular, if* G *is a non-abelian simple group, then* $\widehat{G}$ *is quasisimple.*

Proof This is 33.2, 33.4 and 33.10 in [As00]. $\qquad\qquad\qquad\qquad\qquad$ □

Lemma B.5 *(a) Let* (H, α) *and* (K, β) *be central extensions of* G, *with* K *perfect. Let* $\gamma : (H, \alpha) \to (K, \beta)$ *a morphism of central extensions. Then* (H, γ) *is a central extension of* K. *In particular,* γ *is surjective.*

(b) Let (H, α) *be a central extension of* G, *and let* (K, β) *be a central extension of* H *with* K *perfect. Then* $(K, \beta\alpha)$ *is a central extension of* G.

Proof We have that $\gamma\beta = \alpha$. If $\gamma(h) = 1$, then $\alpha(h) = \beta(\gamma(h)) = 1$, and $h \in \ker(\alpha) \subseteq \mathbf{Z}(H)$. So we only need to show that γ is surjective. Let $k \in K$. Then $\beta(k) = \alpha(h)$ for some $h \in H$. Thus $\beta(k) = \beta(\gamma(h))$. This implies that $K = \gamma(H)\ker(\beta)$. Since $\ker(\beta)$ is central, then this implies that $\gamma(H) \trianglelefteq K$ has an abelian quotient. Since K is perfect, $\gamma(H) = K$, and γ is surjective.

For (b), we have that α and β are surjective, so $\beta\alpha$ is surjective. We only need to show that $U = \ker(\beta\alpha)$ is contained in $\mathbf{Z}(K)$. Let $x \in U$. Thus $\alpha(\beta(x)) = 1$, and $\beta(x) \in \ker(\alpha)$. Therefore $[\beta(x), \beta(y)] = 1$ for all $y \in K$. Hence $[x, y] \in \ker(\beta) \subseteq \mathbf{Z}(K)$, and $[x, y, z] = 1$ for all $y, z \in K$. Thus $[U, K, K] = 1$. By the Three Subgroups Lemma (Corollary 4.10 of [Is08]), $[K, K, U] = 1$, and since K is perfect, we have that $U \subseteq \mathbf{Z}(K)$. $\qquad$ □

Lemma B.6 *Let* $(\widehat{G}, \pi)$ *be a universal central extension of* G, *and let* (H, α) *be a central extension of* $\widehat{G}$ *with* H *perfect. Then* α *is an isomorphism. In particular,* $(\widehat{G}, 1_{\widehat{G}})$ *is a universal central extension of* $\widehat{G}$.

Proof By Lemma B.5(b), we have that $(H, \alpha\pi)$ is a perfect central extension of G. By the universal property, there exists a group homomorphism $\beta : \widehat{G} \to$

H such that $\beta\alpha\pi = \pi$. By the uniqueness property, we have that $\beta\alpha = 1$. In particular, β is injective. Now, notice that $\beta : (\widehat{G}, 1_{\widehat{G}}) \to (H, \alpha)$ is a morphism of central extensions. By Lemma B.5(a), we have that β is surjective. So β is bijective, and so it is $\alpha = \beta^{-1}$. □

Theorem B.7 *Let G_1 and G_2 be perfect groups, and let (G_1, α) be a central extension of G_2. Let $(\widehat{G}_1, \pi_1)$ and $(\widehat{G}_2, \pi_2)$ be universal central extensions of G_1 and G_2, respectively. Then there exists a unique isomorphism $\hat{\alpha} : \widehat{G}_1 \to \widehat{G}_2$ such that $\hat{\alpha}\pi_2 = \pi_1\alpha$.*

Proof By Lemma B.5(b), we have that $(\widehat{G}_1, \pi_1\alpha)$ is a central extension of G_2. So by the definition of universal central extension, there exists a unique $\beta : \widehat{G}_2 \to \widehat{G}_1$ such that $\beta\pi_1\alpha = \pi_2$. By Lemma B.5(a), we have that $(\widehat{G}_2, \beta)$ is a central extension of $\widehat{G}_1$, and by Lemma B.6, we have that β is an isomorphism. Set $\hat{\alpha} = \beta^{-1}$. The uniqueness of $\hat{\alpha}$ follows from the uniqueness of β. □

The following result was used in the proof of Theorem 10.25.

Corollary B.8 *Suppose that K is a perfect group, and let $(\widehat{K}, \pi)$ be a universal central extension of K. Let $Z = \ker(\pi)$.*

(a) If $\alpha \in \operatorname{Aut}(K)$, then there exists a unique $\hat{\alpha} \in \operatorname{Aut}(\widehat{K})$ such that $\pi\alpha = \hat{\alpha}\pi$.
(b) If $\operatorname{Aut}(\widehat{K})_Z$ is the subgroup of automorphisms $\beta \in \operatorname{Aut}(\widehat{K})$ such that $\beta(Z) = Z$, then the map

$$\hat{} : \operatorname{Aut}(K) \to \operatorname{Aut}(\widehat{K})_Z$$

given by $\alpha \mapsto \hat{\alpha}$ is a group isomorphism.

Proof If $\alpha \in \operatorname{Aut}(K)$, notice that (K, α) is a central extension of K. Then the existence and uniqueness of $\hat{\alpha}$ follows from Theorem B.7. If $z \in Z$, we prove that $\hat{\alpha}(z) \in Z$. But this is clear since $\pi(\hat{\alpha}(z)) = \alpha(\pi(z)) = 1$. Using the uniqueness, it is straightforward to check that $\hat{}$ is a group homomorphism. It is also clear that it is injective, for $\hat{\alpha} = 1$ easily implies that $\alpha = 1$ by using that $\pi\alpha = \hat{\alpha}\pi$ and that π is surjective. Finally, if $\delta \in \operatorname{Aut}(\widehat{K})_Z$, then we define $\alpha \in \operatorname{Aut}(K)$ in the following way. For $k \in K$, we choose any $\hat{k} \in \widehat{K}$ such that $\pi(\hat{k}) = k$, and set $\alpha(k) = \pi(\delta(\hat{k}))$. Using that $\delta(Z) = Z$, it is straightforward to check that α is a well-defined automorphism of K such that $\hat{\alpha} = \delta$. □

Corollary B.9 *Let G be perfect. Let $(\widehat{G}, \pi)$ be a universal central extension of G, and let (H, δ) be a central extension of G with H perfect. Then there* *exists a unique surjective homomorphism* $\alpha \colon \widehat{G} \to H$ *such that* $\pi = \alpha \delta$. *Also,* $(\widehat{G}, \alpha)$ *is a universal central extension of* H.

Proof The first part follows from the definition of universal covering together with Lemma B.5(a). Now, suppose that $(\widehat{H}, \pi_1)$ is a universal central extension of H. By Theorem B.7, there is a unique isomorphism $\hat{\delta} \colon \widehat{H} \to \widehat{G}$ such that $\hat{\delta}\pi = \pi_1\delta$. By uniqueness, $\alpha = \hat{\delta}^{-1}\pi_1$. Using that $\hat{\delta}$ is an isomorphism and that $(\widehat{H}, \pi_1)$ is a universal central extension of H, it is straightforward to check that $(\widehat{G}, \alpha)$ is a universal central extension of H.   □

The following result was also used in the proof of Theorem 10.25.

Corollary B.10 *Suppose that* K *is a perfect group such that* $K/\mathbf{Z}(K) \cong S^n$, *where* S *is a non-abelian finite simple group. Let* $\widehat{S}$ *be a universal covering group for* S. *Then* $\widehat{S}^n$ *is a universal covering group for* K.

Proof We know that $\widehat{S}^n$ is a universal covering group for S^n. (See [As00], Exercise 2, Chapter 11.) Now, K is a perfect central extension of S^n. We apply the second part of Corollary B.9.   □

Appendix C

In the proof of the Howlett–Isaacs theorem, we have used Theorem 8.12(b) in which it was claimed that, given a non-abelian simple group S, there exists a prime p such that p divides $|S|$ and whenever $Y/\mathbf{Z}(Y) \cong S^n$, then p does not divide $|Y' \cap \mathbf{Z}(Y)|$. We are going to prove now that this is equivalent to showing that there exists a prime p such that p divides $|S|$ but p does not divide $|\mathbf{Z}(\widehat{S})|$, where $\widehat{S}$ is a universal covering group of S.

Theorem C.1 *Let S be a non-abelian simple group, and let $\widehat{S}$ be a universal covering group of S. Then the following conditions are equivalent.*

(a) There exists a prime p such that p divides $|S|$ and whenever Y is a finite group with $Y/\mathbf{Z}(Y) \cong S^n$ for some $n \geq 1$, then p does not divide $|Y' \cap \mathbf{Z}(Y)|$.

(b) There exists a prime p such that p divides $|S|$ and p does not divide $|\mathbf{Z}(\widehat{S})|$.

Proof Since $\widehat{S}$ is perfect and $\widehat{S}/\mathbf{Z}(\widehat{S}) \cong S$, it is clear that (a) implies (b), using (a) for $n = 1$. Assume now that p does not divide $|\mathbf{Z}(\widehat{S})|$. Suppose that $Y/\mathbf{Z}(Y) \cong S^n$ for some $n \geq 1$. Then $Y/\mathbf{Z}(Y)$ is a perfect group. Let $K = Y'$. Then $K\mathbf{Z}(Y) = Y$ and $K \cap \mathbf{Z}(Y) = \mathbf{Z}(K)$, because $Y/\mathbf{Z}(Y)$ has no nontrivial abelian normal subgroups nor factor groups. Now, $K/\mathbf{Z}(K) \cong Y/\mathbf{Z}(Y) \cong S^n$, and by the same argument $K'\mathbf{Z}(K) = K$. Therefore $K'\mathbf{Z}(Y) = Y$ and Y/K' is abelian. Hence, $Y' \subseteq K'$ and $K' = K$ is perfect. Recall that we want to show that p does not divide $|\mathbf{Z}(K)|$. By Corollary B.10, we have that $\widehat{S}^n$ is a universal covering group of K. Let $\pi: \widehat{S}^n \to K$ be an onto homomorphism with central kernel Z. Then $Z \subseteq \mathbf{Z}(\widehat{S}^n)$. Now, notice that the largest normal solvable subgroup of $\widehat{S}^n$ is $\mathbf{Z}(\widehat{S}^n) = \mathbf{Z}(\widehat{S})^n$, because $\widehat{S}^n/\mathbf{Z}(\widehat{S}^n) \cong S^n$. Since $\mathbf{Z}(\widehat{S})^n$ has order not divisible by p, we conclude that $\mathbf{Z}(\widehat{S}^n/Z) \cong \mathbf{Z}(K)$ has order not divisible by p. $\square$

Bibliographic Notes

Chapter 1. The source for character theory is the book of I. M. Isaacs [Is06]. Some notation, such us $\mathrm{Irr}(G|\theta)$, was developed after Isaacs' book was published. Our definition of irreducible representation assumes Maschke's theorem. In general, a representation is irreducible if it is not similar to a representation in block upper triangular form. Problems 1.3 and 1.13 are in Proposition 1.1 of the celebrated paper on the $k(GV)$-conjecture by R. Knörr [Kn84]. Some parts of it also appear in [Ch86], [Mas86], and [FI89]. Problem 1.4 comes from [Ch86]. Problem 1.12 appeared in [So61]. Problem 1.15 is a well-known theorem of G. Higman in [Hi40].

Chapter 2. The approach that I have followed for the Glauberman correspondence was published in [Na89]. (I thank A. Turull for pointing out a mistake in a previous version of that paper.) The approach for the Glauberman correspondence followed here is good enough for many applications, but not for all of them. For instance, it is a theorem of I. M. Isaacs and mine that the Glauberman correspondence commutes with irreducible induction and restriction of characters ([IN91]), but to show this fact one needs deeper connections between the character values of Glauberman correspondents. Glauberman's lemma appears in [Gl64].

Chapter 3. In the proof of Theorem 3.9, I have simply followed the clever proof of R. Brauer in [Br64].

Chapter 4. Gallagher's generalization of Burnside's theorem of 1903 [Bu03] appears in [Gal66]. The fact that a nonlinear character has a zero on an element of prime power order is the main result in [MNO00]. The Brauer–Nesbitt theorem was first proven in [BN41] using modular representation theory. The other standard known proof by Gallagher uses Brauer's characterization of characters [Gal66]. Our proof follows [Le00]. The main idea of the proof can also be found in Proposition 3 of [Ku94]. Knörr's Theorem 4.8 and its consequence are Theorem 2.4 and Corollary 2.11 in [Kn89]. The proof presented here is an unpublished proof by J. Murray, in which he has removed the need for modular representation theory. I thank him for allowing the use of his proof in this book. For the proof of Frobenius' Theorem 4.9, I have taken ideas from [Ro09]. Strunkov's theorem appears in [St90], although the bit about modular representation theory in the proof has been removed. Brauer's Problem 19 of his celebrated list is in [Br63], and its solution by Robinson appeared in [Ro83]. Another list of open problems in representation theory is given in [Fe82a]. Theorem 4.13 is from [AH85], and Harada's corollary (Corollary 4.15) from [Ha08]. The study of nonvanishing elements starts in [INW99]. The proof of Theorem 4.18 is a variation of a proof of G. R. Robinson from which the modular representation theory has been removed. This theorem appeared first in [Mi12], where N was assumed elementary abelian, and later this hypothesis was removed in [Brou16]. Robinson's Theorem 4.23 is an unpublished result of Robinson, privately communicated to me, for which I thank him. Problem 4.4 appeared in [MNO00], while Problem 4.6 asks for some of the easy parts in [Ga83]. Problem 4.7 is group theoretical and comes from [AH85]. Problem 4.9 appears in [FI89].

Chapter 5. The concepts of a character triple and of character triple isomorphisms are due to I. M. Isaacs, who introduced them in [Is73]. I have mainly followed this paper. Gallagher's count in Theorem 5.16 is in [Gal70]. He does not use character triple isomorphisms, but a nice trick. However, for applications of this theorem, the fact that there is an invertible matrix associated has had some relevance. Theorems 5.19 and 5.20 are my own. Knörr's theorem appeared in [Kn06] with a totally different proof. Turull's Brauer–Clifford group is defined in [Tu09] and gives *character isomorphisms* which also respect fields of values. (See also [La16].) Problem 5.5 appears in [NTV18], although most of it is in Theorem 15.9 of [MW93]. Problem 5.6 is an unpublished result of this author. Problem 5.8 appears in [Gal66] with a different type of proof.

Chapter 6. P. X. Gallagher's Theorem 6.1 appeared in [Gal62]. Theorem 6.5 is my own. The solution of Gow's conjecture is in [NT08]. Lemma 6.8 is essentially Lemma 4.1 of [Is84]. Riese's theorem (Theorem 6.15), appeared in [Ri98]. Problems 6.2 and 6.3 are my own. Problem 6.6 appeared in [NT08] and Problem 6.7 in [RS98].

Chapter 7. The Itô–Michler theorem appeared in [Mi86]. Theorem 7.3 appeared in [MN12]. It turns out that this was a conjecture by A. I. Saksonov of 1987. Isaacs' theorem (Theorem 7.4) appeared in [Is86], Thompson's in [Tho70], Gow–Humphreys' in [GH75], and Berkovich's theorem (Theorem 7.7) in [Be89]. Cossey–Hawkes' theorem (Theorem 7.10) appeared in [CH93]. Higman's theorem (Theorem 7.15) was published in [Hi71], Problem 7.1 in [Na04b], and Problem 7.3 is a special case of [IR92]. To prove Higman's theorem, Brauer's induction theorem is needed. This theorem of Brauer, of 1946, solved a conjecture of the time: every irreducible character of a finite group G can be afforded by a representation with entries in the cyclotomic field $\mathbb{Q}_{|G|}$. This has consequences in number theory on Artin L-functions. Although Isaacs' proof of Brauer's theorem in [Is06] is only three pages of clever (but basic) character theory, it is not fair to present Brauer's theorem in Chapter 1 on the basics. There is a great deal of literature about Brauer induction theorem, including a theorem of R. Boltje [Bo90] about a canonical way of realizing the induction theorem.

Chapter 8. The Iwahori–Matsumoto conjecture was proposed in [IM64]. The proof that I present of the Howlett–Isaacs theorem is largely based on their proof in [HI82], although I have simplified several parts somewhat, removing the need for system normalizers. The deep character triple isomorphism that is mentioned in Section 3, can be found in [Tu08b], assuming Dade's classification of endo-permutation modules of p-groups. The Okuyama–Wajima argument appeared in [OW79]. (See also Lemma 15.8 of [MW93] and Theorem 6.10 in [Is18].)

Chapter 9. Theorem 9.4 appeared in [Na03]. The case $p = 3$ for non-solvable groups appeared in [NTV14]. The fact that non-solvable groups do not have self-normalizing Sylow subgroups for $p > 3$ was published in [GMN04]. Other canonical correspondences for the primes 2 and 3 have been studied in [INOT17], [GKNT17], [Gi17], and [GTT18]. The idea of the relative McKay conjecture appears in [Wo78b] and [Wo90]. The Isaacs-Navarro conjecture appears in [IN02]; the McKay–Galois refinement in [Na04a], and Turull's refinement in [Tu08a]. Another generalization of the McKay conjecture which implies [IN02] was given by A. Evseev in [Ev13]. Theorem 9.9 appears in [NTT07], and Theorem 9.11 in [Sf17], using work in [ST17]. I have followed [KR89] for most of sections 3 and 4, although the sophisticated arguments using the Green correspondence that were necessary to deal with blocks have been removed. This has not been a triviality, and I owe I. M. Isaacs thanks for an argument in Theorem 9.19 and G. R. Robinson for many conversations on this. I only prove the character theoretical consequence in Webb's result Corollary 9.20. I have stated AWC and Dade's ordinary conjecture in block-free form. After all, this book is on ordinary character theory. The fact that Dade's ordinary conjecture implies AWC is in [Da92]. I am not aware of references for the fact that Dade's ordinary conjecture implies McKay's. The connection between Dade's ordinary conjecture and Brauer's results in Theorem 9.29 was also noticed in [HH98]. The remark about radical chains was made in [KR89], and a proof is presented in [Da92]. Problem 9.4 is taken from [Is90]. The proof of the various versions of the counting conjectures for several families of groups occupies many pages in many journals. I apologize for not being able to cite every single work on this. (See the survey article [Ma17].)

Chapter 10. In the course of these years, B. Späth has been reformulating the initial inductive McKay condition to the statement in Definition 10.23. She showed that Alperin's weight conjecture and Dade's projective conjecture also admit similar reformulations. Although central isomorphisms were introduced in [NS14] to reduce Brauer's height zero conjecture, the treatment on how to use central isomorphisms to reformulate the initial inductive McKay condition in [IMN07] is due to Späth, and I would like to thank her for many conversations on this subject. The surveys [Sp17b] and [Sp18] contain many of the ideas in this chapter. In the definition of the inductive McKay condition, I have eliminated the *Malle intermediate subgroup* to avoid technicalities. I believe that the proof of the reduction is complicated enough, and my goal is only to introduce the subject to the reader. To bring Brauer blocks to the reductions is, however, far from being a technicality or a triviality. This is again the work of Späth in [Sp13a], [Sp13b], and [Sp17a]. Corollary 10.5 appears in [NT16a].

References

[Al75]   J. L. Alperin, The main problem of block theory, *Proc. of Conf. on Finite Groups*, New York: Academic Press (1975), 341–356.

[Al76]   J. L. Alperin, Isomorphic blocks, *J. Algebra* **43** (1976), 694–698.

[Al87]   J. L. Alperin, Weights for finite groups, *Proc. Sympos. Pure Math. Amer. Math. Soc.*, **47** (1987), 369–379.

[AF90]   J. L. Alperin and P. Fong, Weights for symmetric and general linear groups, *J. Algebra* **131** (1990), 2–22.

[AC86]   G. Amit and D. Chillag, On a question of Feit concerning character values of finite solvable groups, *Pacific J. Math.* **122** (1986), 257–261.

[AH85]   Z. Arad and M. Herzog (eds.), *Products of Conjugacy Classes in Groups*, Lecture Notes in Math. 1112, New York: Springer-Verlag, (1985).

[As00]   M. Aschbacher, *Finite Group Theory*, Cambridge: Cambridge University Press, (2000).

[Betal16] A. Beltrán, M. J. Felipe, G. Malle, A. Moretó, G. Navarro, L. Sanus, R. Solomon and Pham Huu Tiep, Nilpotent and abelian Hall subgroups in finite groups, *Trans. Amer. Math. Soc.* **368** (2016), 2497–2513.

[Be89]   Ya. G. Berkovich, Degrees of irreducible characters and normal p-complements, *Proc. Amer. Math. Soc.* **106** (1989), 33–35.

[BTZ17] C. Bessenrodt, H. P. Tong-Viet and P. J. Zhang, Huppert's conjecture for alternating groups, *J. Algebra* **470** (2017), 353–378.

[Bo90]   R. Boltje, A canonical Brauer induction formula, *Asterisque* **181–182** (1990), 31–59.

[Bo03]   R. Boltje, Alperin's weight conjecture and chain complexes, *J. London Math. Soc.* **68** (2003), 83–101.

[BN41]   R. Brauer and C. Nesbitt, On the modular characters of groups, *Ann. of Math.* **42** (1941), 556–590.

[Br63]   R. Brauer, Representations of finite groups, in: *Lectures on Modern Mathematics*, vol. I, ed. by T. Saaty, John Wiley & Sons (1963), 133–175.

[Br64]   R. Brauer, A note on theorems of Burnside and Blichfeldt, *Proc. Amer. Math. Soc.* **15** (1964), 31–34.

[Bro88] M. Broué, Blocs, isométries parfaites, catégories dérivées, *C. R. Acad. Sci. Paris* **307** (1998), 13–18.

[Brou16] J. Brough, Non-vanishing elements in finite groups, *J. Algebra* **460** (2016), 387–391.

[Bu03] W. Burnside, On an arithmetical theorem connected with roots of unity and its application to group characteristics, *Proc. London Math. Soc.* **1** (1903), 112–116.

[Bu55] W. Burnside, *Theory of Groups of Finite Order*, Dover (1955).

[Ca13] M. Cabanes, Two remarks on the reduction of Alperin's weight conjecture, *Bull. London Math. Soc.* **45** (2013), 895–906.

[CS17a] M. Cabanes and B. Späth, Equivariant character correspondences and inductive McKay condition for type A, *J. Reine Angew. Math.* **728** (2017), 153–194.

[CS17b] M. Cabanes and B. Späth, Inductive McKay condition for finite simple groups of type C, *Represent. Theory* **21** (2017), 61–81.

[CH80] A. R. Camina and M. Herzog, Character tables determine abelian Sylow 2-subgroups, *Proc. Amer. Math. Soc.* **80** (1980), 533–535.

[Car89] R. W. Carter, *Finite Groups of Lie Type. Conjugacy Classes and Complex Characters*, Wiley (1985).

[Ch86] D. Chillag, Character values of finite groups as eigenvalues of nonnegative integer matrices, *Proc. Amer. Math. Soc.* **97** (1986), 565–567.

[ATLAS] J. H. Conway, R. T. Curtis, S. P. Norton, R. A. Parker, and R. A. Wilson, *An ATLAS of Finite Groups*, Oxford: Clarendon Press, (1985).

[CH93] J. Cossey and T. Hawkes, Computing the order of the nilpotent residual of a finite group from knowledge of its group algebra, *Arch. Math.* **60** (1993), 115–120.

[Cr08] D. A. Craven, Symmetric group character degrees and hook numbers, *Proc. London Math. Soc.* (3) **96** (2008), 26–50.

[Da77] E. C. Dade, Remarks on isomorphic blocks, *J. Algebra* **45** (1977), 254–258.

[Da92] E. C. Dade, Counting characters in blocks I, *Invent. Math.* **109** (1992), 187–210.

[Da94] E. C. Dade, Counting characters in blocks II, *J. Reine Angew. Math.* **448** (1994), 97–190.

[Da97] E. C. Dade, Counting characters in blocks 2.9, in *Proceedings of the 1995 Conference on Representation Theory at Ohio State University.* ed. R. Solomon (1997), 45–59.

[DNPST10] S. Dolfi, G. Navarro, E. Pacifici, L. Sanus, and Pham Huu Tiep, Non-vanishing elements of finite groups, *J. Algebra* **323** (2010), 540–545.

[DPSS09] S. Dolfi, E. Pacifici, L. Sanus, and P. Spiga, On the orders of zeros of irreducible characters, *J. Algebra* **321** (2009), 345–352.

[ER02] C. Eaton and G. R. Robinson, On a minimal counterexample to Dade's projective conjecture, *J. Algebra* **249** (2002), 453–468.

[Ev13] A. Evseev, The McKay conjecture and Brauer's induction theorem *Proc. London Math. Soc.* (3) **106** (2013), 1248–1290.

[Fe80] W. Feit, Some consequences of the classification of finite simple groups, The Santa-Cruz conference on finite groups, *Proc. of Symposia in Pure Math 37*, Amer. Math. Soc. (1980), 175–181.

[Fe82a] W. Feit, *The Representation Theory of Finite Groups*, Amsterdam: North-Holland (1982).

[Fe82b] W. Feit, Some properties of group characters, *Bull. London Math. Soc.* **14** (1982), 129–132.

[FS88] W. Feit and G. Seitz, On finite rational groups and related topics, *Illinois J. Math.* **33** (1988), 103–131.

[FI89] P. Ferguson and I. M. Isaacs, Induced characters which are multiples of irreducibles, *J. Algebra* **124** (1989), 149–157.

[FT86] P. Ferguson and A. Turull, On a question of Feit, *Proc. Amer. Math. Soc.* **97** (1986), 21–22.

[Ga74] S. M. Gagola, Characters fully ramified over a normal subgroup, *Pacific J. Math.* **55** (1974), 107–126.

[Ga83] S. M. Gagola, Characters vanishing on all but two conjugacy classes, *Pacific J. Math.* **109** (1983), 363–385.

[Gal62] P. X. Gallagher, Group characters and normal Hall subgroups, *Nagoya Math. J.* **21** (1962), 223–230.

[Gal66] P. X. Gallagher, Zeros of characters of finite groups, *J. Algebra* **4** (1966), 42–45.

[Gal70] P. X. Gallagher, The number of conjugacy classes in a finite group, *Math. Z.* **118** (1970), 175–179.

[GAP] The GAP group, GAP – *Groups, Algorithms, and Programming*, Version 4.8.7 (2017), www.gap-system.org.

[Gar76] S. Garrison, Determining the Frattini subgroup from the character table, *Canad. J. Math.* **28** (1976), 560–567.

[Gi17] E. Giannelli, Characters of odd degree of symmetric groups, *J. London Math. Soc.* (2) **96** (2017), 1–14.

[GKNT17] E. Giannelli, A. Kleschev, G. Navarro, and Pham Huu Tiep, Restriction of odd degree characters and natural correspondences, *Inter. Math. Research Not.*, **20** (2017), 6089–6118.

[GTT18] E. Giannelli, Pham Huu Tiep, and J. Tent, Irreducible characters of 3'-degree of finite symmetric, general linear and unitary groups, to appear in J. Pure Applied Algebra (2017).

[Gl64] G. Glauberman, Fixed points in groups with operator groups, *Math. Z.* **84** (1964), 120–125.

[Gl68] G. Glauberman, Correspondences of characters for relatively prime operator groups, *Can. J. Math.* **20** (1968), 1465–1488.

[Gl85] D. Gluck, The largest irreducible character degree of a finite group, *Canad. J. Math.* **37** (1985), 442–451.

[GH75] R. Gow and J. F. Humphreys, Normal *p*-complements and irreducible representations, *J. London Math.* **11** (1975), 308–312.

[Go76] R. Gow, Groups whose characters are rational valued, *J. Algebra* **40** (1976), 280–299.

[Go81] R. Gow, Character values of groups of odd order and a question of Feit, *J. Algebra* **68** (1981), 75–78.

[GO96] A. Granville and K. Ono, Defect zero *p*-blocks for finite simple groups, *Trans. Amer. Math. Soc.* **384** (1996), 331–347.

[Gu82] R. M. Guralnick, Commutators and commutator subgroups, *Adv. Math.* **45** (1982), 319–330.

[GMN04] R. M. Guralnick, G. Malle and G. Navarro, Self-normalizing Sylow subgroups, *Proc. Amer. Math. Soc.* **132** (2004), 973–979.

[Ha08] K. Harada, On a theorem of Brauer and Wielandt, *Proc. Amer. Math. Soc.* **136** (2008), 3825–3829.

[Ha17] K. Harada, Revisiting character theory of finite groups, Proceedings of the Conference in Finite Groups and Vertex Algebras, *Bulletin of the Institute of Mathematics, Academia Sinica*, New Series (2017).

[Hi11] R. J. Higgs, Finite groups with irreducible projective representations of large degree, *Comm. in Alg.* **39** (2011), 3897–3904.

[HT94] B. Hartley and A. Turull, On characters of coprime operator groups and the Glauberman character correspondence, *J. Reine Angew. Math.* **451** (1994), 175–219.

[HH98] N. M. Hassan and E. Horváth, Some remarks on Dade's conjecture, *Mathematika Pannonica* **9** (1998), 181–194.

[He05] P. Hedegüs, Structure of solvable rational groups, *Proc. London Math. Soc.* (2) **90** (2005), 439–471.

[Hi40] G. Higman, The units of group rings, *Proc. London Math. Soc.* (2) **46** (1940), 231–248.

[Hi71] G. Higman, Construction of simple groups from character tables, In: *Finite Simple Groups*, Eds: Powell and Higman, London: Academic Press, 205–214, 1971.

[HI82] R. B. Howlett and I. M. Isaacs, On groups of central type, *Math. Z.* **179** (1982), 555–569.

[Hu00] B. Huppert, Some simple groups which are determined by the set of their character degrees I, *Illinois J. Math.* **44** (2000), 828–842.

[Is73] I. M Isaacs, Characters of solvable and symplectic groups, *Amer. J. Math.* **85** (1973), 594–635.

[Is84] I. M. Isaacs, Characters of π-separable groups, *J. Algebra* **86** (1984), 98–128.

[Is86] I. M. Isaacs, Recovering information about a group from its complex group algebra, *Arch. Math.* **47** (1986), 293–295.

[Is90] I. M. Isaacs, Hall subgroup normalizers and character correspondences in M-groups, *Proc. Amer. Math. Soc.* **109** (1990), 647–651.

[Is06] I. M. Isaacs, *Character Theory of Finite Groups*, Providence, RI: AMS Chelsea Publishing, 2006.

[Is08] I. M. Isaacs, *Finite Group Theory*, Graduate Studies in Mathematics, **92**, Providence, RI: Amer. Math. Soc., 2008.

[Is18] I. M. Isaacs, *Characters of Solvable Groups*, Providence, RI: Amer. Math. Soc., 2018.

[IMN07] I. M. Isaacs, G. Malle and G. Navarro, A reduction theorem for the McKay conjecture, *Invent. Math.* **170** (2007), 33–101.

[IN91] I. M. Isaacs and G. Navarro, Character correspondences and irreducible induction and restriction, *J. Algebra* **140** (1991), 131–140.

[IN02] I. M. Isaacs and G. Navarro, New refinements of the McKay conjecture for arbitrary finite groups, *Ann. of Math.* (2) **156** (2002), 333–344.

[IN12] I. M. Isaacs and G. Navarro, Sylow 2-subgroups of rational solvable groups, *Math. Z.* **272** (2012), 937–945.

[INOT17] I. M. Isaacs, G. Navarro, J. Olsson, and Pham Huu Tiep, Character Restrictions and Multiplicities in Symmetric Groups, *J. Algebra* **478** (2017), 271–282.

[INW99] I. M. Isaacs, G. Navarro and T. R. Wolf, Finite group elements where no irreducible character vanishes, *J. Algebra* **222** (1999), 413–423.

[IR92] I. M. Isaacs and G. R. Robinson, On a theorem of Frobenius: Solutions of $x^n = 1$ in finite groups, *Amer. Math. Monthly* **99** (1992), 352–354.

[IM64] N. Iwahori and H. Matsumoto, Several remarks on projective representations of finite groups, *Fac. Sci. Univ. Tokyo Sect. IA Math.* **10** (1964), 129–146.

[JK81] G. James and A. Kerber, *The Representation Theory of the Symmetric Group*, Reading, MA: Addison-Wesley, 1981.

[KM13] R. Kessar and G. Malle, Quasi-isolated blocks and Brauer's height zero conjecture, *Ann. of Math.* **178** (2013), 321–384.

[KS95] W. Kimmerle and R. Sandling, Group theoretic determination of certain Sylow and Hall subgroups and the resolution of a question of R. Brauer, *J. Algebra* **171** (1995), 329–346.

[Kn84] R. Knörr, On the number of characters in a p-block of a p-solvable group, *Illinois J. Math* **28** (1984), 181–210.

[Kn89] R. Knörr, Virtually irreducible lattices, *Proc. London Math. Soc.* (3) **59** (1989), 99–132.

[Kn06] R. Knörr, Partial inner products, *J. Algebra* **304** (2006), 304–310.

[KR89] R. Knörr and G. R. Robinson, Some remarks on a conjecture of Alperin, *J. London Math. Soc.* (2) **39** (1989), 48–60.

[Ku94] B. Külshammer, Central idempotents in p-adic group rings, *J. Austral. Math. Soc. Ser. A* **56** (1994), 278–289.

[La16] F. Ladisch, On Clifford theory with Galois action, *J. Algebra* **457** (2016), 45–72.

[Le00] M. Leitz, Elementary proof of Brauer's and Nesbitt's theorem on zeros of characters of finite groups, *Proc. Amer. Math. Soc.* **128** (2000), 3149–3152.

[LOST10] M. W. Liebeck, E. A. O'Brien, A. Shalev, and Pham Huu Tiep, The Ore Conjecture, *J. Eur. Math. Soc.* **12** (2010), 939–1008.

[LY79] R. A. Liebler and J. E. Yellen, In search of nonsolvable groups of central type, *Pacific J. Math.* **82** (1979), 485–492.

[Li05] M. Linckelmann, Alperin's weight conjecture in terms of equivariant Bredon cohomology, *Math. Z.* **250** (2005), 495–513.

[MAGMA] Wieb Bosma, John Cannon, and Catherine Playoust, The Magma algebra system. I. The user language, *J. Symbolic Comput.* **24** (1997), 235–265.

[Ma08] G. Malle, The inductive McKay condition for simple groups not of Lie type, *Comm. Algebra* **36** (2008), 455–463.

[Ma12] G. Malle, The proof of Ore's conjecture (after Ellers-Gordeev and Liebeck-O'Brien-Shalev-Tiep), *Astérisque* **361** (2014), 325–348.

[Ma17] G. Malle, Local–global conjectures in the representation theory of finite groups, Representation Theory–current trends and perspectives, 519–539, EMS Ser. Congr. Rep., Eur. Math. Soc., Zurich, 2017.

[MN12] G. Malle and G. Navarro, Characterizing normal Sylow p-subgroups by character degrees, *J. Algebra* **370** (2012), 402–406.

[MNO00] G. Malle, G. Navarro, and J. B. Olsson, Zeros of characters of finite groups, *J. Group Theory* **3** (2000), 353–368.

[MS16] G. Malle and B. Späth, Characters of odd degree, *Ann. of Math.* **184** (2016), 869–908.

[MW93] O. Manz and T. R. Wolf, *Representations of solvable groups*, Cambridge: Cambridge University Press, 1993.

[Mas86] G. Mason, Some applications of quasi-invertible characters, *J. London Math. Soc.* (2) **33** (1986), 40–48.

[Mat92] S. Mattarei, Character tables and metabelian groups, *J. London Math. Soc.* (2) **46** (1992), 92–100.

[McK71] J. McKay, Abstract 71T-A31, *Notices Amer. Math. Soc.* **18** (1971), 397.

[McK72] J. McKay, Irreducible representations of odd degree, *J. Algebra* **20** (1972), 416–418.

[Mi86] G. Michler, A finite simple group of Lie type has p-blocks with different defects if $p \neq 2$, *J. Algebra* **104** (1986), 220–230.

[Mi12] M. Miyamoto, Non-vanishing elements in finite groups, *J. Algebra* **364** (2012), 88–89.

[Mo07] A. Moretó, Complex group algebras of finite groups: Brauer's Problem 1, *Adv. in Math.* **208** (2007), 236–248.

[Na89] G. Navarro, On the Glauberman correspondence, *Proc. Amer. Math. Soc.* **105** (1989), 52–54.

[Na98] G. Navarro, *Characters and blocks of finite groups*, Cambridge: Cambridge University Press (1998).

[Na03] G. Navarro, Linear characters of Sylow subgroups, *J. Algebra* **269** (2003), 589–598.

[Na04a] G. Navarro, The McKay conjecture and Galois automorphisms, *Ann. of Math.* (2) **160** (2004), 1129–1140.

[Na04b] G. Navarro, Problems on Characters and Sylow Subgroups, Finite Groups, Berlin: de Gruyter (2004), 275–281.

[NRi14] G. Navarro and N. Rizo, Nilpotent and perfect groups with the same set of character degrees, *J. Algebra and Appl.* **13** (2014), 1450061-3.

[NRi16] G. Navarro and N. Rizo, A Brauer-Wielandt formula (with an application to character tables), *Proc. Amer. Math. Soc.* **144** (2016), 4199–4204.

[NR12] G. Navarro and G. R. Robinson, Irreducible characters taking root of unity values on p-singular elements, *Proc. Amer. Math. Soc.* **140** (2012), 3785–3792.

[NST15] G. Navarro, R. M. Solomon, and Pham Huu Tiep, Abelian Sylow subgroups in a finite group, II, *J. Algebra* **421** (2015), 3–11.

[NS14] G. Navarro and B. Späth, On Brauer's height zero conjecture, *J. Eur. Math. Soc.* **16**, (2014), 695–747.

[NT08] G. Navarro and Pham Huu Tiep, Rational irreducible characters and rational conjugacy classes in finite groups, *Trans. Amer. Math. Soc.* **360** (2008), 2443–2465.

[NT11] G. Navarro and Pham Huu Tiep, A reduction theorem for the Alperin weight conjecture, *Invent. Math.* **184** (2011), 529–565.

[NT14] G. Navarro and Pham Huu Tiep, Abelian Sylow subgroups in a finite group. I, *J. Algebra* **398** (2014), 519–526.

[NT16a] G. Navarro and Pham Huu Tiep, Representations of odd degree, Math. Ann. **365** (3) (2016), 1155–1185.

[NT16b] G. Navarro and Pham Huu Tiep, Real groups and Sylow 2-subgroups, *Adv. in Math.* **299** (2016), 331–360.

[NTT07] G. Navarro, Pham Huu Tiep, and A. Turull, p-rational characters and self-normalizing Sylow p-subgroups, *Represent. Theory* **11** (2007), 84–94.

[NTV14] G. Navarro, Pham Huu Tiep, and C. Vallejo, McKay natural correspondences of characters, *Algebra and Number Theory* **8** (2014), 1839–1856.

[NTV18] G. Navarro, Pham Huu Tiep, and C. Vallejo, Local blocks with one simple module, to appear in *Trans. Amer. Math. Soc.*

[Ok81] T. Okuyama, Module correspondence in finite groups, *Hokkaido Math. J.* **10** (1981), 299–318.

[OW79] T. Okuyama and M. Wajima, Irreducible characters of p-solvable groups, *Proc. Japan Acad. Ser. A* **55** (1979), 309–312.

[Ol76] J. B. Olsson, McKay numbers and heights of characters, *Math. Scand.* **38** (1976), 25–42.

[OU95] J. B. Olsson and K. Uno, Dade's conjecture for symmetric groups, *J. Algebra.* **176** (1995), 534–560.

[Ri85] I. M. Richards, Characters of groups with quotients of odd order, *J. Algebra* **96** (1986), 45-47.

[Ri98] U. Riese, A subnormality criterion in finite groups related to character degrees, *J. Algebra* **201** (1998), 357–362.

[RS98] U. Riese and P. Schmid, Characters induced from Sylow subgroups, *J. Algebra* **207** (1998), 682–694.

[Ro83] G. R. Robinson, The number of blocks with a given defect group, *J. Algebra* **84** (1983), 495–502.

[Ro00] G. R. Robinson, Dade's projective conjecture for p-solvable groups, *J. Algebra* **229** (2000), 234–248.

[Ro09] G. R. Robinson, Characters and the commutator map, *J. Algebra* **321** (2009), 3521–3526.

[Ro11] G. R. Robinson, Generalized characters whose values on non-identity elements are roots of unity, *J. Algebra* **333** (2011), 458–464.

[Sa67] A. I. Saksonov, An answer to a question of R. Brauer, *Vesci Akad. Navuk BSSR, Ser. Fiz.-Mat. Navuk* **1** (1967), 129–130.

[Sf17] A. Schaeffer-Fry, Galois automorphisms on Harish-Chandra series and Navarro's self-normalizing Sylow 2-subgroup conjecture, https://arxiv.org/pdf/1707.03923.pdf

[ST17] A. Schaeffer-Fry and J. Taylor, On self-normalizing Sylow 2-subgroups in type A, https://arxiv.org/pdf/1701.00272.pdf

[So61] L. Solomon, On the sum of the elements in the character table of a finite group, *Proc. Amer. Math. Soc.* **12** (1961), 962–963.

[Sp12] B. Späth, Inductive McKay condition in defining characteristic, *Bull. London Math. Soc.* **44** (2012), 426–438.

[Sp13a] B. Späth, A reduction theorem for the Alperin-McKay conjecture, *J. reine angew. Math.* **680** (2013), 153–189.

[Sp13b] B. Späth, A reduction theorem for the blockwise Alperin weight conjecture, *J. Group Theory* **16** (2013), 159–220.

[Sp17a] B. Späth, A reduction theorem for Dade's projective conjecture, *J. Eur. Math. Soc.* **19** (2017), 1071–1126.

[Sp17b] B. Späth, Inductive conditions for counting conjectures via character triples, Representation theory–current trends and perspectives, 665–680, EMS Ser. Congr. Rep., *Eur. Math. Soc.*, Zurich (2017).

[Sp18] B. Späth, Reduction theorems for some counting conjectures, to appear.

[St90] S. P. Strunkov, On blocks of defect 0 in finite groups, *Math. USSR Izvestiya* **34** (1990), 677–683.

[Th92] J. Thévenaz, Locally determined functions and Alperin's Weight Conjecture, *J. London Math. Soc.* (2) **45** (1992), 446–468.

[Th93] J. Thévenaz, Equivariant K-theory and Alperin's conjecture, *J. Pure App. Algebra* **85** (1993), 185–202.

[Tho70] J. G. Thompson, Normal p-complements and irreducible characters, *J. Algebra* **14** (1970), 129–134.

[Tho08] J. G. Thompson, Composition factors of rational solvable groups, *J. Algebra* **319** (2008), 558–594.

[Tu08a] A. Turull, Strengthening the McKay conjecture to include local fields and local Schur indices, *J. Algebra* **319** (2008), 4853–4868.

[Tu08b] A. Turull, Above the Glauberman correspondence, *Adv. Math.* **217** (2008), 2170–2205.

[Tu09] A. Turull, The Brauer–Clifford group, *J. Algebra* **321** (2009), 3620–3642.

[Tu13] A. Turull, The strengthened Alperin–McKay conjecture for p-solvable groups, *J. Algebra* **394** (2013), 79–91.

[Tu14] A. Turull, The strengthened Alperin weight conjecture for p-solvable groups, *J. Algebra* **398** (2014), 469–480.

[Tu17] A. Turull, Refinements of Dade's projective conjecture for p-solvable groups, *J. Algebra* **474** (2017), 424–465.

[We86] P. J. Webb, Subgroup complexes, *Proceedings of Symposia in Pure Mathematics* **47**, *Amer. Math. Soc.*, Providence, (1987).

[Wi06] T. Wilde, Orders of elements and zeros and heights of characters in a finite group (2006), https://arxiv.org/abs/math/0604337

[Wil88] W. Willems, Blocks of defect zero in finite simple groups of Lie type, *J. Algebra* **113** (1988), 511–522.

[Wils98] R. A. Wilson, The McKay conjecture is true for the sporadic simple groups, *J. Algebra* **207** (1998), 294–305.

[Wo78a]   T. R. Wolf, Character correspondences in solvable groups,   *Illinois J. Math.* **22** (1978), 327–340.

[Wo78b]   T. R. Wolf, Characters of p'-degree in solvable groups, *Pacific J. Math.* **74** (1978), 267–271.

[Wo90]   T. R. Wolf, Variations on McKay's character degree conjecture, *J. Algebra* **135** (1990), 123–138.

Index